The U.S. Nuclear Arsenal

A HISTORY OF WEAPONS AND DELIVERY SYSTEMS SINCE 1945

The U.S. Nuclear Arsenal

A HISTORY OF WEAPONS AND DELIVERY SYSTEMS SINCE 1945

The U.S. Nuclear Arsenal

A HISTORY OF WEAPONS AND DELIVERY SYSTEMS SINCE 1945

Norman Polmar and
Robert S. Norris

NAVAL INSTITUTE PRESS
Annapolis, Maryland

Naval Institute Press
291 Wood Road
Annapolis, MD 21402

Library of Congress Cataloging-in-Publication Data
Polmar, Norman.
The U.S. nuclear arsenal : a history of weapons and delivery systems since 1945 / Norman Polmar and Robert S. Norris.
p. cm.
Includes bibliographical references and index.
ISBN 978-1-55750-681-8 (alk. paper)
1. Nuclear weapons—United States—History. 2. Ballistic missiles—United States—History. 3. Guided missiles—United States—History. 4. Tactical missiles—United States—History. 5. Bombers—United States—History. 6. United States—Military policy. 7. Deterrence (Strategy) I. Norris, Robert S. (Robert Stan) II. Title. III. Title: United States nuclear arsenal.
U264.3.P65 2009
358'.39820973—dc22

2008054725

Printed in the United States of America on acid-free paper

15 14 13 12 11 10 09 9 8 7 6 5 4 3 2
First printing

Frontispiece: The first use of a nuclear weapon occurred on 6 August 1945, when a Little Boy atomic bomb was detonated over the Japanese city of Hiroshima. The bomb devastated the city. Three days later a Fat Man atomic bomb similarly destroyed Nagasaki. Although nuclear weapons were "almost" used several times since 1945, these were the only ones ever employed in combat. (U.S. Army Air Forces)

For Beverly and Myriam
for always

Thus, the first and most vital step in any American security program for the age of atomic bombs is to take measures to guarantee to ourselves in case of attack the possibility of retaliation in kind. The writer in making that statement is not for the moment concerned about who will win the next war in which atomic bombs are used. Thus far the chief purpose of our military establishment has been to win wars. From now its chief purpose must be to avert them. It can have almost no other useful purpose.

—Bernard Brodie
The Absolute Weapon (1946)

Contents

Foreword

The advent of atomic weapons dramatically transformed global security and the nature of big-power rivalry. Over time, as an increasing number of nations gained access to relevant knowledge and technology, the atomic arena steadily expanded, stimulated by the East-West rivalry of the Cold War and the instabilities of the Middle East and East Asia. Now, two decades after the fall of the Berlin Wall, the Cold War–rooted fears of long-standing speculative fiction—a world of irresponsible national leaders, irredeemable mistakes, shadowy substate actors, vicious criminal cartels, and ruthless terrorists, captured in motion pictures and novels such as Stanley Kubrick's *Dr. Strangelove*, Eugene Burdick's and Harvey Wheeler's *Fail-Safe*, Nevil Shute's *On the Beach*, and the wilder scenarios of Ian Fleming's *Goldfinger* or *Thunderball*—have become disturbing contemporary reality, adding further levels of complexity to the nuclear debate.

This unstable world is at once more unpredictable, chaotic, and dangerous than the nuclear era that preceded it. Ironically, when the Wall fell, the world seemed poised on a far-safer era, one in which the world's atomic powers would turn away from such destructive weaponry, embarking on a new world characterized by widespread stability and peace. Indeed, a widespread nuclear disarmament did occur, and one rogue nation—South Africa—turned its back on nuclear weapons entirely. But, otherwise, the expectations of the post–Cold War era did not come to pass. Instead, within a year, America's anticipated "peace dividend" had evaporated amidst Iraq's aggression against neighboring Kuwait; by January 1991, 500,000 soldiers, sailors, airmen, and Marines (and their allies) were poised to wrest it back from Saddam Hussein. In the many other conflicts that followed, any hopes of a new and more rational era of human affairs withered. Today, a variety of technologies and systems exist to make possible ever-more deadly and destructive nuclear weapons, giving a wider range of nations and interest groups access to their unprecedented power, and raising a specter of ever-more frightening and foreboding scenarios. That growing threat, coupled with concerns over the safety and security of America's own nuclear arsenal following some well-publicized lapses in security procedures involving critical data and transportation of weapons, make even more imperative that decision-makers and public-minded citizens alike have access to reliable, authoritative, and unbiased information regarding the nuclear world built since the advent of the Manhattan Project.

Accordingly, it is good to have this impressively detailed and thoughtful book. Authors Norman Polmar and Robert S. Norris provide a remarkably thorough and encyclopedic look at the investment that the United States has made in atomic weaponry. They cogently present the history of American nuclear strategy and development, analyze the various technical streams of research and the doctrinal evolution that led to the weapons and systems government sponsored and fielded, and assess the impact of those developments upon the larger history of the Cold War and its aftermath. Eminently readable and the product of decades of research and thought, The *U.S. Nuclear Arsenal* is certain to become a standard reference for think tanks, decision-makers, and students of nuclear affairs. Hopefully, as well, it will stimulate creative thought about the future of nuclear weaponry, and how to best confront its threat to the world today.

—Dr. Richard P. Hallion
Former Senior Advisor for Air and Space Issues,
Directorate for Security, Counterintelligence, and
Special Programs Oversight, and The Air Force
Historian, Headquarters Air Force, Washington, D.C.

Preface

Atomic bombs became the centerpiece of American military strategy after they ended the war against Japan in August 1945. For the next 45 years the relative strength of the principal protagonists of the Cold War was measured by their nuclear arsenals. And, in the post–Cold War era nuclear weapons continue as a major factor in U.S. security strategy; they are viewed as a deterrent—and hence potential retaliation—to an enemy's use of weapons of mass destruction (chemical, biological, nuclear) against the United States.

In the late 1940s the American debate over whether the atomic bomb—delivered by long-range manned bombers—was the ultimate arbitrator of international differences led to the infamous carrier-versus-B-36 bomber controversy. American schoolchildren in the 1950s practiced "duck and cover" because an atomic attack was feared against American cities, and trillions of dollars were spent to develop vast fleets of bombers and missiles to strike the Soviet Union and the rest of the communist world.

At the battlefield level the U.S. Army developed the 280-mm atomic cannon, man-carried atomic demolitions, and the infamous Davy Crockett "atomic grenade" launcher—the last intended to give battalion commanders their own nuclear arsenal. Similarly, the U.S. Navy entered the atomic era, developing carrier-based nuclear strike aircraft to compete with the U.S. Air Force in strategic deterrence, and then a vast variety of strategic and tactical nuclear weapons, including 16-inch nuclear projectiles for battleships and unmanned (drone) helicopters that could carry a nuclear depth bomb.

This volume focuses on the nuclear attributes of U.S. weapons development from the beginning of the "atomic age" through today. Details of the nuclear aspects of weapon development, the warheads, their deployment, and dismantlement are provided to the extent possible. While some secrets are necessary to safeguard current technology and deployments of nuclear weapons, the U.S. government continues to overclassify and keep secret significant facts and figures about the role of nuclear weapons in the Cold War, even the early aspects of that conflict that began more than a half-century ago.

Other reference books, such as the distinguished Jane's publications, are for the most part totally silent on the nuclear aspects of the weapons that they describe, overlooking in many respects the raison d'être for the aircraft, missile, rocket, bomb, or gun having been developed. Thus, the authors have sought to provide a single-volume reference to U.S. nuclear weapons and their delivery systems.

Both authors have long worked in fields related to nuclear weapons development and have written extensively on the subject. We have benefited from the efforts of several of our colleagues in these efforts, which were the basis for this volume. We are grateful to many of them, especially William Arkin, Thomas Cochran, Hans Kristensen, and the late Chuck Hansen.

The following individuals have been especially helpful in producing this book: Dana Bell, aviation historian and author; Dr. Richard Hallion, former historian of the Air Force; Dr. John C. Lonnquest, chief historian, Army Corps of Engineers; and Dr. David A. Rosenberg, historian. Several individuals at the Naval Institute Press were especially helpful in this effort—Susan Brook, Susan Corrado, Chris Onrubia, and Patricia Bower, and the press director, Richard Russell, who encouraged us to write this book. Also, James Caiella of the Naval Institute *Proceedings* was most helpful, as was Maryam Rostamian, book designer par excellance.

Abbreviations and Acronyms

AAF	Army Air Forces (1941–1947)
AAM	air-to-air missile
ABM	anti-ballistic missile
ACM	advanced cruise missile
ADM	atomic demolition munition
AEC	Atomic Energy Commission (1946–1974)
AFAP	artillery fired atomic projectile
AFB	Air Force base
ALCM	air-launched cruise missile
AMSA	advanced manned strategic aircraft (B-1)
AN/	prefix for U.S. military electronic equipment
ANG	Air National Guard
ASM	air-to-surface missile
ASROC	anti-submarine rocket (nuclear or conventional warhead)
ASTOR	anti-submarine torpedo (Mk 45)
ASW	anti-submarine warfare
ATB	advanced technology bomber (B-2)
BOAR	bombardment aircraft rocket
BOAR	Bureau of Ordnance atomic rocket
CAT	category (of PAL)
CEP	circular error probability
EC	emergency capability
ECM	electronic countermeasures
ELINT	electronic intelligence
ER	enhanced radiation
GAM	guided aircraft missile
GAR	guided aircraft rocket
GLCM	ground-launched cruise missile
HE	high explosive
hp	horsepower
ICBM	intercontinental ballistic missile (range in excess of 3,400 statute miles)
IHE	insensitive high explosive
IM	interceptor missile
INF	intermediate-range nuclear forces
IOC	initial operational capability
IRBM	intermediate-range ballistic missile (range approx. 1,860 to 3,400 statute miles)
JSTPS	Joint Strategic Target Planning Staff
KT	kiloton (thousands of tons of TNT)
LABS	low altitude bombing system
lbst	pounds static thrust
LTV	Ling-Temco-Vought
MADM	medium atomic demolition munition
MaRV	maneuvering reentry vehicle
MiG	Soviet aircraft developed by the Mikoyan–Gurevich design bureau
MIRV	multiple independently targeted reentry vehicle
Mk	mark
Mod	modification
MRBM	medium-range ballistic missile (range approx. 620 to 1,860 statute miles)
MRV	multiple reentry vehicle

MT	megaton (millions of tons of TNT)
NATO	North Atlantic Treaty Organization (established 1949)
n.mile	nautical mile
NUCAP	nuclear cannon artillery projectile
PAL	permissive action link
psi	pounds-per-square-inch
QRA	quick-reaction alert
RAP	rocket-assisted projectile
SABMIS	sea-based anti-ballistic missile intercept system
SAC	Strategic Air Command (1946–1992)
SADM	special atomic demolition munition
SALT	Strategic Arms Limitation Talks
SAM	surface-to-air missile
SCAD	subsonic cruise armed decoy
SCM	strategic cruise missile
SIOP	single integrated operational plan
SLBM	submarine-launched ballistic missile
SLCM	ship-launched cruise missile
SRAM	short-range attack missile
SSBN	strategic missile submarine (nuclear propelled)
SSN	torpedo-attack submarine (nuclear propelled)
START	Strategic Arms Reduction Talks
SUBROC	submarine rocket
TAC	Tactical Air Command (1946–1992)
tacnuke	tactical nuclear
TEL	transporter-erector
TERCOM	terrain contour matching
TLAM(N)	Tomahawk land-attack missile (Nuclear); also TLAM-N
TM	tactical missile
ULMS	underwater long-range missile system (i.e., Trident)
USAF	U.S. Air Force (established 1947)
USMC	U.S. Marine Corps
USN	U.S. Navy
USNR	U.S. Naval Reserve
VAH	heavy attack squadron (Navy)
VC	composite squadron (Navy)
WS	weapons system (U.S. Air Force)

CHAPTER ONE

Nuclear Weapons

The mass destruction resulting from the use of nuclear weapons by the United States in August 1945 convinced political and military leaders throughout the world that these weapons could have a decisive role in the postwar world. Only one bomber carrying a nuclear weapon—if it could reach its target—would inflict the same damage that previously required thousands of bomber sorties. Nuclear weapons appeared at last to validate the words of Italian air-power advocate General Giulio Douhet, written more than two decades earlier:

> The complete destruction of the objective has moral and material effects, the repercussions of which may be tremendous. To give us some idea of the extent of these repercussions, we need only envision what would go on among the civilian population of congested cities once the enemy announced that he would bomb such centers relentlessly, making no distinction between military and nonmilitary objectives.[1]

BEFORE HIROSHIMA

The beginning of the U.S. atomic bomb program is generally dated as August 1939, when noted physicist Albert Einstein agreed to sign a letter prepared by Hungarian physicists Leo Szilard, Eugene Wigner, and Edward Teller. The letter was sent to President Franklin D. Roosevelt to attract his attention to the potentially dangerous possibility of the military use of nuclear energy by Nazi Germany.

President Roosevelt read the letter and told an aide "it required action." That action was the establishment of the Advisory Committee on Uranium consisting of the director of the Bureau of Standards and a representative each from the Army and the Navy. But the committee's report of November 1939, while briefed to the president, remained in a file, moot and inactive well into 1940. It was only when the British government informed the Roosevelt administration in the summer of 1941 of the conclusions of the British Cabinet's Maud Committee reports that described the military applications of nuclear energy and the means to achieve them that the U.S. government took the issue seriously.

In June 1940 the U.S. National Defense Research Committee (NDRC), headed by Vannevar Bush, was established to study, among other things, the feasibility of military applications of nuclear energy. The studies were conducted through November 1941. (In a reorganization, the Office of Scientific Research and Development was established in late June 1941, subsuming the NDRC and the Committee on Uranium.)

As the United States entered World War II in December 1941, President Roosevelt made the decision to initiate a program to develop and produce an atomic bomb. After lengthy discussions, on 17 June 1942, responsibility for the atomic bomb project was assigned to the U.S. Army's Corps of Engineers, and the Manhattan Engineer District was established to direct all U.S. nuclear weapons development. That September, Army Colonel Leslie R. Groves was appointed to direct the "Manhattan Project."[2]

Groves, with Dr. Robert Oppenheimer as his scientific director, managed the massive Army–Navy–civilian effort that produced the first nuclear weapon in July 1945.

---

1 Gen. Giulio Douhet, *The Command of the Air* (Washington, D.C.: Office of Air Force History, 1983), p. 20. This is a reprint/translation of Douhet's book by the same title published in 1921 under the auspices of the Italian Ministry of Defense.

2 See Lt. Gen. Leslie R. Groves, *Now It Can Be Told: The Story of the Manhattan Project* (New York: Harper, 1962); Robert S. Norris, *Racing for the Bomb: General Leslie R. Groves, the Manhattan Project's Indispensable Man* (South Royalton, Vt.: Steerforth, 2002); and Richard Rhodes, *The Making of the Atomic Bomb* (New York: Simon and Schuster, 1988).

Lieutenant General Leslie R. Groves, USA. (U.S. Army)

That "gadget"—a prototype Fat Man atomic bomb—was successfully detonated at Alamogordo, New Mexico, on 16 July 1945, in history's first man-made nuclear detonation.

Under the direction of Groves, the preparation of special bases was already under way on Tinian in the Marianas. In the United States, the 509th Composite Group flying specially modified Boeing B-29 Superfortress bombers was preparing for the delivery of atomic bombs on Japan. That unit had been activated on 17 December 1944, under the command of Lieutenant Colonel Paul W. Tibbets.[3]

THE FIRST ATOMIC WEAPONS

The beginning of the atomic age was dominated by the fact that only the United States had nuclear weapons and the means to deliver them. The initial delivery method was by four-engine B-29 Superfortress bombers, the only delivery platform that could carry such heavy weapons (approximately five tons for the early bombs) to targets more than a thousand miles from friendly bases. Thus, the initial responsibility for strategic warfare within the U.S. armed forces fell within the province of the Army Air Forces (after 1947, the U.S. Air Force).

The U.S. nuclear laboratory at Los Alamos, New Mexico, produced a Fat Man (plutonium) bomb in July 1945, which was test detonated atop a tower that month at Alamogordo, New Mexico. Subsequently, the laboratory produced one Little Boy (uranium) and two Fat Man bombs for use against Japan during August; the Little Boy was dropped on Hiroshima by the B-29 *Enola Gay* on 6 August and a Fat Man on Nagasaki by the B-29 *Bockscar* on 9 August. Components for the second Fat Man bomb, ready for overseas shipment, was retained in the United States when, in mid-August, it appeared that Japan was about to surrender.[4]

The U.S. atomic bomb availability was estimated to be three weapons in September 1945, with the production increasing to seven or more weapons per month in December 1945.[5] However, with the end of hostilities, atomic bomb production slowed precipitously, in part because of the rapid demobilization of civilian and military scientific personnel and the general postwar cutbacks.

At the time, the B-29 was the only aircraft capable of carrying a Fat Man or Little Boy atomic bomb, but only a few of the hundreds of B-29s were nuclear capable. When the 509th Composite Group deployed to Tinian in May 1945, the unit had 15 B-29s that had been extensively modified under the "Silverplate" program. After the war additional B-29s were modified for nuclear weapons delivery under this program. In January 1946, the AAF designated the 58th Very Heavy Bombardment Wing at March Field, California, as its atomic striking force: the wing consisted of three bomb groups—the 40th (March

Colonel Paul W. Tibbets, USA. (U.S. Army Air Forces)

3 The date was the 41st anniversary of the Wright brothers' first flight. The term "composite" indicated that the group would be self-sufficient, that is, capable of carrying out its mission without additional support. After World War II, the unit's designation was changed to 509th Bomb Group.

4 That bomb was scheduled for drop on a Japanese city after 17 or 18 August 1945, the exact date to be determined by weather conditions. Maj. Gen. Leslie R. Groves, USA, Memorandum to Chief of Staff, U.S. Army, 10 August 1945.

5 Adjutant General, War Department, message to Secretary of War Henry Stimson, 24 July 1945, No. WAR-37350. At the time Mr. Stimson was with President Truman, attending the Tripartite Conference in Babelsberg, Germany.

The B-29 *Enola Gay* returning to its base at Tinian after the Hiroshima mission. The "circle R" insignia was that of the 6th Bombardment Group; the 509th Composite Group's insignia was an arrowhead in a circle, but the planes usually flew over Japan with the insignia of other units. The name "Enola Gay"—for Paul Tibbets' mother—was painted on the aircraft's left side the day before the mission. (U.S. Army Air Forces)

Field), 444th (Castle Field, California), and 509th (Roswell Field, New Mexico). At that time the total bomber force of the Strategic Air Command (SAC) was 148 B-29s.

The number of B-29s modified for nuclear weapons delivery has been estimated at

Sep 1945	46 aircraft
Jan 1946	29 aircraft
Dec 1946	23 aircraft
Mar 1947	35 aircraft
Apr 1947	46 aircraft
June 1948	32 aircraft
Dec 1948	38 aircraft
Jan 1949	66 aircraft
Jan 1950	95 aircraft

Official sources differ on exact numbers, and not all aircraft were fully operational (i.e., only 20 of those on the AAF roster in April 1946). And there were never fully trained crews available for all of these aircraft nor were nuclear weapons available for all of these aircraft.

During the late 1940s, several new bomber aircraft were introduced into the U.S. strategic arsenal. First came the B-50, an improved version of the piston-engine B-29, and then the long-range Convair B-36 Peacemaker, a piston-engine giant with a takeoff weight of more than 300,000 pounds that could, under ideal conditions, fly a round-trip of more than 8,000 miles with a single early nuclear weapon. The maximum bomb load of a B-36 was 86,000 pounds, which reduced its range to 3,400 miles.[6]

A large force of nuclear-capable B-29, B-50, and B-36 bombers was in service by 1950, and U.S. perceptions of this striking power were considerable. In May 1947, for example, a force of 101 B-29s flew over New York City in a "maximum mission effort" to demonstrate the striking power of the U.S. nuclear arsenal. This was one-third of the existing strategic bomber strength; another 30 B-29s planned for the mission remained at their bases because of maintenance and supply problems. In a comparably impressive demonstration, a single B-50 flew the first non-stop flight around the world in March 1949 without landing; it was refueled four times in flight by B-29 tanker aircraft. These flights, as well as overseas flights by small

6 Convair was derived from the Consolidated Vultee Aircraft Corporation; the firm became the Convair Division of the General Dynamics Corporation in April 1954.

The most comprehensive source of data on post–World War II bomber aircraft is Marcelle Size Knaack, *Post–World War II Bombers 1945–1973*, vol. II in *Encyclopedia of U.S. Air Force Aircraft and Missile Systems* (Washington, D.C.: Office of Air Force History, 1988). Force levels are based primarily on Office of the Historian, Headquarters, Strategic Air Command, *Alert Operations and the Strategic Air Command 1957–1991* (Offutt AFB, Neb. [n.d.]). See Appendix C.

7 SAC was established as an Air Force command on 21 March 1946; subsequently, it additionally became a specified command of the Department of Defense.

detachments of B-29s, were intended for training as well as to demonstrate capabilities of the SAC, which operated the U.S. strategic bombers.[7]

The first direct confrontation with the Soviet Union in which the SAC was directed to prepare for a nuclear strike was the Berlin crisis of June 1948, when the Soviets interdicted Allied ground and canal transit to the city. One B-29 squadron was already in West Germany on a training flight; a short time later, additional B-29 squadrons were flown into West German and British air bases. Some 90 B-29s were forward-based in Europe during the crisis period. The B-29s sent to Europe in 1948 were *not* nuclear capable. At that time SAC had only 32 B-29s capable of delivering nuclear weapons. All of them were assigned to 509th Bomb Group, and none was deployed to Britain until the summer of 1949.

At the height of the Berlin crisis—in September 1948—there were about 50 Mk 3 weapons available, those being the only U.S. atomic bombs produced from 1946 to 1948. (Nonnuclear components of atomic bombs were not sent to Britain until 1950; nuclear bombs were not stored in Britain until 1954.)

Had the decision been made to strike the Soviet Union or East Germany with nuclear weapons, several hours would have been required to assemble the bombs, and several more to load and ready the aircraft at bases in the eastern United States. The B-29s would then have to stage through bases in Britain to be refueled before going on to their targets. This was a ponderous process but still constituted the world's only nuclear strike capability. Airlift rather than atomic bombs became the Allied strategy for this "Battle for Berlin." Consideration of U.S. nuclear strike options did, however, point up a number of limitations in SAC capabilities, among them: (1) the limited ranges of aircraft, (2) the inability to store nuclear weapons overseas (even in Britain), (3) the lack of ready overseas bases, and (4) the time and specialized personnel required to assemble and load nuclear weapons.

The U.S. monopoly of nuclear weapons ended on 29 August 1949, when the Soviet Union detonated its first atomic bomb, an implosion device based largely on the U.S. Fat Man design. This event was a milestone on the Soviet path to becoming a superpower, and set the stage for a highly energetic arms race that would continue for four decades. But it would be several years before the Soviet Union had nuclear weapons in large numbers or the means to deliver them against the United States.

The first Soviet aircraft capable of carrying an atomic bomb was the Tupolev Tu-4 (NATO code name Bull). Three U.S. B-29 Superfortress bombers had landed in Soviet Siberia in 1944 after bombing raids against Manchuria and Japan. These aircraft were copied by the Soviets in remarkably short time, and the Soviet-built Tu-4 first flew late in 1946. The aircraft was soon in series production, and some 1,200 Bulls were delivered during the next few years.

The Tu-4 provided a strategic bomber to *Aviatsiya Dalnovo Deistviya* or Long-Range Aviation (LRA). With a bomb load of 5½ tons, the Bull could fly 3,000 miles at a cruising speed of some 225 mph. This meant that by taking off from bases in northern Russia and Siberia, the planes could reach some targets in the northern United States on a 13-hour, one-way flight. The limited number of Soviet bombers available through the 1950s, the duration of their flight, the system of U.S. radar warning stations that stretched across the Canadian Arctic and into the Atlantic and Pacific oceans, and the manned interceptor aircraft of the United States and Canada reduced the potential effectiveness of a Soviet nuclear strike capability against the United States during the decade of the 1950s.

But this limited Soviet nuclear strike capability and the subsequent Soviet detonation of a thermonuclear device in 1953 spurred additional U.S. strategic weapons developments. The United States exploded the first nuclear fusion or thermonuclear device—the precursor of a deliverable hydrogen bomb—on 1 November 1952, during Operation Ivy. The experimental thermonuclear device detonated in event Mike, a 10.4 megaton (MT) explosive force or yield. The Mike detonation vaporized the island of Elugelab in Eniwetok atoll in the Marshall Islands.

The device was a "proto-bomb" contained in a cylindrical steel structure about 20 feet high, varying from 8 to 10 feet in diameter, and weighing 82 tons. It was a two-stage "weapon" using a TX-5 fission bomb as the primary stage with the secondary stage consisting of liquid deuterium fusion fuel stored in a cylindrical Thermos-like flask. An estimated 77 percent of the yield (8 MT) was due to the natural uranium pusher/tamper, with the remainder (2.4 MT) coming directly from fusion of the deuterium fuel. (The Soviet Union detonated its first thermonuclear device on 12 August 1953; although not a modern two-stage design, it was a bomber-deliverable bomb.)

Mastery of fusion technology would have far-reaching implications in the arithmetic of nuclear arms. The significance of the thermonuclear or hydrogen bomb lies in

This was the 31 October 1952 detonation of the world's first "H bomb"—known as the Mike device. The nuclear fusion or thermonuclear device had an explosive force or "yield" of 10.4 megatons. But the device was not a bomb; the Soviet thermonuclear weapon that was detonated on 12 August 1953, was a bomber-deliverable weapon. (U.S. Air Force)

the enormous energy-releasing efficiency of fusion as compared with fission. Whereas fission weapons are measured in terms of kilotons (KT), equivalent to thousands of tons of TNT, larger fusion weapons could be discussed only in terms of megatons (MT), equivalent to millions of tons of TNT. A 1-MT thermonuclear warhead possesses about 50 times the explosive power of the A-bomb used at Nagasaki but weighs only half as much and is more compact.

JET-AGE BOMBERS

In 1951 the first large B-36s became operational in SAC (ten years after the first contract and three years after first aircraft delivery) and the first jet-propelled B-47s were being delivered. The B-36 was big and slow (440 mph maximum), but could reach the Soviet Union from U.S. bases without in-flight refueling. It was a large airplane (almost 360,000 pounds gross for the D model), powered by six piston engines, later supplemented by four jet engines in wing pods. Its size and cost made it the most-publicized bomber of the immediate postwar era. Its development was predictable, inasmuch as the Air Force continually sought larger bombers that could lift greater payloads of bombs and fuel. The B-36 was also heavily armed, fitted with 16 20-mm cannon in twin, remote-controlled turrets.

This heavy gun armament personified Douhet's declaration,

> What determines victory in aerial warfare is fire power. Speed serves only to come to grips with the foe or to flee from him, no more. A slower, heavily armed plane, able to clear its way with its own armament, can always get the best of the faster pursuit plane. A unit of combat composed of slower, heavily armed planes is in a position to stand up to the fire of enemy pursuit planes and carry out its mission successfully.[8]

The six-turbojet Boeing B-47 could accelerate to 630 mph carrying a nuclear weapon, but lacked the range to reach the Soviet Union from bases in the United States. Accordingly, several SAC bases were established in Britain, Spain, Libya, and French Morocco in 1950–51, and large numbers of KC-97 tanker aircraft were procured to provide the B-47s with in-flight refueling.

8 Douhet, *The Command of the Air*, p. 44. The B-17 Flying Fortress had up to 13 .50-caliber machine guns, while the B-29 Superfortress had up to 12 .50-caliber machine guns and 1 20-mm cannon.

The sleek B-47 propelled the U.S. Air Force into the jet bomber age. Designed by the Boeing Aircraft Company, which had produced the famed B-17 Flying Fortress and B-29 Superfortress bombers, the new aircraft was one of several designs considered at the end of World War II for the initial generation of jet-propelled bombers. After studying a number of configurations, Boeing engineers took advantage of German research into swept-back wings to produce the final B-47 design. The pilot and copilot sat in tandem in the narrow fuselage with a navigator-bombardier in the nose, a total of 3 men compared to 10 for the B-17 and 11 or 12 for the B-29, both smaller aircraft. Similarly, the B-47 had only two 20-mm tail guns for self-defense. Space and weight were instead allocated to fuel, with the B-47 carrying almost ten times the B-17's fuel load and close to three times that of the B-29. The distinctive swept wings of the B-47 carried six turbojet engines in a twin and single pod on each side. They could push the B-47E to just over 600 mph (Mach 0.84 at 16,300 feet), and the plane could carry two nuclear weapons (10,000 pounds) to a combat radius of 2,000 miles, and farther with in-flight refueling.

The B-47s began joining SAC squadrons in 1951, with the ultimate B-47E version making its first flight in 1953. B-47 production totaled 2,060 aircraft, including several hundred EB-47 electronic warfare and RB-47 reconnaissance variants. The B-47 was the most numerous aircraft in SAC during the early 1960s and was produced in greater numbers than any jet-propelled bomber except possibly the Soviet Tu-16 (NATO code name Badger).[9] In quantity and quality, the B-47 was a milestone in the development of strategic bombers and gave the United States a potent nuclear strike capability if overseas bases and sufficient aerial tankers were available.

The Soviets, ever defensively oriented and fully aware of U.S. reliance on nuclear strike aircraft, in the same period revealed a 500-mph jet fighter-interceptor that could operate at altitudes up to 50,000 feet. This was the MiG-15 (NATO code name Fagot), a swept-wing fighter that could make bombing missions by the B-36s and B-47s, which flew at lower altitudes, extremely hazardous. Beginning in 1947, some 15,000 to 18,000 MiG-15s were built, more than twice the number of F-86 Sabre fighters produced, which was the most numerous Western jet fighter design.

In the immediate postwar era the U.S. Navy was without a "strategic" mission. The Navy had emerged from World War II as the unquestioned ruler of the seas. No nation, friend or late foe, had any warships capable of seriously interfering with U.S. use of the sea. The "potential" enemy, the Soviet Union, had neither a fleet nor a reason to use the sea. Not until Korea could the U.S. Navy demonstrate that it still was needed in the atomic age, albeit in support and projection missions (rather than for control of the sea). If future national strategy was to be based on nuclear weapons, those in existence were too large for carrier-based planes, let alone for long-range naval guns, the largest of which was only able to shoot a projectile 16 inches in diameter some 23 miles.

In searching for a role in the nuclear era, the Navy sought the means to deliver nuclear weapons against the Soviet Union or other future enemies. Shipboard and submarine test-launchings were conducted with captured German V-2 rockets and V-1 jet-propelled missiles. But these efforts yielded no practical result until the mid-1950s. More immediately, experiments were undertaken with the P2V Neptune, a twin-engine, long-range patrol/reconnaissance bomber. Neptunes were used in test launches from carrier flight decks, and 12 aircraft were modified to the P2V-3C configuration to carry nuclear weapons. The Neptunes could be launched from aircraft carriers but could not land aboard them.

As previously noted, the Korean War, which began in June 1950, shattered the illusions of the nuclear-capable bomber as the final arbiter of modern war. On 19 October 1949, General Omar N. Bradley, chairman of the U.S. Joint Chiefs of Staff, had addressed the question of conventional-versus-strategic actions in these words: "I am wondering whether we shall ever have another large-scale amphibious operation. Frankly, the atomic bomb, properly delivered, almost precludes such a possibility."[10]

9 Approximately 2,000 Badgers were built in the Soviet Union through the mid-1960s plus more than 100 produced in China designated H-6D. The Tu-16 Badger was a swept-wing bomber roughly comparable in size, role, and performance to the B-47. The Badger, which first flew in late 1952, has only two engines, each developing an estimated 19,180 pounds of thrust, as compared to 7,200 pounds thrust for each of six engines in the B-47E—a manifestation of the Soviet tendency to "build big."

10 Gen. Omar N. Bradley, USA, Chairman, Joint Chiefs of Staff, testimony before the Committee on Armed Services, House of Representatives, 19 October 1949 (Unification and Strategy hearings).

In-flight refueling of U.S. strategic bombers to enable them to reach the Soviet Union and return to base or, at least, to neutral nations. Here a KC-97 Stratofreighter—the first purpose-built U.S. tanker aircraft—refuels a SAC B-47B Stratojet bomber. Note the provision for rocket boosters in the fuselage of the B-47 (just forward of the national insignia). (Boeing Airplane Co.)

The Korean War nevertheless included several amphibious landings. (The invasion force at the September 1950 amphibious Inchon landing consisted of a U.S. Marine division, an Army infantry division, most of a Marine aircraft wing plus supporting troops, all carried and supported by an armada of more than 230 ships.) As conventional war raged on the Korean peninsula, the theater commander, General of the Army Douglas MacArthur, asked that nuclear weapons be employed to halt the masses of Chinese "volunteers" that entered North Korea in the winter of 1950–1951. President Truman (and other U.S. leaders) were against using nuclear weapons. There was a fear in Washington that American use of nuclear weapons would widen the war, possibly bringing Soviet forces directly into the conflict or precipitating Soviet action in Europe. Other factors included the repeated use of nuclear weapons against an Asian people, the credibility of the war as a United Nations action if a weapon that only the United States possessed were employed, the difficulty of determining targets that warranted the devastation of nuclear weapons, and the possibly limited number of nuclear weapons in the U.S. arsenal. Reportedly, General MacArthur argued for employing between 30 and 50 bombs. That would have been a significant portion of the nuclear weapons available at the time.

During the Korean War, U.S. striking power in Europe was increased against fear that Korea was merely a communist diversion in Asia, while the main blow was about to fall in Europe. SAC began rotational training flights of squadrons to overseas bases in order to provide forward deployments of bombers. Most of these flights were sent to Britain, but SAC bombers also operated from bases in French Morocco, Libya, Guam, and Japan. (B-29s participating in the Korean War, which were not nuclear capable, flew from bases in Japan and Okinawa.)

STRIKE FROM THE SEA

By the early 1950s the U.S. Navy was ready to make an initial contribution to the nation's nuclear strike capability. In February 1951—eight months after the Korean War began—six of the new North American AJ-1 Savage (piston) and three Lockheed P2V-3C Neptune (piston) bombers flew across the Atlantic to Port Lyautey (now Kenitra), Morocco. Beginning in March 1951, the Savages periodically flew from two of the large carriers of the *Midway* (CVA 41) class that were then operating in the Mediterranean.

Both of the aircraft carriers had nuclear weapons on board, having been the first ships modified for handling atomic bombs. The Savages could operate from carriers, but the larger Neptunes had to remain ashore because they required dockside loading by large cranes. Thus, if nuclear war appeared imminent, the Savages would land on the carriers, refuel, load nuclear weapons, and take off when the carriers came within range (1,600 miles on a one-way trip) of Soviet targets. The bombs were assembled aboard the carriers in special, highly restricted spaces by large numbers of technicians who required about 24 hours to put the bombs together. Still, the aircraft carriers were sovereign U.S. territory, fully independent of the complex considerations involved in a nuclear-armed aircraft taking off from a land base on foreign soil.

Subsequently, additional aircraft carriers, including the smaller ships of the *Essex* (CVA 9) class, were refitted to handle nuclear weapons. The AJ Savage, and later A3D Skywarrior (jet) attack planes, were added to their normal air groups, up to a dozen planes on the larger carriers, and four planes on each of the *Essex*-class ships.

The Korean War was not a communist diversionary tactic, and no direct U.S.-Soviet confrontation occurred during the conflict, although large numbers of Soviet-piloted MiG-15s participated in the Korean War. U.S. military planners did consider using atomic bombs in the Far East during the winter of 1950–1951, when communist Chinese troops threatened to destroy the U.S. ground forces in North Korea, and again in the spring of 1954, when communist Viet Minh units surrounded French ground forces at Dien Bien Phu in Indochina.

B-29s or B-50s based on American-controlled Okinawa would have been used to strike Chinese troops massing in Manchuria or entering Korea across the Yalu River. The plan for strikes in Indochina presented more complex problems. Initial proposals involved sending large numbers of B-29s from the Philippines or carrier-based planes to pound Viet Minh positions around Dien Bien Phu with conventional bombs. As the situation worsened for the French, consideration was given to employing nuclear weapons. The sovereignty of the Philippines prevented loading nuclear weapons at U.S. bases in that island nation. Okinawa and Guam were too far away for the effective use of the B-50s. On the other hand, at least two of the U.S. Navy's eight aircraft carriers in the western Pacific had nuclear weapons on board, and the squadron of AJ Savages then based in Japan could have rapidly flown out to the carriers to provide a nuclear strike capability.

For a number of reasons, the U.S. government decided against direct intervention with either conventional or nuclear bombs. The reasons were similar to those that led to the decision not to use nuclear weapons in Korea. Dien Bien Phu fell in May 1954, and the First Indochina War soon came to an end, creating the conditions for U.S. entry into the Vietnam conflict in the 1960s. Thus, U.S. military planners did consider using nuclear weapons on three occasions during the periods of U.S. monopoly and overwhelming superiority in nuclear weapons: the Berlin crisis of 1948, the Korean War, and Dien Bien Phu. During

The U.S. Navy developed carrier-based nuclear strike aircraft beginning in the late 1940s. Here 12 AJ-1 Savages of Heavy Attack Squadron (VAH) 9 crowd the flight deck of the carrier *Coral Sea* (CVB 43). The subsequent adoption of the British-developed angled flight deck made operations by such large as well as faster jet-propelled aircraft more feasible and much safer. (U.S. Navy)

those years, the U.S. nuclear strike capability increased from a handful of bombs deliverable by piston-powered B-29s to—by the end of 1955—just over 1,300 nuclear-capable strategic bombers:

205	B-36 Peacemakers
1,086	B-47 Stratojets
18	B-52 Stratfortresses

Added to this force was the Navy's small but growing carrier-based nuclear strike capability. The Navy had lost in its effort to construct a huge "supercarrier" in 1949. Secretary of Defense Louis Johnson and, to a lesser extent, General Dwight D. Eisenhower (then president of Columbia University) had the decisive influence on a split Joint Chiefs of Staff on the issue of whether to build the ship or alternatively, by implication, to emphasize land-based strategic bombers. The Navy continued to argue for a sea-based nuclear strike capability beyond the few AJ Savages aboard its existing carriers. Navy spokesmen contended that "we cannot safely place reliance on any single weapon or weapon system, but must carry a relatively 'full bag'—must keep them versatile and adaptable to any situation." The words were similar to those that would be used two decades later by the proponents of the so-called Triad strategic weapons concept, who were arguing against increased reliance on sea-based strategic weapons.

The Korean War demonstrated that navies and, especially, aircraft carriers were still needed in the atomic age, and a program of large carrier construction was accelerated, initially at the rate of one supercarrier per year. The first, the USS *Forrestal* (CVA 59), was completed in 1955. The *Forrestal* and later large or supercarriers further increased U.S. forward-deployed nuclear strike capabilities. Also, numerous fighter and attack aircraft embarked in those ships, including the outstanding AD Skyraider single-engine (piston) aircraft, were configured to deliver nuclear weapons. The AD Skyraider was unquestionably the best piston-engine attack aircraft ever produced by any nation. Vice Admiral Gerald E. Miller later wrote that in the nuclear strike role

> the propeller-driven [Skyraider] had a high probability of achieving the desired results—better than most. It was a reliable airplane, and it traveled at low level to the target, giving it a high penetration reliability. In addition, the targets to which it was assigned were generally elements of early warning air defense systems such as radars, which were on the periphery of the vast target area.[11]

U.S. NUCLEAR SUPERIORITY

When President Dwight D. Eisenhower entered the White House in January 1953, the Cold War had been under way for several years. The threat of Soviet aggression against western Europe, still recovering from the destruction of World War II, was of widespread concern, having led to the establishment of the North Atlantic Treaty Organization (NATO) in April 1949.

Eisenhower had served as the first NATO Supreme Allied Commander in Europe. His planners told "Ike" that to resist a Soviet onslaught into western Europe, the NATO allies must field 96 divisions. (By comparison, at the peak of World War II the U.S. Army had 89 divisions and the Marine Corps had 6.) The situation was much the same when Eisenhower became president and, knowing that NATO could not build large conventional ground forces, he initiated the "new look" strategy that stressed large nuclear bombs, the production of large numbers of tactical nuclear weapons, and extensive covert operations by the Central Intelligence Agency against the Soviets and their allies. The emphasis on nuclear weapons was not only reflected in the number of weapons produced over the next few years but also in the proliferation of different nuclear weapon systems in the U.S. armed forces and the beginning of a sustained airborne bomber alert by the SAC—Operation Chrome Dome.

The advent of smaller fission and fusion weapons led to the development of U.S. Air Force and Navy fighter/attack-type aircraft that could deliver "tactical" nuclear weapons. The first such weapon was the B5 lightweight bomb, which became available in 1952. In the Air Force, these new nuclear-capable tactical aircraft began with the F-100 Super Sabre "fighter," while the Navy's first new nuclear plane was the piston-engine AD Skyraider and the diminutive (11-ton gross weight) A4D Skyhawk. Although these fighter/attack aircraft were considered as "tactical" in U.S. military semantics, an F-100 based in

11 Vice Adm. Jerry Miller, USN (Ret.), *Nuclear Weapons and Aircraft Carriers: How the Bomb Saved Naval Aviation* (Washington, D.C.: Smithsonian Institution Press, 2001), pp. 214–15.

West Germany or an A4D aboard an aircraft carrier in the eastern Mediterranean could deliver a nuclear weapon against cities in the Soviet Union, thus blurring efforts to make clear distinctions between tactical and strategic nuclear delivery systems.[12]

The U.S. Navy operated 14 to 16 attack carriers in the post-Korea period, each provided with 50 or more nuclear strike aircraft: up to one squadron of large Douglas A3D Skywarrior or, later, the North American A3J Vigilante attack aircraft, and two or three squadrons of smaller Douglas-built AD Skyraider and A4D Skyhawk bombers. For a brief period, most carrier-based fighter aircraft could also deliver nuclear weapons. During periods of crisis, the carriers in the Mediterranean and Western Pacific would off-load some of their fighters and embark additional strike aircraft. For example, in 1960–1961 the carrier *Coral Sea* (CVA 43) operated in the Western Pacific with 83 attack aircraft capable of delivering nuclear weapons (plus 3 electronic warning aircraft and 1 cargo aircraft).

Possibly more significant on a long-term basis, the advent of smaller and lighter nuclear warheads meant that the delivery of weapons of mass destruction was feasible with missiles. In the United States, this potential was partially fulfilled with the development and limited deployment of low-altitude, air-breathing "cruise" or guided missiles. This concept was an outgrowth of the German V-1 or "buzz bomb," with the missiles following a preset flight path to a fixed target; the lack of terminal accuracy in these weapons was compensated for by the high destructive force of the nuclear warhead.

These air-breathing cruise missiles included the Air Force's Matador, with a 600-nautical-mile range (which was renamed Mace in later variants, with a range of 1,200 miles in the final Mace-B missile), and Snark, with a 5,000-nautical-mile range; and the Navy's submarine-launched Regulus I missile, with a range of 500 nautical miles. The Matador/Mace missiles were forward-based in Europe and in the western Pacific, while the limited number of Snark missiles procured were based—for a matter of months—at Presque Isle, Maine. The Navy adapted two World War II–built submarines to carry two Regulus missiles each, and then built three additional submarines, one of which was nuclear powered, to each carry four or five missiles. With only one or two submarines (four or five missiles) at sea at any given time, however, the Navy's Regulus force added comparatively little megatonnage to the nation's nuclear striking forces.

The development of small nuclear warheads also permitted the U.S. Army to deploy "tactical" nuclear weapons. These included nuclear rounds for 280-mm and 8-inch (203-mm) guns, the Honest John and Redstone battlefield missiles, and the very short-range Davy Crockett, a small, battalion-level nuclear weapon with a range measured in yards and a W54 warhead producing a one-fourth kiloton of explosive power. Small nuclear warheads were also employed in land mines—atomic demolition munitions (ADM) developed for use by the Army and Marine Corps. Their primary use was to detonate the small nuclear devices in mountain passes and to destroy massive bridges to stop the advance of Soviet troops into western Europe, or to force them into advancing on routes where they would be more vulnerable to NATO forces. Some of the ADMs were man-portable. (The British Army also sponsored the development of a nuclear land mine—Blue Peacock—beginning in 1954; the project was abandoned after about four years of development effort.)

U.S. Army interest in nuclear weapons transcended the battlefield, leading to the Jupiter intermediate-range ballistic missile (IRBM).[13] This was a true rocket-propelled missile, developed like the Redstone, employing the genius of Wernher Von Braun and other former German missile scientists who had developed the V-2 antecedent to these missiles. The Jupiter—with a range of 1,500 miles—was intended for emplacement in NATO nations for strikes against the Soviet Union; it was in fact a strategic weapon.

Simultaneously, the U.S. Air Force was belatedly developing an interest in ballistic missiles and initiated the Thor IRBM (1,500 miles) and the Atlas and Titan intercontinental ballistic missiles (ICBM), with initial ranges of 5,500 and 6,300 miles, respectively. Interservice rivalry led to a secretary of Defense decision on 26 November 1955, that the Air Force would operate all IRBMs and

12 The most comprehensive source of data on post–World War II fighter aircraft is Marcelle Size Knaack, *Post–World War II Fighters 1945–1973*, vol. I in *Encyclopedia of U.S. Air Force Aircraft and Missile Systems* (Washington, D.C.: Office of Air Force History, 1978).

13 A ballistic missile relies on the initial thrust of its engine to launch it into a ballistic trajectory; propulsion is not sustained in flight. A ballistic missile does not have wings for lift and relies on tail fins or engine thrust for directional control.

The brilliant aircraft designer Ed Heinemann produced the largest carrier-based aircraft, the A3D Skywarrior (later A-3), and the smallest, the A4D Skyhawk (later A-4). Both were designed specifically for the nuclear strike role and both served in several roles in conventional conflicts. This A4D-2, carrying a nuclear "shape" and two drop tanks, is about to be catapulted from the carrier *Franklin D. Roosevelt* (CVB 42). (U.S. Navy)

ICBMs, ending Army participation in long-range missile programs. The Army would have operational control only of weapons with a range of 200 miles or less.

The first six Atlas ICBMs became operational in 1959 in the United States, while the Thor IRBMs were installed in Great Britain under joint UK-U.S. control, and later the Jupiter IRBMs in Italy and Turkey. These were all above-ground installations, vulnerable to enemy air or missile attack.

In retrospect, the decision to assign IRBMs to the Air Force, which was already developing the longer-range ICBMs as well as the Thor IRBM, cannot be argued. The IRBM and ICBM missions—from the American perspective—were the same and they shared the characteristics of being land-based weapons with similar command, control, and basing issues. The Navy was charged with developing sea-based ballistic missiles, with the early efforts to be collaborative with the liquid-propellant Jupiter IRBM intended for launching from surface ships and possibly from submarines. However, the Navy abhorred the use of liquid-propellant missiles, especially in submarines.

TACNUKES AT SEA

With the advent of small nuclear warheads, the Navy also put to sea an array of tactical nuclear weapons for anti-submarine warfare (ASW) and for air defense against Soviet bomber strikes on carrier task groups. In the ASW role, nuclear weapons could compensate for the uncertainty of a submarine's precise location, which was needed for some weapons, and the use of decoys and jamming devices that could be used against ASW homing torpedoes. In the anti-air role a single nuclear weapon could destroy multiple attacking aircraft or cruise missiles.

The first nuclear ASW weapons were air-dropped nuclear depth bombs: the "Betty" with the Mk 7 warhead, which became available in 1955; the "Lulu" with the W34 warhead (1958); and the B57 bomb (1963). They could be carried by a variety of carrier-based and land-based ASW aircraft, including helicopters. These nuclear depth bombs were also prepositioned in several NATO countries—under U.S. control—for use by allied ASW aircraft.

In addition, to provide surface warships with a "long-range" delivery system, the *unmanned* DASH—drone anti-submarine helicopter—could carry a B57 bomb as well as conventional torpedoes.[14] Deployed in the 1960s, the DASH could be carried aboard cruisers, destroyers, and frigates considered too small for manned helicopters.[15] The radius of DASH was some 30 nautical miles, limited by the ship's tracking radar. The helicopter carried no sensors, relying on guidance to the target from the ship and being told when to release either an ASW homing torpedo or the B57 bomb.

14 The DASH vehicle was formally designated DSN and, after 1962, QH-50.

15 In the same period the Royal Navy was deploying the relatively large Sea King ASW helicopter in destroyers.

An Atlas E intercontinental ballistic missile is raised from its protective magazine during an exercise. Liquid oxygen is seen venting from the missile. Although the United States lagged behind the Soviet Union in the development and testing of ICBMs, and especially in throw weight, the U.S. weapons proved to be more robust and more reliable. (U.S. Air Force)

U.S. warships were also fitted with ASROC (anti-submarine rocket), a ship-launched weapon of shorter range—about seven miles. This unguided, ballistic-trajectory weapon delivered either a rocket-boosted, conventional homing torpedo or a W44 nuclear depth bomb. Obviously, both DASH and ASROC had ranges far shorter than Soviet anti-ship cruise missiles that were at sea in the 1960s.

U.S. submarines carried the Mk 45 ASTOR (anti-submarine torpedo) beginning in 1958. This was the only nuclear torpedo developed in the West (the Soviets developed several torpedoes with nuclear warheads). The Mk 45 was wire-guided by the launching submarine with the torpedo having no homing guidance; the warhead was detonated by (wire) command from the submarine, having no contact or influence exploder. The ASTOR carried a W34 nuclear warhead. In a form of gallows humor, some naval officers estimated that it had a probability of kill (Pk) of 1.5: The one was the enemy submarine—the half was the launching submarine; a few believed that the ASTOR's Pk was closer to 2—both submarines![16]

The U.S. Navy deployed two nuclear surface-to-air missiles, the Terrier and Talos, both of which also had conventional warheads. Intended for use against Soviet bomber formations, the Talos was a large missile, some 7,000 pounds, that became operational in 1959 and was fitted in seven cruisers. It had a range of 70 miles and could reach approximately 85,000 feet carrying a W30 warhead with yields up to 5 KT.

The smaller Terrier went aboard scores of cruisers and destroyers, the BTN—beam-riding Terrier nuclear—variant from 1962. It carried a W45 warhead with yields up to 5 KT and was credited with a range of 23 miles and an altitude of 80,000 feet. Ship-launched Talos and Terrier missiles were both used in the Vietnam War with conventional warheads.

The U.S. Navy considered but did not develop a nuclear sea mine. (The Soviet Navy did develop at least one type of nuclear mine; up to four of those weapons were planned for shipment to Cuba prior to the missile crisis of October 1962. In 1954 the British Naval Staff proposed development of a nuclear mine—code-named Cudgel—

16 Two Mk 45 torpedoes were on board the U.S. nuclear-propelled attack submarine *Scorpion* (SSN 589) when she sank in the eastern Atlantic on 27 May 1968.

to be used by midget submarines to attack Soviet harbors, especially Kronshtadt in the Gulf of Finland, a major base of the Soviet Baltic Fleet. The nuclear mine was to be adopted from the Red Beard aerial bomb weighing 1,750 pounds with a 15-KT yield.)

A major breakthrough in nuclear weapons at sea—for the land-attack role—occurred in 1987 when the Tomahawk land-attack missile/nuclear (TLAM/N) became operational. The Tomahawk cruise missile went to sea in 1983 in the anti-ship configuration, followed by the conventional TLAM version in 1984. With a range of almost 1,000 miles in the nuclear configuration carrying a W80 warhead, the Tomahawk could be launched by nonspecialized, torpedo-attack submarines (SSN), destroyers, cruisers, and battleships. (The conventional versions have a 1,000-pound warhead.) Initially fitted with contour-matching and then with more advanced guidance systems, the missile provided the Navy with a widely distributed, affordable, long-range, and accurate strike weapon.

EMPHASIS ON BOMBERS

Despite the variety of cruise and ballistic missile programs undertaken by the U.S. military services during the 1950s, primary U.S. attention was focused on strategic bombers. The U.S. Air Force had reached a peak strategic bomber strength in 1959 with 1,366 B-47s and 488 B-52s, in addition to 174 RB-47 reconnaissance aircraft, which were supported by more than 1,000 KC-97 and KC-135 tanker aircraft. These SAC bombers were the mainstay of U.S. strategic forces, and the nation's available cruise and ballistic missiles and carrier-based bombers paled in comparisons of numbers and megatonnage of the weapons carried in the bombers. For example, the six Atlas D missiles operational at the end of 1959 could each lift an estimated 3-MT warhead, which was only one-half the megatonnage that a single B-52 could deliver carrying bombs and short-range nuclear attack missiles.

Through 1957, U.S. production lines built 2,041 B-47s. These planes, in the bomber configuration, could deliver one or two nuclear weapons and had a maximum range of 4,000 miles. Based overseas and requiring in-flight refueling, they could reach Soviet targets. They were then to fly on to neutral territory or return to base. Beginning in June 1955, the SAC received the larger, eight-engine B-52. This aircraft, which first flew in 1952, would become the "ultimate" strategic bomber produced in quantity by the United States.

The early Boeing B-52s, with a takeoff weight of some 450,000 pounds, could carry four H-bombs some 3,500 miles, strike a target, and return to base without refueling. Maximum speed of this eight-engine giant was 600 mph. B-52 production reached 744 aircraft when deliveries ceased in late 1962. And, like the B-47, the B-52s carried only tail guns—four remote-control .50-caliber machine guns in the early models.

Two subsequent strategic bomber programs were initiated in the 1950s, the B-58 Hustler and the B-70 Valkyrie. The Convair-built B-58 was developed following Boeing and Convair studies of delta-wing bombers that dated back to 1946. Convair ultimately offered a very small aircraft to provide higher speeds and to reduce radar reflectivity (radar cross section). The B-58, which flew in 1956, was a "small" aircraft (160,000 pounds gross compared to 488,000 pounds for the later model B-52s). It was to streak over targets at twice the speed of sound. A sleek-looking plane with four turbojet pods slung under its delta wing, the B-58 had a limited range, and only 116 aircraft were built, including some configured as TB-58 trainers.

The North American B-70 program was essentially stillborn. This aircraft, which was to have been successor to the B-52, was a giant (500,000 pounds gross) six-turbojet bomber with intercontinental ranges and a service altitude of over 70,000 feet. Top speed was to be more than three times the speed of sound (in excess of 2,000 mph). The B-70 aircraft was cancelled in 1961 by the incoming Kennedy administration whose Secretary of Defense, Robert S. McNamara, preferred the development of strategic missiles rather than bombers. McNamara cited the B-70 as being "unnecessary and economically unjustifiable." Only three XB-70 aircraft were produced for research purposes.

However, with the loss of a U-2 reconnaissance aircraft to an SA-2 missile over the Soviet Union on 1 May 1960, American military leaders realized that the effectiveness of the SA-2 Guideline (Soviet V-75 Dvina) surface-to-air missile. This weapon forced SAC bombers to change to low-level flight "profiles," which considerably reduced their range.[17]

17 The Soviet Union had also deployed a massive anti-aircraft artillery defense; a U.S. Army study in 1958 concluded that Moscow had the densest anti-aircraft artillery defenses of any location in the world and greater than the German concentration around Berlin in World War II.

General Thomas S. Power, USAF. (U.S. Air Force)

The high-flying B-70 would undoubtedly have been vulnerable to the advanced air defense missiles that were being deployed by the Soviets. U.S. concern for Soviet offensive and defensive capabilities in the late 1950s led SAC and the Navy to revise bombing procedures to provide for low-level attacks that were hoped to be less vulnerable to Soviet radar detection and interception. Low-level attack profiles could reduce the probability of aircraft detection and interception, but greatly increased fuel consumption and increased the stress placed on the aircraft structures. This was the period of major Soviet reorganization for defense of the homeland and the development of advanced fighter-interceptor aircraft, ground radar warning and intercept control stations, and the wide installation of surface-to-air missiles.

In 1959 SAC began an "airborne alert" on a test basis. As a means of increasing the readiness and survivability of SAC bombers, especially against the threat of Soviet ICBM attacks on bomber bases, General Thomas S. "Tommy" Power conducted an exercise in 1959 in which one of the new B-52 Stratofortress bombers took off from Westover AFB in Massachusetts. The seven-man crew, with a mattress and stack of box lunches, remained aloft for about 25 hours. In the plane's bomb bay were two B39 thermonuclear bombs. Subsequently, six B-52s were kept aloft on a sustained basis, proving that SAC could keep bombers in the air on a continuous basis—24 hours a day, 365 days a year. This was the beginning of Operation Chrome Dome.[18]

The airborne bombers, flying proscribed routes, had target folders on board, and upon receipt of an encoded message—a "go code" authenticated by two or more crew members—the plane would pass though a "fail safe" line and proceed toward its preassigned target. This procedure meant that a percentage of SAC's bombers would streak toward the communist world even if there was a surprise attack on SAC bases by Soviet intercontinental missiles, submarine-launched missiles, or even saboteurs.

The airborne alert was expensive. It required an entire wing of 30 to 45 B-52s to keep 6 bombers airborne. Each airborne B-52 had to be refueled two to four times during the 24-hour flight, placing a strain on the SAC aerial tanker fleet and consuming vast quantities of aviation fuel. Still, the airborne alert effort was continued, and at the time of the Cuban missile crisis in the fall of 1962, about 60 of SAC's 600-plus B-52s were on continuous airborne alert for a brief period, while one-half of the remainder of the B-52s and one-half of the command's almost 900 B-47 Stratojet bombers were on ground alert, armed with nuclear weapons, and ready to begin takeoff 15 minutes after being "scrambled." (At the time SAC operated just over 1,000 tanker aircraft.)[19]

General Power, testifying before Congress, explained:

> We in the Strategic Air Command have developed a system known as airborne alert where we maintain airplanes in the air 24 hours a day, loaded with bombs, on station, ready to go to the target. . . . We must impress Mr. Khrushchev that we have it, and that he cannot strike this country with impunity.[20]

Soviet leader Nikita Khrushchev, although far less belligerent and truculent than his predecessor, Josef Stalin, did pursue an intensive missile development and space program in the late 1950s, as evidenced by the world's first full-range ICBM test flight and the launch of the world's first artificial satellite—*Sputnik 1*—both in 1957.[21] These

18 Two fictional accounts of Operation Chrome Dome have become classics. The first, based on Peter Bryan George's novel *Red Alert* (1958), is Stanley Kubrick's outstanding film *Dr. Strangelove: Or, How I Learned to Stop Worrying and Love the Bomb* (1964); the second is the Eugene Burdick and Harvey Wheeler's novel *Fail-Safe* (1962) and the film of the same name (1964).

19 At the end of 1962 SAC operated 503 KC-97 piston-engine tankers and 515 jet-propelled tankers.

20 SAC, Office of the Historian, *The Development of Strategic Air Command* (Offutt AFB, Neb., 1 September 1986), p. 83.

21 The missile was the SS-6 Sapwood (the Soviet designation was *Semyorka*—Number 7). The SS-6 was a massive, primitive missile; only six SS-6 missiles were operationally deployed.

accomplishments, coupled with Soviet press reports and Khrushchev's bluster, led some American military and political leaders to demand a massive increase in U.S. missile programs to counter the perceived "missile gap." Earlier, Soviet aerial demonstrations revealing advanced bomber aircraft had led to political cries of a "bomber gap" between the United States and Soviet Union.

In reality, neither a bomber gap nor missile gap existed. In the late 1950s the strategic balance of power was significantly in favor of the United States. This was evident to those with the highest security clearances—from the intelligence acquired from 23 overflights of the Soviet Union by U-2 spyplanes from 1956 to 1960, several peripheral flights by U.S. spyplanes, and other sources, mostly Soviet defectors and the trickle of émigrés.[22]

U.S. SECOND-GENERATION WEAPONS

The deployment of 1950s-era U.S. strategic weapons—bombers and land-based missiles—was coming to an end when the Kennedy administration entered the White House in January 1961. Amid cries of a "missile gap" by Democrat leaders, the new administration accelerated both strategic and conventional weapon programs. These weapons, particularly the Minuteman ICBM and Polaris SLBM, had been initiated earlier under the Eisenhower administration.

Simultaneously, the concept of systems analysis was brought forward to a previously unprecedented degree in an effort to develop the optimum "mix" of strategic forces. The nature of the future strategic mix was discussed in these terms by Secretary of Defense McNamara:

> The introduction of ballistic missiles is already exerting a major impact on the size, composition, and deployment of the manned bomber force, and this impact will become greater in the years ahead. As the number of . . . ballistic missiles increases, requirements for strategic aircraft will be gradually reduced. Simultaneously, the growing enemy missile capability will make grounded aircraft more vulnerable to sudden attack.

During the early 1960s the remaining U.S. first-generation ICBMs became operational: the Atlas E in 1960, and the Atlas F and Titan I in 1962, all with storable-liquid propellants and capable of reaching the Soviet Union from launch pads in the United States. These weapons were followed into service by the second-generation ICBMs, the Minuteman I and Titan II. The Minuteman I (of which there were actually two versions developed simultaneously) had solid propellant and could deliver a one-MT warhead against targets 6,300 n. miles away. The Minuteman ICBMs were installed in underground launch silos to provide protection against preemptive enemy missile or bomber attacks on U.S. deterrent forces. The first ten Minuteman ICBMs went on alert on 27 October 1962 (at the height of the Cuban missile crisis).

Whereas the Minuteman was a "light" ICBM, the Titan II was the largest ICBM developed by the United States, weighing almost five times as much as the Minuteman and carrying a W53 warhead of nine MT. It had a range of 6,300 nautical miles.

As the Minuteman and Titan production got under way, their proponents advocated large deployments to ensure numerical superiority over Soviet strategic forces—the deployment of up to 2,500 Minuteman ICBMs was discussed. Actual planning in the early 1960s provided for 1,200 Minuteman missiles and 120 Titan II weapons as the second-generation SAC missile force. General Power, the head of SAC, spoke in terms of 10,000 Minuteman missiles.[23]

Although Secretary McNamara favored missiles over bombers, he restricted the numbers of ICBMs deployed to what he considered realistic force levels. In addition, the later Minuteman II (operational in 1966) would have increased accuracy over the earlier Minuteman missiles.

Accordingly, the eventual force levels were established at 1,000 Minuteman and 54 Titan II missiles. At the same time, Secretary McNamara accelerated phasing out the earlier strategic missiles. The Snark long-range cruise missile was discarded in June 1961, less than four months after 30 of these weapons had been declared combat ready. During the next few years, all of the nation's first-generation ICBMs were phased out: the 30 Atlas D, 33

22 Francis Gary Powers was making the 24th overflight of the Soviet Union when his aircraft was shot down by an SA-2 missile on 1 May 1960. No further overflights of the Soviet Union were made by U.S. spyplanes.

23 The various proposals for ICBM force levels are discussed in detail in Desmond Ball, *Politics and Force Levels: The Strategic Missile Program of the Kennedy Administration* (Berkeley, Calif.: University of California Press, 1980). The official Air Force (vice SAC) position on Minuteman ICBMs fluctuated in the 1,000 to 1,500 range.

Atlas E, 80 Atlas F, and 63 Titan I missiles. Thus, for much of the remaining Cold War the SAC would operate approximately 1,000 Minuteman missiles (of various models) and 54 Titan II missiles.

A factor in U.S. Minuteman-Titan ICBM effectiveness was the question of reliability. Of the three components of the Triad, the land-based ICBMs were the only force that was not extensively tested. Bombers regularly took off, flew missions, and dropped bombs; prior to test-ban agreements, bombers dropped nuclear weapons (and dummy bombs) in full-system tests. Similarly, submarines regularly fired unarmed ballistic missiles—sans warheads—on test ranges; and on 6 May 1962, the USS *Ethan Allen* (SSBN 608) fired a Polaris A-1 missile almost 1,200 nautical miles in the Pacific with a nuclear detonation. This was the only full-system test of a U.S. nuclear-armed ICBM/IRBM/SLBM missile from an "operational silo."

No nuclear-armed ICBM has been launched from an operational silo. Periodically, the silo crews fired various ICBMs from test facilities at the Vandenberg Air Force Base in California and from Cape Kennedy in Florida under highly controlled conditions. But even periodic efforts to launch an ICBM with reduced fuel and no warhead from an operational silo have failed, and Congress has refused approval of full-range test firings from an operational silo that would take even an unarmed missile over urban areas.

The exposed Thor IRBMs in Britain, operated by the Royal Air Force, were discarded in 1963, as were the U.S.-controlled Jupiter missiles in Italy. The few Jupiter missiles in Turkey, which had been the catalyst for the Cuban missile crisis in the fall of 1962, were similarly withdrawn. (Ironically, the first Jupiter had become operational in Turkey in October 1962, at the height of the missile crisis.)[24]

The loss of the joint-controlled missiles in Britain was compensated, in part, by U.S. proposals to provide stand-off nuclear missiles for British bombers. That program—the Skybolt—was cancelled and instead the British received submarine-launched Polaris A-3 and, later, Trident D-5 missiles, both fitted with British warheads. To help overcome Turkish opposition to removal of its Jupiter missiles, the U.S. Navy deployed nuclear-propelled Polaris submarines to the Mediterranean in 1963. The USS *Sam Houston* (SSBN 609), the first Polaris submarine to enter the "Med," ensured that all interested parties knew of the Polaris presence when, on 14 April, the submarine entered the port of Izmir, Turkey.[25]

With respect to manned strategic bombers, the Kennedy administration began phasing out the entire B-47 medium bomber force. SAC bomber strength had reached a peak of 1,854 aircraft in late 1959: 488 of the intercontinental B-52s, and the remainder medium-range B-47s. The phaseout of older B-47s had begun in 1959 and was accelerated after the Kennedy administration took office in January 1961. But the Berlin crisis of 1961 and the Cuban crisis of 1962 briefly delayed the B-47 retirements. The deliveries of B-52 and B-58 jet bombers also briefly offset the B-47 reductions.

The strike capabilities of the B-52 aircraft were improved beginning in 1961 with deployment of the Hound Dog air-to-ground missile fitted with the W28 nuclear warhead. A B-52 could carry two Hound Dog missiles under each wing. The turbojet missile had a Mach 2+ speed and a range of more than 500 miles for attacking enemy defenses or heavily defended targets. (The B-52s also carried a large Quail decoy in the rear bomb bay; also turbojet powered, it had a high subsonic speed and a range of over 200 miles. The Quails were to be released by B-52s to simulate bomber images on enemy radars to aid bombers in penetrating to their targets.)

The last B-52s and B-58s were delivered to SAC in 1962, and for the first time since before World War II there was no strategic bomber aircraft being developed or produced in the United States. SAC bomber strength fell to 591 B-52s and 83 B-58s in late 1966, when the last B-47 bombers were retired, although 16 reconnaissance RB-47s were held for a few months longer. Meanwhile, phaseout of the B-58s had begun, and all of those high-speed, short-range aircraft were gone by 1970.

As the number of manned bombers declined, the highly expensive airborne alert program was also cancelled.

24 The first Jupiter missile was formally turned over to the Turkish Air Force for operation and maintenance on 22 October 1962 (the nuclear warheads remaining under U.S. custody). Although the move was publicized in Turkey and undoubtedly detected by Soviet intelligence, President Kennedy and his confidants apparently were not aware of the event.

25 This was the first Polaris SSBN visit to a foreign port other than Holy Loch, Scotland, which had served as a forward base for U.S. Polaris submarines since March 1961. Subsequently, Rota, Spain, became a Polaris support base, as did Guam in the Mariana Islands.

A B-52D Stratofortress releases conventional bombs over an enemy position in South Vietnam during the lengthy conflict in Southeast Asia. B-29 Superfortresses were used as conventional bombers in the Korean War, but apparently none was configured to carry nuclear weapons to prevent any nuclear-related equipment from falling into communist hands. Thus, B-52s were the first U.S. nuclear-configured bombers to be sent into combat—and several were shot down. (U.S. Air Force)

Contributing to the demise of the airborne alert were crashes of bomb-laden B-52s off Palomares, Spain, in 1966, and near Thule, Greenland, in 1968.[26] The increasing number of dispersed Minuteman ICBMs in underground silos, and the completion of early warning radar installations that could provide at least 15 minutes of warning of a Soviet missile attack, alleviated the need for the airborne alert. Still, a large percentage of the strategic bomber force was kept on 15-minute runway alert. On 21 April 1964, the number of U.S. ICBMs on alert was equal to the number of SAC manned bombers on alert.

STRATEGIC BOMBERS AT WAR

Also during the 1960s, B-52 strategic bombers were employed in a conventional warfare role against Viet Cong positions in South Vietnamese jungles as well as targets in North Vietnam and Laos in Operation Arc Light. This was the first U.S. employment of strategic bomber aircraft in conventional strike roles following the use of non-nuclear-capable B-29 Superfortress bombers in the Korean conflict. On 18 June 1965, 27 B-52F aircraft flying from Guam initiated strategic bomber participation in the Vietnam War. Subsequent B-52 raids conducted during

26 On 17 January 1966, a B-52 and a KC-135 tanker collided over Palomares during a routine aerial refueling operation. Both aircraft crashed. One bomb was recovered on the ground and one was recovered from the sea after extensive search and recovery efforts. The two other bombs came down on land and their high-explosive components exploded on impact, releasing some radioactive materials. On 21 January 1968, a B-52 crashed and burned some seven miles southwest of the runway at Thule Air Base. The bomber carried four nuclear weapons, all of which were destroyed by fire with some radioactive contamination occurring in the area of the crash, which was on the sea ice.

the next seven years were considered highly effective in support of ground operations as the bombers were undetectable by Viet Cong troops in South Vietnam until their bombs were falling. Older B-52 aircraft were also committed to the Vietnam War.

In June 1964, 28 B-52F aircraft were fitted with external racks to carry 24 750-pound bombs in addition to their internal bomb bays under Project South Bay. Their success led to another 46 aircraft being modified under Project Sun Bath. Starting in late 1965 the Air Force modified most of its B-52D aircraft from an internal capacity of 27 500-pound or 750-pound bombs to 84 500-pound weapons or 72 750-pound weapons, in addition to 24 bombs of either size carried on wing racks. Thus, Project Big Belly provided a conventional bomb load of 54,000 pounds when carrying 108 of the 500-pound bombs or 49,500 pounds with 750-pound weapons. These planes entered combat over Vietnam in April 1966, replacing the B-52Ds in the conventional strike role. (Both the B-52D and B-52F retained the ability to carry four nuclear bombs.)

When B-52s were used over North Vietnam they suffered significant losses to Soviet-provided SA-2 missiles, the first being a B-52D lost in a raid over Vinh on 22 November 1972. President Richard M. Nixon ordered a maximum bombing effort aimed at Hanoi and Haiphong in late 1972 as the North Vietnamese delayed peace negotiations. On 18 December the first strikes of operation Linebacker II were flown with waves of B-52s and F-111 strike aircraft attacking Hanoi, as well as nearby airfields and missile sites. A defensive curtain of flak, SAMs, and MiGs met the attackers. Three B-52s were shot down on the first night. (The tail-gunner of a B-52D claimed the first fighter kill ever by a jet-propelled strategic bomber, a MiG-21 destroyed by the plane's quad .50-caliber gun turret. A second MiG was destroyed by another B-52 gunner on 24 December.)

The Linebacker II strikes continued through 24 December—Christmas Eve—when there was a 36-hour stand-down over the holiday. Then, on the night of 26–27 December, an elaborately planned strike hit Hanoi and Haiphong in a one-two-three stream of planes that proved to be the peak of the 11-day Linebacker II operation. Within a 15-minute period a force of 120 B-52s struck the cities from three directions, overtaxing and confusing the North Vietnamese defenses. The Air Force, Navy, and Marine Corps flew 113 aircraft in support of the B-52s. Two B-52s were shot down that night.[27]

In all, Linebacker II cost 17 B-52s lost to SA-2 missiles in a period of 11 days. Several Air Force and Navy aircraft, mostly F-111s and A-6 Intruders that were attacking SAM sites in support of the bombers, were also lost.[28] The last of 126,615 B-52 sorties over North and South Vietnam was flown on 15 August 1973. During the conflict the Air Force also suffered 14 B-52s lost to operational causes, including 2 in a collision. In total, 73 crewmen were killed and 33 became prisoners of war.[29]

During the Vietnam War period (1964–1973) the B-52 force was reduced from some 625 aircraft to 450, although there were qualitative improvements to the bombers through modernization. In the midst of this bomber reduction, Secretary McNamara, without warning, introduced a new "strategic" manned bomber. On 8 December 1965, he announced that the controversial swing-wing TFX "fighter" would be procured by SAC as a follow-on manned bomber. The F-111 program had been initiated by the Air Force to provide a successor, the F-105 Thunderchief, a long-range, tactical nuclear strike aircraft. The program was redirected by McNamara in 1961—and relabeled TFX for tactical fighter, experimental—to provide a single aircraft design for tactical strike, close air support, fleet air defense, and battlefield interdiction. He envisioned the Air Force, Navy, and Marine Corps all procuring TFX variants. Although the Air Force and Navy–Marine Corps had differing requirements for a tactical aircraft, the senior admirals initially endorsed the concept. However, the Navy, abetted by supporters in Congress, soon withdrew from the program (as did the British).

27 The 26–27 December mission was the largest B-52 operation ever flown with 78 taking off from Anderson AFB on Guam and 42 from U Tapao Royal Thai Navy Air Base in Thailand; all 118 survivors returned to Guam. The Guam-launched bombers flew an 18-hour mission.

28 North Vietnamese SA-2 missiles shot down between 150 to 160 U.S. aircraft from 1965 to 1973. More than 9,000 missiles were fired for a kill rate of 1.7 percent. Despite the low aircraft loss rate, the vast number of SAMs caused many aircraft to jettison their bombs short of their targets, required numerous missions by anti-SAM and radar jamming aircraft, and often forced attacking planes to adopt tactics that made them more vulnerable to anti-aircraft guns.

29 The combat losses were 11 B-52D and 6 B-52G aircraft. B-52 missions and losses are detailed in Chris Hobson, *Vietnam Air Losses: United States Air Force Navy and Marine Corps Fixed-Wing Aircraft Losses in Southeast Asia 1961–1973* (Hinckley, England: Midland Publishing, 2001).

U.S. submarine-launched ballistic missiles also lagged behind Soviet developments. But the Polaris was a far superior weapon when it went to sea in 1960. It was the first U.S. solid-propellant strategic missile and, in the A-3 variant, the first U.S. multiple-warhead weapon. Here the USS *Henry Clay* (SSBN 625) makes a rare surface launch of a Polaris A-2 missile. The tall mast is a temporary telemetry antenna used for test launches. (U.S. Navy)

In an effort to increase F-111 production, Secretary McNamara directed that 210 FB-111 bomber variants be produced by General Dynamics. The FB-111, like the B-58 Hustler and B-52 Stratofortress that it was to replace, would have to penetrate Soviet air defenses. In the end, only 76 of the FB-111 strategic bombers were built.

THE STRATEGIC TRIAD

The third major U.S. strategic system developed during the 1960s was the Polaris SLBM fleet. Although the Navy was initially opposed to large-scale Polaris development, a forceful Chief of Naval Operations, Admiral Arleigh Burke, and his full endorsement of his project manager, Rear Admiral William F. "Red" Raborn, got Polaris to sea far in advance of early schedules. The Navy was able to separate its missile development from the Army's Jupiter IRBM and employ sold-propellant motors, making the Polaris suitable for submarine use, both because of its small size (initially 28 feet long and 54 inches in diameter), and the relatively safe fuel. The first Polaris submarine, the USS *George Washington* (SSBN 598), went to sea on 15 November 1960, carrying 16 nuclear-tipped Polaris missiles on the first two-month nuclear deterrent patrol. This A-1 missile had a range of 1,200 nautical miles and was fitted with a W47 warhead of 600-KT explosive force.

By the early 1960s the Navy was proposing a force of 45 missile submarines—five squadrons of nine submarines each. Secretary McNamara, however, cut the planned force to 41 submarines (carrying 656 missiles), and the last Polaris submarine went to sea in April 1967, seven and a half years after the first. Thirteen nuclear-propelled Polaris submarines were commissioned in the peak construction year of 1964, in addition to four nuclear torpedo-attack submarines.[30] In 1961, McNamara described the Polaris as having

> the most survival potential in the wartime environment of any of our long range nuclear delivery systems. Polaris missiles do not have to be launched early in the war, they can be held in reserve and used in a controlled and deliberate way to achieve our wartime objectives. For example, Polaris is ideal for counter-city retaliation.[31]

30 At that time the United States had seven shipyards building nuclear-propelled submarines, both ballistic missile (SSBN) and torpedo-attack (SSN). There were two government and five private shipyards; today there are two yards, both private, building only SSNs.

31 John P. Glennon, Edward C. Keefer, and David W. Mabon (eds.), *National Security Policy*, vol. VIII, *Foreign Relations of the United States, 1958–1960* (Washington, D.C.: Department of State, 1996), p. 151.

Polaris became one "leg" of the U.S. Triad—the term coined by the U.S. Air Force to rationalize the "need" for three U.S. strategic offensive forces: manned land-based bombers, land-based ICBMs, and submarine-launched ballistic missiles. The term "Triad" is believed to have been coined about 1970 by Major General Glenn Kent, a leading Air Force strategic planner. The first public use of the term on record was Air Force chief of staff General John D. Ryan using it on 30 October 1970, in a speech before the Los Angeles Chamber of Commerce. The use of Triad was quickly encouraged "throughout the Air Force to help explain the continuing need for a manned strategic bomber."[32]

While initially opposing the Navy's development of the Polaris program, the Air Force subsequently attempted to embrace it under the operational control of the Strategic Air Command. The Air Force's rationale was that strategic missiles—whether launched from land or sea—were targeted against an enemy's homeland, and that such strikes and their targets must be under centralized control. This move was strongly resisted by the Navy. Instead of a single strategic command, in 1960 Secretary of Defense Thomas S. Gates established the U.S. Joint Strategic Target Planning Staff (JSTPS)—headed by the commander in chief of SAC with a Navy deputy—with the responsibility for the targeting of all U.S. strategic nuclear weapons and coordinating their use.

The JSTPS developed the National Strategic Target List that enumerated the targets in the Soviet Union, eastern Europe, North Korea, and China. It also designed the Single Integrated Operational Plan (SIOP) for attacking the targets on the target list with specific aircraft and missiles. Secretary Gates approved the first target list and SIOP in December 1960, less than four months after the JSTPS was established. Soon the JSTPS challenge was how to most efficiently use the large numbers of strategic weapons available in the U.S. arsenal, Between 1955 and 1960 the U.S. inventory grew from about 1,000 to more than 10,000 nuclear weapons of all sizes.

In the early 1960s, as the Polaris submarines joined the fleet, the Navy's forward-deployed aircraft carriers were relieved of their primary nuclear strike role. The aircraft carriers did retain a nuclear strike capability, however, and continued to be a part of the nuclear attack plans until the end of the Cold War.

The components of the U.S. strategic offensive force, developed during the 1960s with differing origins, effectiveness, and operational concepts, became linked together by the term "Triad" after the fact of their existence. The decisions of the 1960s with respect to number of platforms dictated the size of U.S. strategic offensive forces in the 1970s and 1980s: 1,054 ICBMs, approximately 450 land-based manned bombers, and 41 submarines with 656 missiles. In addition, some forward-based tactical aircraft in Europe and forward-deployed aircraft carriers could reach the Soviet Union with "tactical" nuclear weapons.

Significantly, Soviet nuclear strategy also changed in the late 1960s. Nikita Khrushchev, blamed for the "defeat" in the Cuban missile crisis of 1962 and subsequent Soviet economic problems, was replaced in October 1964 as head of the Communist Party by his protégé Leonid Brezhnev. Almost immediately, Brezhnev initiated a major military buildup of conventional forces as well as new strategic bomber and missile programs. Thus, during the 1960s both the United States and, subsequently, the Soviet Union deployed large numbers of strategic missile submarines and land-based strategic missiles, while the Soviets initiated the development and production of a new generation of long-range bombers.

OFFENSE VERSUS DEFENSE

Two factors began to dominate the strategic arms race during the later 1960s: ballistic missile defense and multiple warheads. The two factors were closely related in that each factor was a counter to the other. With the proliferation of ICBMs and SLBMs by the United States and Soviet Union in the 1960s, and the subsequent fitting of multiple warheads to those missiles, the amount of destruction that each side could inflict became enormous.

Both superpowers had large proportions of their population and industrial capacity concentrated in cities (see Table 1-1). In February 1965, Secretary of Defense McNamara announced a strategy of "assured destruction" against the Soviet Union in the event of a nuclear exchange. According to McNamara, the destruction of 200 American cities could kill more than one-half of the nation's population and eliminate three-quarters of its

32 Col. Charles D. Cooper, USAF, Office of Public Affairs, Department of the Air Force, letter to N. Polmar, 5 February 1981.

TABLE 1-1. Distribution of Population and Industrial Capacity, 1970

	United States		Soviet Union	
Number of cities	Percent of population	Percent of industrial capacity	Percent of population	Percent of industrial capacity
1	25	33	8	25
50	42	55	20	40
100	48	65	25	50
200	55	75	34	62
400	60	82	40	72
1,000	63	86	47	82

industrial capacity. The loss of 400 Soviet cities would mean that almost one-half of that nation's population and three-quarters of its industry could be destroyed. The numbers of cities were counted in the hundreds, while the available nuclear warheads soon numbered in the thousands because of the technology enabling multiple warheads to be delivered by a single missile.

At the same time, technologies were becoming available that gave promise of an active ballistic missile defense whereby incoming ICBM warheads could be intercepted by anti-ballistic missile (ABM) systems. An active missile defense system consists principally of radars to detect and track incoming reentry vehicles (RV) and interceptor missiles to destroy these RVs. Formal U.S. efforts to develop an ABM system began with the Army's Nike-Zeus system, initiated in 1956. This was a development of the Army's Nike-Ajax and Nike-Hercules anti-aircraft missiles, the latter fitted with the W31 nuclear warhead.

The Nike-Zeus ABM system could have become operational by 1963–1964, but would—according to subsequent estimates—have been obsolete by that time because of Soviet development of advanced reentry vehicles. Other ABM concepts were put forward in the United States, including one for Polaris-armed missile submarines to fire their weapons in either an offensive or ABM defensive mode. That proposal was not pursued.

Soviet ABM efforts were moving rapidly and would impact U.S. strategic missile development. As early as 1960, deployments began around Leningrad (now St. Petersburg) of what appeared to be an enlarged version of the SA-2 missile. This missile—given the NATO designation Griffon (possibly RZ-25 by the Soviets)—could have been, in the words of Soviet weapons expert Steven J. Zaloga, "evisioned as the first step on the ABM ladder."[33] He continued, "it was of very conventional design, and rather archaic for an ABM missile. It may have been developed as an exercise before the initiation of the main ABM effort, or it may have been a technological hold-over from an earlier programme." By 1962 there were 30 launch sites around Leningrad, the second-largest city of the USSR. Each launcher was in a hardened bunker with reload missiles. In October 1961, Minister of Defense Rodion Ya. Malinovskiy, said that "the problem of destroying missiles in flight has been successfully overcome"; in July 1962, Khrushchev boasted that Soviet air defenses could "hit a fly" in space. Despite this bravado, the Soviet Union did not have a realistic ballistic missile defense capability.

In the early 1960s there was another new missile deployment, this one around the Estonian capital of Tallinn. This was the SA-5 Gammon (Soviet S-200), which, with massive associated radars, was initially considered by U.S. intelligence to be an ABM system. While the SA-5 remains a subject of controversy, it was in fact an advanced anti-bomber defense. Rather, the successor to the abortive Griffon as an ABM system would be the ABM-1 Galosh (Soviet UR-96). This was a large interceptor missile, first publicly displayed in its launch canister at a Moscow parade in November 1964. The ABM-1 was a three-stage missile that could reach a speed in excess of Mach 3. It carried a 2- to 3-MT warhead.

The initial Galosh ABM system reached a limited operational capability with a few interceptors around Moscow in November 1967. This Moscow deployment consisted of 64 reloadable launchers for interceptor missiles,

33 Steven J. Zaloga, *Soviet Air Defense Missiles: Design, Development and Tactics* (London: Jane's, 1989), pp. 120–21.

each carrying a warhead with a 2- to 3-MT yield. (In addition, Moscow was surrounded by rings of SA-1 Guild surface-to-air missiles with 48 multilauncher sites that could launch either nuclear or conventional missiles against attacking bombers.)

The ABM-1 deployment was limited. Several variations of the missile and its accompanying radar systems. Beginning in the mid-1970s the Soviet ABM upgrades provided a counter to a "limited" U.S. ballistic missile attack or to a Chinese ballistic missile attack. This upgrade was believed to be intended to stop about 100 incoming warheads. After an initial reduction, by 1989 the Moscow ABM system was increased to 100 ABM launchers for improved missiles, the ABM-3 Gazelle and ABM-4 Gorgon. However, significant problems delayed the improved system from becoming fully operational until 1995—after the Cold War had ended.

Simultaneously, beginning in the late 1950s, thousands of SA-2 Guideline launchers were deployed throughout the Soviet Union to intercept attacking bombers and reconnaissance aircraft, with an SA-2 missile shooting down the U-2 spyplane in 1960.[34] Most SA-2 missiles, which reached a peak deployment of some 4,800 launchers in the USSR in 1968, had conventional warheads. However, the SA-2E variant had a 15-KT warhead.

With the early evidence of a Soviet ABM deployment, the United States initiated a counterdeployment of weapons to overcome ballistic missile defenses. The principal means to counter an ABM system were to saturate the defensive radars and associated electronic equipment used to track incoming reentry vehicles, or to launch sufficient attacking missiles to exhaust the ABM interceptor missiles. Other methods, such as a preliminary attack on ABM radars (possibly by bombers), were less certain to be effective.

To overcome an ABM system by increasing the number of offensive land-based or submarine missiles meant more launch equipment, more crews, more submarines (for the SLBMs), more real estate (for the ICBMs), and so on. Instead, another technological approach to saturating ABM interceptors was undertaken—the multiple-warhead missile. With multiple warheads, a single missile could carry aloft several RVs that would be released in flight to come down separately against one or more targets. A simplistic analogy is the single, large bullet compared to a buckshot cartridge. The development of more accurate controls for rocket motors, computers, miniaturized electronics, inertial navigation, and smaller warheads made multiple warheads feasible.

MULTIPLE WARHEAD DEPLOYMENT

The United States began deployment of the first multiple warhead on the Polaris A-3 submarine-launched missile, which became operational in 1964. With a range of 2,500 nautical miles the A-3 missile greatly increased the ocean areas in which Polaris submarines could operate and remain within missile range of Moscow and other Soviet targets. The A-3 variant carried a multiple reentry vehicle (MRV) payload. As each Polaris missile streaked aloft, the warhead separated into three separate RVs, or "bomblets," that could strike a single target. Whereas the basic Polaris A-1 and A-2 warheads were generally described as being in the 1-MT range, the A-3 missile's three RVs had an explosive force of only 200 KT each because of the additional weight and space required for the separation apparatus (the PBV, or post-boost vehicle). Although the sum of the three RVs was less than the Polaris A-1 and A-2 payload, the three RVs would impact on a "soft" target—a city or installation—in a pattern that could result in greater overall destruction. The first Polaris submarine armed with multiple-warhead A-3 missiles, the USS *Daniel Webster* (SSBN 626), went to sea in 1964. Subsequently, all of the Navy's 41 ballistic missile submarines were rearmed with the A-3 missile.

But this missile, with its three MRVs, was only an interim weapon. Development was already under way on multiple independently targeted reentry vehicle (MIRV) warheads. Whereas the MRV warheads were "shotgunned" down onto a single target, with MIRV warheads each reentry vehicle or "bomblet" can be directed to a separate target. This separation is accomplished by the rocket boosters carrying aloft a "bus" containing the individual RVs. After the boosters burn out and fall away from the bus, the latter continues toward enemy territory, releasing the RVs in sequence. After each release, the orientation and velocity of the bus is changed by a preset program to aim the next RV release. Obviously, the area of land or "footprint" in which the separate RVs can be aimed is limited, and must fall within the overall range of the missile, greatly complicating both missile design and targeting.

34 A Soviet-launched SA-2 missile shot an American U-2 over Cuba in October 1962, and Chinese-launched SA-2 missiles shot down five U-2s flown by Taiwanese pilots over China in the 1960s.

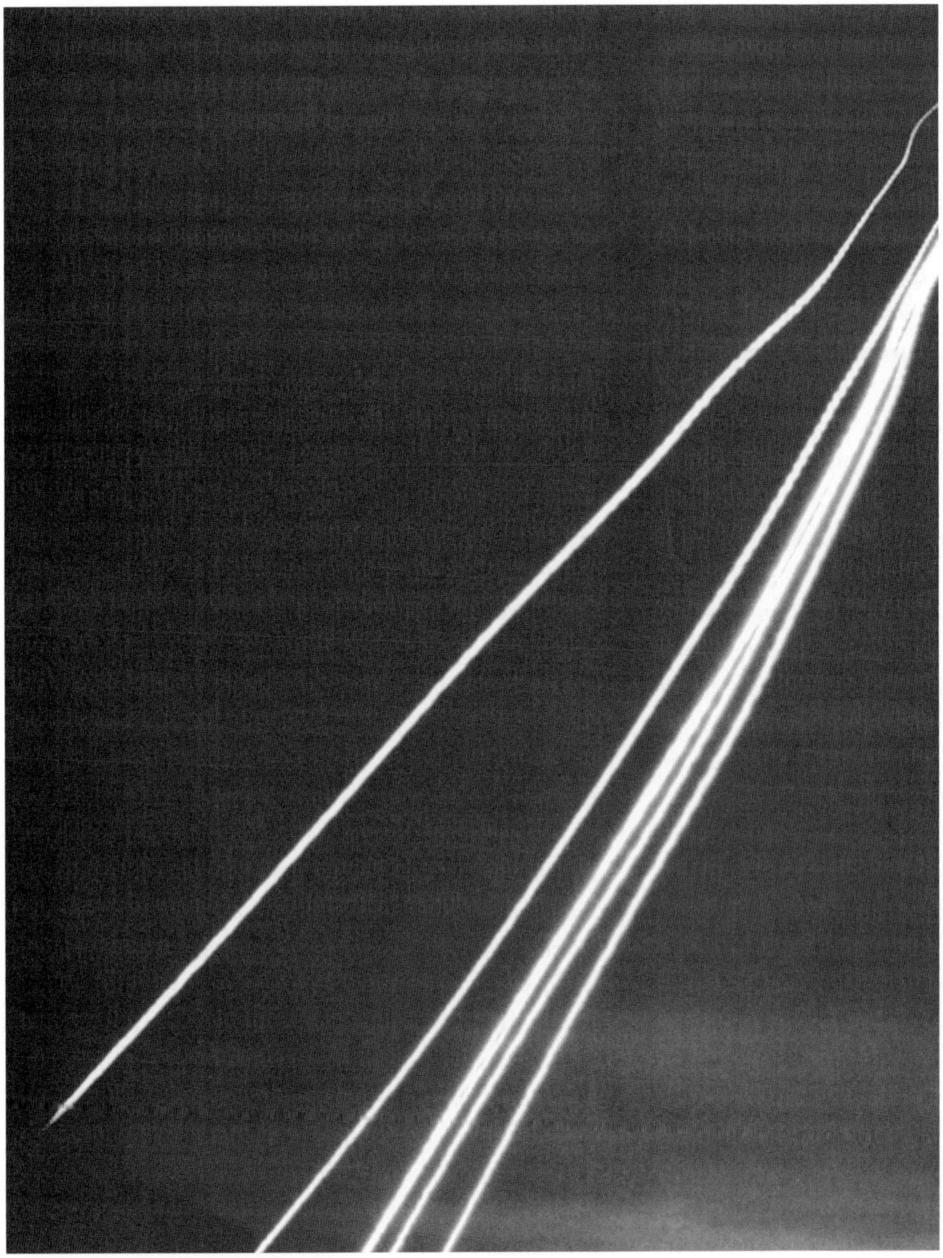

Multiple independently-targeted warheads on strategic missiles totally changed the "strategic equation" between the Soviet Union and the United States. The Minuteman III ICBM and Poseidon C-4 SLBM introduced MIRVs to the U.S. arsenal. The Soviet Union soon followed. This photo shows seven reentry vehicles from a Peacekeeper ICBM smashing through the atmosphere on a Pacific test range. (U.S. Air Force)

The first U.S. tests of an operational MIRV system began in 1968 with the Minuteman III, an ICBM with a nominal range of 6,000 miles. Whereas the earlier Minuteman I and II missiles carried a single warhead of about one MT, the Minuteman III had three RVs with W62 warheads, each with a 170-KT yield. Between 1970 and 1975, 550 of the earlier Minuteman ICBMs in the SAC arsenal were replaced by Minuteman III missiles fitted with MIRV warheads.

MIRV technology was also applied to the SLBM program, and in 1970 the first submarine firings were conducted with the Poseidon C-4 missile. This weapon, successor to the Polaris A-3 missile, had a range of 2,500 nautical miles and with a reduced range could deliver up to 14 RVs, each with a yield of 50 KT. The Navy converted 31 Polaris submarines to each launch 16 Poseidon MIRV missiles, each generally carrying 10 RVs. (The 10 oldest Polaris submarines were not suitable for modification to launch the newer weapon.)

The Minuteman and Poseidon MIRV programs greatly increased the number of reentry vehicles of the U.S. strategic offensive forces, easily providing the capability of overcoming any foreseeable Soviet ABM system and still destroying at least 400 cities. Furthermore, the "bounty" of multiple warheads on ICBMs and SLBMs enhanced the survivability of the strategic bomber fleet by providing RVs that could be used to destroy Soviet air-defense radars, command centers, and surface-to-air missile sites that were not previously targeted (see Table 1-2).

In the Soviet arsenal the giant SS-9 Scarp (R-36) was the first strategic missile to carry multiple payloads. The missile became operational in 1967 with a single 25-MT warhead, and testing of the SS-9 with a MRV warhead began in August 1968. Western intelligence estimates soon credited the missile with the ability to carry three reentry vehicles of 5 MT each. Thus, each RV dispensed from the SS-9 was larger than any U.S. ICBM warhead except for the single-warhead Titan II.

The SS-9 was also tested as the launch platform for the fractional orbital bombardment system (FOBS). In the FOBS configuration the missile was fired in a low trajectory in the opposite direction from the target, entered a partial Earth orbit, and then was launched onto the target. The tactic increased the problems of defending a target, increased the range of the weapon, and provided considerably less warning time than an ICBM following a high ballistic trajectory. In this configuration, however, the SS-9 was less accurate and carried a smaller warhead.

In the same period the Soviets began deploying the SS-11 (RS-10) in 1966 and SS-13 (RS-12) in 1969. These ICBMs were smaller weapons, the SS-11 having a storable-liquid propellant and able to strike targets in the United States with a 1-MT warhead; the SS-13, which introduced solid-fuel propellant to Soviet ICBMs, had a range of some 5,500 nautical miles and also carried a single, 1-MT warhead. Subsequently Soviet land-based and sea-based strategic missiles were fitted with MRV and MIRV warheads.

MULTIPLE WARHEADS AND FIRST STRIKE

Soviet multiple warhead development meant that the proposed U.S. missile-defense systems could theoretically be overcome through RV saturation. Even discounting other Soviet ICBMs and possibly SLBMs, the Soviets could add RVs to the SS-9 force at much lower costs than the United States could add interceptor missiles to an ABM system. With its massive payload capacity, the SS-9 might ultimately be developed to carry as many as 25 RVs per missile.

Beginning in the late 1950s intercontinental missiles were viewed primarily for attacking opposing strategic offensive forces in a concept known as "counterforce."

TABLE 1-2. U.S. Strategic Missile Warheads, 1970s

Pre-MIRV		Post-MIRV	
Titan	54	Titan	54
Minuteman	1,000	Minuteman II	450
		Minuteman III	1,650
Polaris[a]	656	Polaris A-3[b]	160
		Poseidon C-4[c]	4,960
Total	1,710	Total	7,274

a In 41 submarines.
b In 10 submarines.
c In 31 submarines.

However, the effectiveness of ICBMs carrying a single warhead would be limited against an enemy strategic force of about the same size. For example, ICBMs are susceptible to failures; for an ICBM to function perfectly, hundreds of different components must work within limited tolerances; although the percentage possibilities of individual failure are small, in the aggregate they significantly reduce the effectiveness of an entire ICBM force.

In the late 1960s, Secretary of Defense Melvin Laird judged Soviet ICBMs to have an effectiveness factor of 80 percent. This meant that—on average—for every ten missile firings, eight missiles would launch properly, travel to their target, come down within expected accuracy, and detonate properly. Hence, if 1,000 Soviet ICBMs were fired against 1,000 ICBMs in the United States in a preemptive strike, 800 of the Soviet weapons could be expected to destroy their targets. This would leave the United States with about 200 ICBMs ready to launch (assuming no U.S. missiles were launched until after the Soviet attack). Accepting the same 80 percent effectiveness for U.S. missiles, after absorbing a preemptive attack the U.S. force could still destroy 160 cities in the Soviet Union ($200 \times 0.8 = 160$). This second-strike capability would destroy almost one-third of the population and more than one-half of the industry of the Soviet Union.

The same arguments given above, but with the roles of aggressor and victim reversed, would also serve to deter the United States from launching a preemptive missile attack against the Soviet ICBM force. Thus, one nation could always deter an enemy first strike against its ICBM force by maintaining the same number of ICBMs as, or a larger number than, the opponent.

Multiple warhead deployments changed the counterforce equations. With MIRV technology, the Soviets could conceivably arm 500 large ICBMs with multiple RVs, with each RV having a yield of 5 MT and a quarter-mile accuracy. Soviet leaders could plan a preemptive strike launching the 500 multiple-warhead ICBMs against the 1,000 U.S. ICBMs. With an 80 percent reliability, some 1,200 RVs could be expected to strike their targets ($500 \times 3 \times 0.8 = 1,200$), enough to pose a serious threat to the survivability of Minuteman, in the words of Secretary Laird. A surprise first strike of this magnitude could destroy almost all of the U.S. land-based ICBMs; possibly only ten or a score of U.S. missiles would survive. But the Soviet Union would still retain some 500 missiles with which to threaten U.S. cities, in the event that the United States should retaliate against the Soviet counterforce strike which—because of the fact that the ICBMs are emplaced in Midwestern states—would have killed relatively few Americans.

Because of these equations and potential scenarios, multiple warheads were a severely destabilizing influence on strategic weapons development. With MIRV technology, the Soviets could effectively make U.S. nonmobile, undefended, land-based ICBMs highly vulnerable to preemptive attack. In such an environment, the principal utility of nonmobile, land-based systems would be a spasmodic retaliatory strike. In a MIRV world, sea-based systems or mobile land-based systems appeared to be the best means available to ensure a stable assured destruction capability.

At that time, the U.S. Navy was developing the Poseidon SLBM to succeed the Polaris missile while the Air Force was promoting a very large ICBM, at the time known as WS-120A. The Poseidon C-3 missile went to sea in existing Polaris submarines beginning in 1971. Each missile carried 10 MIRVs of 50-KT explosive force. A single Poseidon submarine carried 16 missiles with up to 160 warheads. The large WS-120A would evolve into the MX or Peacekeeper ICBM, a 10-MIRV weapon with long range and a high degree of accuracy for targeting Soviet missile silos and command centers.

THE STRATEGIC SHIFT

The Cold War began with the United States possessing the only nuclear weapons and the only effective, long-range means of delivering them. Not until the early 1950s did the Soviet Union possess either, and even then the Tu-4 bomber—a copy of the B-29 Superfortress—could reach U.S. cities only on one-way missions.

As the USSR developed more advanced long-range bombers and, subsequently, ballistic missiles, the United States retained a strategic advantage in all measures of strategic force effectiveness well into the 1960s. Building on the technology base developed under his predecessor, Party secretary Leonid Brezhnev initiated the mass production of advanced strategic systems while continuing development into advanced weapon and warhead concepts. The following list chronicles the shift in strategic force effectiveness from the U.S. advantage to Soviet advantage.[35]

1965	ICBM throw weight
1967	ICBM equivalent megatonnage
1969	ICBM launchers
1973	Nuclear-propelled ballistic missile submarines[36]
1973	SLBM launchers on nuclear-propelled submarines
1974	SLBM equivalent megatonnage
1976	Independently targeted ICBM warheads
1981	Intercontinental strategic bombers *if* Tu-22M Backfires are included (without including Backfires the United States retained the advantage)

Because of the Poseidon missile—which could carry up to 14 MIRV warheads—and the follow-on Trident missiles, the United States always held the advantage of independently targetable SLBM warheads. In addition, the United States had more accurate ICBMs and SLBMs. However, because of its greater number of RVs and equivalent megatonnage, the Soviet Union retained the advantage in hard-target kill capability (i.e., destroying a hardened ICBM silo or hardened command structures for ICBMs). The United States did retain the advantage in hard-target kill capability with SLBMs.

Throughout the Cold War the Soviet Union had the advantage—qualitative as well as quantitative—in air defense interceptor aircraft and surface-to-air missiles. By the early 1960s Soviet air defenses included more than 4,000 jet interceptor aircraft and almost 8,000 surface-to-air missiles, with a continuing modernization program. And, of course, the Soviet Union did deploy and retain limited ABM defenses, albeit of questionable effectiveness.

35 Herbert Hoppe, Norman Polmar, and A. C. Trapold, *Measures and Trends U.S. and USSR Strategic Force Effectiveness* (Alexandria, Va.: Santa Fe Corp., March 1978); report prepared under contract to the Defense Nuclear Agency, Washington, D.C.

36 The Soviet Union also operated diesel-propelled ballistic missile submarines, a strategic platform not operated by the United States.

In response to Soviet MIRV and ABM developments, coupled with the large numbers of strategic weapons deployed during the Brezhnev era (1964–1982), and the lack of coordination of the U.S. strategic missile programs, Secretary of Defense McNamara established the so-called Strat-X study in late 1966 to

> characterize U.S. alternatives to counter the possible Soviet ABM deployment and the Soviet potential for reducing the U.S. assured-destruction-force effectiveness during the 1970s. It is desired that U.S. [missile] alternatives be considered from a uniform cost-effectiveness base as well as from solution sensitivity to various Soviet alternative actions.[37]

The Strat-X study, which included Air Force and Navy officers as well as civilian scientists and engineers, marked the first time that U.S. strategic weapon requirements had ever been addressed in an integrated and analytical manner. The study took a half year and considered about 125 candidate missile systems. Significantly, only two were sea-based systems.[38] (Some of the land-based options were waterborne, with proposals to mount ICBMs on barges on existing waterways or on a new series of canals to make them mobile and hence more survivable.)

The survivability of U.S. missiles from a Soviet first-strike attack was given particular consideration by the Strat-X participants, hence mobile ICBMs—moved by truck, railroad, and even in aircraft—were particularly attractive.[39] Two of the most expensive methods of enhancing ICBM survivability were to deploy ABM interceptors to defend the weapons and to provide hardened missile silos.

The two sea-based candidates proposed carrying missiles in surface ships and in a new generation of nuclear-propelled submarines. ("Sub-options" of Strat-X looked into fitting the new Poseidon missile in up to 31 of the Polaris submarines plus building 20 to 25 new submarines to carry Poseidon. This probably would have been the least costly method of reaching the Strat-X warhead weight objective with a high degree of weapon survivability.)

The final Strat-X recommendations sent to Secretary McNamara in mid-1967 proposed the consideration of four new missile systems that were determined as much by interservice politics as by analysis. Two were land-based and two were sea-based: (1) ICBMs in hardened underground silos, (2) land-mobile ICBMs, (3) a ship-based long-range missile system (SLMS), and (4) an underwater long-range missile system (ULMS). Thus, rather than providing the *best* strategic options—as McNamara had requested—Strat-X provided a set of options balanced between land-based (Air Force) and sea-going (Navy) alternatives.

The SLMS option consisted of specialized, merchant-type ships carrying long-range missiles that could operate in U.S. coastal waters, at sea in company with other naval ships, or in merchant shipping lanes. They also could launch their missiles while in ports or harbors.[40] Although surface ships were far more vulnerable than were submarines, the argument could be made that they were less vulnerable than fixed, land-based ICBMs, and they were the lowest cost of the proposed four options for attaining a specific amount of missile payload.

Significantly, the ULMS submarine option was the only Strat-X candidate to be fully developed. From its inception, ULMS was envisioned as the successor to the Polaris and Poseidon. At the time, the oldest missile submarines were scheduled to begin being retired in the early 1980s. The ULMS submarine—with missile ranges of more than 6,000 nautical miles—would have a very high at-sea to in-port time ratio, would be designed for rapid missile modernization, and would have enhanced noise-reduction features to increase their survivability against Soviet anti-submarine forces. The last was particularly significant. Although U.S. Navy officials steadfastly denied that the Polaris-Poseidon submarine force was vulnerable to detection or attack by Soviet forces, there was continued

37 Institute for Defense Analyses, *The Strat-X Report* (August 1967), p. 1.

38 See Graham Spinardi, *From Polaris to Trident: The Development of U.S. Fleet Ballistic Missile Technology* (New York: Cambridge University Press, 1994), pp. 113–63, and N. Polmar and Capt. D. A. Paolucci, USN (Ret.), "Sea-Based 'Strategic' Weapons for the 1980s and Beyond," U.S. Naval Institute *Proceedings* (May 1978), pp. 99–113. Also Rear Adm. George H. Miller, USN, one of the two Navy participants in Strat-X, discussions with N. Polmar, Pentagon, Washington, D.C., during 1968–1969. In this period Adm. Miller was Director, Strategic Offensive and Defensive Systems (OP-97); Capt. Paolucci was his deputy.

39 A Minuteman ICBM was test launched from a C-5A transport aircraft on 24 October 1974; the missile was extracted by parachute, and while in descent the solid-propellant rocket was ignited.

40 The original Polaris was proposed for launching from surface ships, both cruisers and specialized merchant-type ships, the latter to be operated by multinational NATO crews.

concern over possible Soviet advances in ASW. Secretary Laird's 1970 statement to Congress noted:

> However, a combination of technological development, and the decision by the Soviets to undertake a world-wide ASW effort, might result in some increased degree of Polaris/Poseidon vulnerability beyond the mid-1970s. I would hope that Polaris would remain invulnerable at least through the 1970s. But, as a defense planner, I would never guarantee the invulnerability of *any* strategic system beyond the reasonably foreseeable future, say 5–7 years.[41]

Thus, the ULMS/Trident submarine program was initiated. Ten existing Polaris-Poseidon submarines were modified to carry the initial Trident I missile, followed by construction of a new class of very large missile submarines. The first of the new submarines, the USS *Ohio* (SSBN 726), went on patrol in October 1982 carrying 24 Trident C-4 missiles. These and the later D-5 missiles each held eight MIRV warheads, either W76 (100 KT) or W88 (450 KT). Trident submarine construction continued until the end of the Cold War with 18 Trident submarines being built (of a program goal of about 24 submarines).[42]

The high cost and large size of the Trident submarine led Secretary of Defense James Schlesinger in the early 1970s to initiate design efforts for a new and less costly SLBM submarine tentatively known as the "Narwhal" program, because of its use of the same kind of nuclear reactor as in the torpedo-attack submarine *Narwhal* (SSN 671). This submarine was to carry 16 Trident C-4 missiles. As envisioned at the time, the *Narwhal* design would complement the Trident submarine, although criticism of the size and cost of the latter led to speculation that the smaller *Narwhal* type could become the successor to the Polaris/Poseidon submarines. Strong opposition from Admiral H. G. Rickover, head of the Navy's nuclear propulsion program, and resulting congressional opposition to developing a second Trident submarine class simultaneous with the first led to cancellation of the *Narwhal* proposal.

The MX/Peacekeeper development was highly convoluted as U.S. administrations wrestled with the basing problem for the ICBM. Like the Minuteman before it, various hardened and mobile concepts were examined.

While the MX was in development, in the mid-1970s consideration was also given to adopting the Trident C-4 SLBM (which became operational in 1979) to the land-based ICBM role. A "common missile" would have provided an advanced ICBM at an earlier date and, presumably, at less cost than a two-missile effort. However, the Air Force fought the proposal and, according to their perspective, "by the time an SLBM missile variant and silo variant were fielded, the cost would have equaled or exceeded the cost of a two-missile program."[43]

Eventually, between 1986 and 1988, 50 MX/Peacekeeper missiles were installed in hardened former Minuteman silos. That deployment was relatively short lived with the missiles discarded in 2002–2005 under the terms of strategic arms limitation treaties. Each of the 50 missiles carried ten W87 warheads (300 KT) with a range of 6,700 miles.

Beyond MIRV, a more advanced warhead was being developed in the 1970s—the maneuvering reentry

---

41 Melvin R. Laird, *Fiscal Year 1971 Defense Program and Budget* (Washington, D.C.: Department of Defense, 25 February 1970), p. 40.

42 The *Ohio*-class submarines were the largest undersea craft to be constructed in the United States. Two Soviet nuclear-propelled submarines were larger, the Project 941 (NATO Typhoon) and Project 949 (NATO Oscar); the Soviet submarines were armed with ballistic missile (SSBN) and cruise missiles (SSGN), respectively.

	Ohio class	Project 941 Typhoon	Project 949 Oscar
IOC	1981	1981	1980
Displacement[a]	18,750 tons	55,000 tons	22,500 tons
Length[b]	560 feet	564¼ feet	472⅓ feet
Beam	42 feet	74¾ feet	59¾ feet
Draft	36¼ feet	36 feet	30⅙ feet

a Submerged
b Overall

43 Brig. Gen. William B. Davitte, USAF, Commentary on "One Missile for the Triad," U.S. Naval Institute *Proceedings* (February 1990), pp. 20, 22. The original article was D. D. Dalgleish and L. Schweikart, "One Missile for the Triad," *Proceedings* (August 1989), pp. 73–77.

vehicle (MaRV). The Mk 500 MaRV effort was initiated to help evade ABM interceptor missiles, although terminal guidance to increase ICBM accuracy was also an option. However, as developed, the Mk 500 actually had less accuracy than contemporary MIRVs. The MaRV program reached the test stage but was neither fully developed nor deployed as Soviet ABM deployments remained limited. The MaRV was configured to be compatible with Trident I/II and possibly the Minuteman III missiles.

CONTINUING ATTEMPTS AT MISSILE DEFENSE

Based on the continuing concern over Soviet strategic missiles and advances in radar and intercept technologies, by 1969 the Department of Defense was advocating a ground-based missile defense system. It was first proposed in 1967 as the Sentinel system to provide a full-fledged defense against Chinese missiles. In 1969 it was realigned to defend the Minuteman ICBMs against a Soviet attack and renamed Safeguard as President Nixon shifted his attention to the increasing Soviet missile threats. In August 1969, the Senate narrowly approved the Safeguard deployment with Vice President Spiro T. Agnew casting the tie-breaking vote.

Under the various ABM constructs the defensive missiles being developed would carry nuclear warheads. These would be the short-range Sprint (W65 and then W66 warhead) and long-range Spartan (W71 warhead).

The ABM concept was changed again by the U.S.-Soviet agreement resulting from the Strategic Arms Limitation Talks (SALT) of 1972. Under this agreement, each nation was allowed to build and operate one ABM site in defense of the national capital and another in defense of an ICBM base. Each site was defined according to area, number of missile launchers, and types of radars. This aspect of SALT permitted the Soviet Union to maintain the Galosh ABM already emplaced around Moscow (which also defends some ICBMs in that area), and the United States to continue the construction of a Safeguard ABM being installed near Grand Forks, North Dakota, to defend 150 Minuteman ICBMs. A protocol of 3 July 1974, reduced the number of ABM deployment areas from two to one, and the number of allowable interceptor missiles from 200 to 100.

Finally, on 1 October 1975, the Safeguard installation at Grand Forks, North Dakota, became operational. The next day the House of Representatives voted to close the Safeguard site because new Soviet multiple warheads could easily overwhelm Safeguard. Beyond the Safeguard's vulnerability to direct attack, technical problems, such as radar blinding by electromagnetic pulse from exploding nuclear warheads, made the system unreliable, and even threatened to damage the Minuteman ICBMs it was intended to protect. On 18 November 1975, the Senate also voted to terminate Safeguard.

In February 1976 the Safeguard system was closed down—it had been operational for 133 days! Some Safeguard radars were incorporated into the North American Air Defense Command's warning and assessment network. In the interim, the Soviet Union had continued to develop and deploy two ABM systems, one around Moscow and one around Leningrad. Subsequent to the 1974 protocol, the Leningrad system was dismantled, followed two decades later by the Moscow system.

Another ABM concept that received attention in the late 1960s was the U.S. Navy's proposal for a sea-based ballistic missile intercept system (SABMIS). This concept provided for deploying SABMIS ships in the North Atlantic and North Pacific where their interceptor missiles could shoot down Soviet- or Chinese-launched ICBMs. The advantages of the forward-deployed ships included early intercept, before multiple warheads fully separated and hence were easier to kill. Moreover, the intercepts would take place over the sea and not the United States and Canada, as with land-based ABM systems, and the launch of interceptor missiles would not interfere with ICBM launches from the United States, as would colocated ABM systems. Studies indicated that the SABMIS ships—working either independently or with terminal defenses in the United States—would be both survivable and effective. But SABMIS was rejected primarily because of U.S. interservice policies that favored maintaining Army leadership in continental air-missile defense; hence, only the (limited) Safeguard ABM program was put forward by the government for consideration.

However, low-level interest in land-based ABM systems continued, with limited research being undertaken. Ballistic missile defense again received impetus when President Ronald Reagan, on 23 March 1983, in a nationwide television speech, asked: "What if free people could live secure in the knowledge that their security did not rest upon the threat of instant U.S. retaliation to deter a Soviet attack, that we could intercept and destroy strategic ballistic missiles before they reached our own soil or that of our allies?" The President continued:

> I know this is a formidable, technical task, one that may not be accomplished before the end of the century. Yet, current technology has attained a level of sophistication where it's reasonable for us to begin this effort. It will take years, probably decades of efforts on many fronts. There will be failures and setbacks, just as there will be successes and breakthroughs. And as we proceed, we must remain constant in preserving the nuclear deterrent and maintaining a solid capability for flexible response. But isn't it worth every investment necessary to free the world from the threat of nuclear war? We know it is.

The press quickly dubbed the president's proposed program as "Star Wars," after the epic science fiction films by George Lucas.

Although President Reagan was unable to garner sufficient public or congressional support for a full-scale ABM research and development program, investment continued at a significant level through the administrations of George H. W. Bush and Bill Clinton. Subsequently, President George W. Bush announced early in his first administration that a limited ABM system would be operational by the winter of 2004, primarily to defend against ICBMs launched from "rogue states," especially North Korea. (At this writing North Korea has demonstrated that it could detonate a nuclear device but not that it possesses a missile with ranges that could reach the continental United States.)

By mid-2008 there were 21 interceptor missiles installed at Fort Greely (near Fairbanks), Alaska, and three interceptors at Vandenberg AFB in California. These were nonnuclear missiles intended to intercept and kill an incoming ICBM warhead in a high-speed collision. However, a series of embarrassing intercept failures delayed the program and raised more questions about the efficacy of ballistic missile defenses.

At the same time, under a joint missile defense office, the Navy's shipboard Aegis anti-air/anti–cruise missile system was being upgraded to provide an ABM capability against ICBMs in their boost phase. The Navy was directed to modify several Aegis cruisers and destroyers for this role, using an enhanced (nonnuclear) Standard-series missile. The Aegis—an advanced radar/fire control system—was the most capable air/missile defense capability in existence, either ashore or afloat. On 20 February 2008, a Standard SM-3 missile fired from the Aegis cruiser *Lake Erie* (CG 70) successfully intercepted an errant U.S. spy satellite that was reentering the atmosphere. The "hard-kill" intercept of the satellite, reentering the atmosphere at hypersonic speed, occurred at an altitude of about 130 miles. Most of the Navy's 22 Aegis cruisers and 62 Aegis destroyers were to be modified for the ballistic missile intercept role in addition to their anti-air/anti–cruise missile capabilities.

And, not to be excluded, the Air Force entered the competition for an advanced ABM system with an airborne laser concept. Despite the obvious limitations of a laser intercept of ballistic missiles, that is, interference from clouds, the need to keep aircraft airborne during a crisis, and kill-range limitations, the Air Force persevered, flying a small-scale test aircraft in 2004, the YAL-1, a reconfigured Boeing 747. Of all of the candidate "Star Wars" systems, the airborne laser was the most complex and offered the least flexibility; accordingly, the Department of Defense decided to pursue the airborne laser as a test program without plans for development of an operational system. Significantly, none of these "Star Wars" ABM systems employed nuclear warheads as a kill mechanism.

MORE STRATEGIC BOMBERS

Strategic bombers survived throughout the Cold War as a part of the Triad. Although Secretary McNamara preferred strategic missiles to bombers, the Air Force was able to initiate studies in 1965 for the advanced manned strategic aircraft (AMSA), a successor to the B-52. The AMSA was to deliver standoff or lay-down weapons; the AMSA would be supersonic at high altitudes and have a high subsonic, low-level penetration speed. The program gained momentum after McNamara stepped down as secretary of Defense on 29 February 1968.

In June 1970 the Air Force selected a Rockwell International design for the AMSA, soon redesignated B-1.[44] It would carry bombs as well as the new SRAM—short-range attack missile—which had a range of 135 miles when launched from high altitudes, and carried the W69 warhead (170–200 KT). The SRAM could also be delivered by the older B-52G/H and FB-111A. But in the 1970s

---

44 The Rockwell International Corp. produced the aircraft at the firm's North American Aircraft Operations at El Segundo and Palmdale, Calif.

three factors argued against manned bombers. First, the B-52 losses in the bomber strikes over North Vietnam and of Israeli tactical aircraft losses to Soviet SAMs in the Middle East led many government officials and bomber opponents to argue that the large and modern Soviet air defense network made strategic bombers obsolete.

Second, U.S. strategic bomber and tanker bases were considered to be increasingly vulnerable to the possibility of prelaunch attacks by modern Soviet SLBM submarines operating off of American coasts. The bomber prelaunch survivability was of special concern should the Soviets develop a depressed-trajectory flight profile. (However, Western intelligence did not observe any Soviet SLBM flights that demonstrated a deliberate attempt to minimize flight time.[45])

A third factor—understood only by "insiders"—was the fallacy of a principal Air Force argument for manned bombers: that they could be "recalled." The argument asserted that while ICBMs and SLBMs, once launched, could not be stopped, manned bombers could be recalled. The argument, however, ignored the fact that a 500-mph bomber flying the Arctic route from the United States to Moscow would be over the offshore area that was part of the Soviet air-defense zone and over Soviet territory for about three hours before reaching Moscow. During that time it would be attacked by Soviet fighters and, subsequently, SAMs. The bomber's recall would be problematical at best. Flight times for ICBMs and SLBMs were 30 minutes or less, in reality giving them a longer "recall" time.

The existing B-52 force was predicted to "wear out" in the 1980s, when the newest of the aircraft would be more than 18 years old. The FB-111 was considered an "interim" strategic bomber, essentially forced on SAC by Secretary McNamara's need to increase the TFX/F-111 production base.

The B-1 strategic bomber first flew in December 1974. It was a large aircraft and featured variable-sweep (swing) wings to provide optimum flight configurations for high-altitude cruise, low-level dash, and landing. Powered by four jet engines, the B-1 had a gross weight of some 390,000 pounds (compared to 480,000 pounds for the B-52G/H and 100,000 pounds for the FB-111A). At high altitudes the B-1 had a Mach 2 speed; for low-level strategic attacks it slowed to about 600 mph. The unrefueled combat radius was reported as 3,000 miles, meaning that the B-1 would require in-flight refueling from tanker aircraft to reach primary targets in the Soviet Union from bases in the United States.

Early Air Force proposals called for 240 B-1 production aircraft. Four prototypes had been produced when the program was cancelled in its entirety in June 1977 by the recently installed Carter administration. In response to the outcries from Air Force leaders as well as Republican critics, President Jimmy Carter hinted that a more advanced bomber was being developed.

The inauguration of President Ronald Reagan in January 1981 brought renewed interest in strategic and naval programs. The aborted bomber program was resuscitated. Two of the prototypes were brought into a B-1B development program. The "new" aircraft was to be a standoff bomber, carrying standoff cruise missiles to attack Soviet targets. Without the penetration requirement, the maximum speed was reduced from Mach 2 to Mach 1.25, which made possible a higher gross weight—477,000 pounds. The aircraft's radar cross section was also reduced.

Although the Reagan administration immediately accelerated the Advanced Technology Bomber (ATB) program, as an interim measure the B-1 program was reinstated in 1981 with the first production aircraft flying on 18 October 1985. One hundred B-1B Lancer aircraft were built through 1988. The B-1B became operational October 1986, and was immediately found to suffer major problems with its electronic systems needed to penetrate Soviet air defense systems. The plane first saw combat as a conventional bomber using unguided "iron bombs" against Iraqi targets in Operation Desert Fox in December 1998. The aircraft subsequently participated in the NATO air strikes in Kosovo and the campaigns against Afghanistan (beginning in 2001) and Iraq (2003).

The ATB, which would fly as the B-2, was a low-observable or stealth aircraft. It was essentially a flying wing based on Jack Northrop's flying wing bombers—the B-35 piston-engine and B-49 turbojet variant—that flew shortly after World War II. The ATB/B-2 was developed in great secrecy, the first public viewing taking place in late 1988 at its "roll out" at the Northrop facility at Palmdale, California. The first flight occurred six months later, on 17 July 1989.

45 Central Intelligence Agency, "Depressed Trajectories: Unlikely Role for Soviet SLBMs" (Intelligence Memorandum, SW 82-10075) (August 1982), p. 6.

The shape of things that go boom in the night: A B-52H Stratofortress and a B-21A Spirit flying in formation—the oldest and newest strategic bomber aircraft in U.S. service. The radical, flying-wing design of the B-2 dates to the efforts of Jack Northrop in the late 1930s, with the Army procuring limited numbers of his B-35 (piston) and B-49 (jet) flying wings in the late 1940s. (U.S. Air Force)

The B-2's flying wing or "bat" shape belies its great size—maximum takeoff weight is 376,000 pounds with a payload of 50,000 pounds of bombs or missiles in two weapon bays. Propelled by four turbofan engines imbedded in the wing, the aircraft has a crew of only two with an unrefueled intercontinental range.

Even as the first aircraft became operational in December 1993, the B-2 was bedeviled by cost growth, technical problems, and the fragile nature of its skin coatings, the last a key factor in stealth technology. The Air Force persevered, and despite the demise of the Soviet Union, the single prototype, which flew on 17 July 1989, was followed by 20 production aircraft. All were assigned to the 509th Bomb Wing—successor to the 509th Composite Group, the world's first nuclear strike force.

As late as 2003, the Pentagon's annual report on operational test and evaluation noted that the B-2's availability was still inadequate, mainly because of maintenance on the aircraft's low-observable materials, and the B-2's defensive avionics also had problems. In December 2003 the B-2 force reached full operational capability.

The B-2s saw combat in the air campaign in Kosovo and the 2003 invasions of Iraq and Afghanistan as did B-52 and B-1 bombers. In some of these operations the B-2s flew missions of about 30 hours from their U.S. bases; the B-52s flew missions of 35 hours, both using multiple in-flight refuelings.

Although not generally considered in the context of strategic aircraft, from the early 1950s through the end of the Cold War the United States deployed "tactical" strike aircraft in Europe and Asia as well as aboard aircraft carriers with nuclear weapons that could strike the Soviet Union, eastern Europe, China, and North Korea. In addition, several NATO air forces flew strike aircraft that could carry U.S. nuclear weapons.

In Europe these aircraft flew from several NATO bases, with nuclear weapons stored on their territory under U.S. military control. In Asia the planes and weapons were located at bases in Okinawa, South Korea, and Guam. (Okinawa was Japanese territory and Guam became an independent territory of the United States in 1950.)

The shape of things to come: The first flight of the F-35A Lightning II on 15 December 2006. This aircraft apparently will have a nuclear weapons capability, but not the closely related F-35B short takeoff/vertical landing nor F-35C carrier-capable variants. Originally known as the joint strike fighter (JSF), the F-35 will also be flown in its various configurations by several other nations. (Lockheed Martin)

The Air Force land-based strike aircraft were primarily the F-84 Thunderstreak/Thunderjet, F-100 Super Sabre, F-105 Thunderchief, F-4 Phantom, and F-111. The F-84 was evaluated with a zero-length launcher for quick-response takeoff of nuclear strike aircraft in forward areas. At various points of the Cold War these tactical strike aircraft at forward bases numbered in the hundreds.

At sea Navy carriers operated a succession of attack aircraft—the AD Skyraider, A3D Skywarrior, A4D Skyhawk, A3J Vigilante, A-7 Corsair—as well as several fighter aircraft—the F7U-3 Cutlass, FJ Fury—with a nuclear strike capability. The later Navy fighters, such as the F8U Crusader and F4H Phantom, did not have a nuclear capability as the sea-based component of the Triad was shifted to submarine-launched missiles, and there were sufficient "attack" planes for tactical/theater nuclear strikes. For most of the Cold War the Navy normally operated two large aircraft carriers in the Mediterranean Sea and two or three in the western Pacific. During crisis periods, and during the Vietnam War, the number of carriers forward deployed was increased to a majority of the carriers in service.

In addition to Air Force and Navy nuclear-strike aircraft, the U.S. Marine Corps has a large air arm, usually three aircraft wings, generally with fighter and attack aircraft of the same type as the Navy operated. Several Marine fighter-attack aircraft were configured to carry nuclear weapons and could operate from land bases or aircraft carriers, with all Marine pilots being trained in carrier operations.

THE NUCLEAR DRAWDOWN

The SALT agreement of 1972 set the course for future strategic arms negotiations between the United States and the Soviet Union, although President Jimmy Carter withdrew a SALT II treaty from the Senate in January 1980 because of the Soviet invasion of Afghanistan. The SALT sessions were followed by the more comprehensive Strategic Arms Reduction Talks (START), which further limited strategic weapons deployments, both offensive and defensive.

Perhaps the most comprehensive agreement between the two superpowers came when President Reagan and First Secretary Mikhail Gorbachev personally negotiated the Intermediate-range Nuclear Force (INF) treaty. Signed in Washington, D.C., on 8 December 1987, it was history's first treaty to eliminate a complete class of weapons, with the more than 7,600 INF nuclear weapons to be eliminated by 31 May 1991. The treaty also had unprecedented rights for both nations to conduct on-site inspections in the United States and Soviet Union.

The U.S. weapons covered by the INF treaty were the Pershing II missile and the ground-launched cruise missile (GLCM) version of the Tomahawk, both of which were in the process of being installed in European NATO nations as a counter to recent Soviet theater missile deployments. Even before the end of the Cold War, the large Titan II ICBMs were taken off strategic alert, the last on 5 May 1987.

When the Cold War ended with the dissolution of the Soviet Union in December 1991, the United States had some 20,000 nuclear weapons in service. President George H. W. Bush, on 27 September 1991, announced a massive cutback in U.S. nuclear weapons—tactical and strategic—a decision that led to the rapid reduction of approximately one-half of the weapons. Rapidly discarded, or at least placed in storage under this unilateral effort, were all ship-based weapons—except for Trident SLBMs in 18 submarines—and all nuclear rounds for Army and Marine Corps artillery.

The START I Treaty entered into force on 5 December 1994, establishing the post–Cold War nuclear arsenals for the United States and Russia:

- 1,600 deployed ICBMs, SLBMs, and "heavy" bombers for each country.
- 6,000 "accountable" warheads on ICBMs, SLBMs, and heavy bombers of which
    - –no more than 4,900 on ICBMs and SLBMs
    - –no more than 1,540 on heavy missiles (the Russian SS-18)
    - –no more than 1,100 on mobile ICBMs (the Russian SS-27 Topol-M).
- Ballistic missile throw weight is limited to 3,600 metric tons for each country.
- START I to continue for 15 years (to December 2009) with an option to extend for successive 5-year periods or indefinitely.

Two subsequent agreements further shaped the U.S. and Russian nuclear arsenals—the START II Treaty (1993) and the Moscow Agreement (2002).

The allowable B-2 and B-52H bomb strength would remain the same, with the B-1 Lancer being configured to carry only conventional bombs and air-to-surface missiles. The MX Peacekeeper ICBM force has been retired—the last missiles taken out of their hardened silos in September 2005. Future plans entail that the 450 surviving Minuteman III missiles will carry 500 warheads—that is, some carrying two warheads—with the downloading to be finished by the end of 2012. Some of the Minuteman III force will be refitted with W87 warheads from the retired MX missiles.

The 4 oldest Trident submarines were converted in 2004–2006 to combination Tomahawk cruise missile/special forces submarines (designated SSGN). The next 4 Trident submarines have been rearmed from the C-4 to the D-5 variant of the Trident missile (i.e., currently 14 submarines carrying 336 missiles). The D-5 missile, in turn, is being downloaded to an average of six MIRVs, still providing the largest number of "weapons" as well as launch systems in the U.S. nuclear arsenal. Still, with the downloading of most of the surviving Minuteman III missiles to a single-warhead configuration and the demise of the MX/Peacekeeper, the Trident SLBM is the only U.S. missile with a multiple warhead except for the few Minuteman III missiles (see Table 1-3).

TABLE 1-3. U.S. Nuclear Weapon Systems, January 2009

	Weapon	Missiles/Aircraft	Yield	Total Warheads
ICBM	Minuteman III	450		
	Mk 12A RV	(250)	1–3 W78 × 335 KT	350
	Mk 21/SERV	(200)	1 W87 × 300 KT	200
SLBM	Trident II D-5	336[a]		
	Mk 4/A		6 W76 × 100 KT	960
	Mk 5		6 W88 × 455 KT	384
Bomber	B-52H Stratofortress	44[b]	ALCM/W80-1 × 5–150 KT	375
Bomber	B-2A Spirit	15[c]	B61-7/11, B83-1 bombs	175
Attack	F-15E Eagle[d]	}	B61-3/4 bombs × 0.3–170 KT	400
Attack	F-16 Fighting Falcon[d]			

a In 14 nuclear-propelled submarines; 2 are normally in overhaul.
b Mission aircraft; total inventory is 93 aircraft with the others employed in research and training activities.
c Mission aircraft; total inventory is 20 aircraft with the others employed in research and training activities.
d Fighter-bomber aircraft based in Europe and the United States.

THE COLD WAR WON

The collapse of the Soviet Union brought about precipitous reductions in the U.S. nuclear arsenal—from some 20,000 weapons in 1991 to some 7,000 nuclear weapons in 2004, to about 4,300 in 2008, and to between 1,700 and 2,200 "operationally deployed" strategic warheads by 2012. Assistant Secretary of Defense (International Security Policy) J. D. Crouch, in briefing the U.S. nuclear posture review in January 2002, said that many of the removed warheads would be kept in reserve as a "responsive force," able to be installed in missiles within weeks or months.

The Cold War, Crouch said, had provided in the Soviet Union

> basically a known, single ideological opponent. . . . We relied not exclusively, but very heavily on our offensive nuclear forces, and we had a threat-based approach to nuclear planning. . . . The focus was on the Soviet Union, which of course is no longer with us, and because everything else was sort of a lesser included case—any other potential contingencies were lesser included cases.

With the end of the Cold War came a new appraisal of the U.S. structure for nuclear war. The surviving nuclear systems—bombers and missiles—were taken off alert status and their targets were deleted from missile guidance systems, although, obviously, target data could be quickly reentered. And the basic Triad structure was changed.

In 2002 the George W. Bush administration proposed a new Triad composed of nuclear forces—ICBMs, SLBMs, and manned bombers—to form a mini-Triad within a broader structure. The reduced U.S. nuclear arsenal was to be augmented in the new Triad by a national missile defense system and highly response conventional forces, all linked by upgraded command-control, intelligence, and planning activities (see Figure 1-1).

The U.S. strategic offensive force, however, retains the "old Triad" flavor: land-based Minuteman III ICBMs, sea-based Trident II SLBMs, and land-based B-2A and B-52H manned bombers. The B-2s carry only gravity "dumb" or guided nuclear (gravity) bombs; the B-52s can carry gravity bombs as well as ALCM nuclear-armed cruise missiles. The nuclear bombs are a mix of B61 and B83 weapons. (Both bombers can also carry a variety of conventional strike weapons.)

In addition, about 200 B61 nonstrategic nuclear bombs are available for F-16C/D Fighting Falcon, F-15E Strike Eagle, and allied Tornado "fighters" located at six bases in five European NATO countries. (The nuclear weapons storage site at RAF Lakenheath, England, was closed in 2008.) About 200 more nonstrategic B61s

FIGURE 1. The New Triad

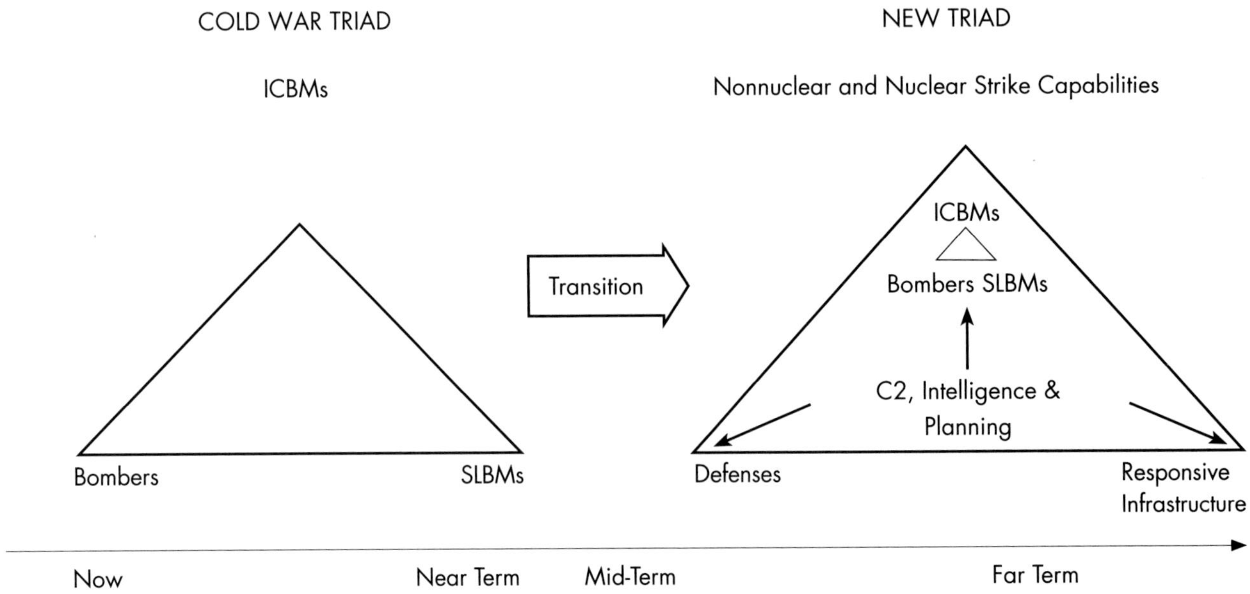

A B-1B Lancer stands ready at an airfield in Southwest Asia. The "bone"—a nickname derived from B-one—failed to meet the requirements for a strategic bomber that could effectively penetrate Soviet air space, but has served credibly as a conventional bomber in post-Cold War conflicts. (U.S. Air Force)

are in reserve in the United States. Several accidents in 2007–2008 forced the grounding of some F-15 aircraft, questioning the viability of U.S. tactical nuclear strike capabilities.

Efforts by both of the George W. Bush administrations to develop follow-on nuclear bombs have been defeated or deferred by Congress. The Air Force has modernized the warheads and guidance of the Minuteman III ICBM force; similarly, the Navy has extended the service life of the 14 remaining Trident submarines to 44 years (i.e., the first to retire in 2029) and has upgraded their missiles.

For the future, the Air Force is looking at the design of a new bomber to carry out both conventional and nuclear strike missions, and a nuclear-capable, air-launched cruise missile. The Navy has requested funding for the initiation of a design of a new strategic missile submarine.

The F-35 Lightning II series joint strike fighter (JSF) may be nuclear capable, at least under current plans. This aircraft will be flown in three principal configurations:

- F-35A conventional variant for U.S. Air Force replacing:
    F-15 Eagle, F-16 Fighting Falcon, F-117A Nighthawk (stealth aircraft)
- F-35B short-takeoff/vertical-landing variant for U.S. Marine Corps and Royal Navy replacing:
    AV-8B Harrier, RAF and RN Harrier
- F-35C carrier variant for U.S. Navy replacing:
    F/A-18C/D Hornet

Most likely only the F-35A will be nuclear configured as neither U.S. nor British aircraft carriers are currently fitted to store nuclear weapons.

One other nuclear system warrants mention in this context: the next-generation bomber. By 2007 the U.S. Department of Defense and Congress were agreed that manned, long-range bombers were important for conventional conflicts, as demonstrated by the use of strategic bombers in the conflicts in Afghanistan and Iraq. The new aircraft—to become operational in 2018—would have a nuclear as well as conventional weapons capability. The aircraft would replace some or all of the surviving B-52H and B-1B long-range bombers (see Chapter 3).

Thus, while the number of nuclear weapons and the number of nuclear delivery systems in the U.S. arsenal are declining, nuclear weapons and their delivery systems will be a component of U.S. national strategy and military forces for the foreseeable future.

CHAPTER TWO

Nuclear Warheads

Following the original Fat Man and Little Boy nuclear weapons—which were not given numerical designations—all U.S. nuclear weapons have been numbered in sequence. (The Mark 3 was essentially a duplicate of the Fat Man weapon.)

All U.S. nuclear weapons have been of the implosion type except for seven that had a gun-assembly configuration: Little Boy, Mk 8, Mk 9, Mk 11, Mk 19, Mk 23, and W33. The Mk 9, Mk 19, Mk 23, and W33 warheads were for artillery shells. (Several other gun-type weapons were initiated but not completed.)

The Atomic Energy Commission (AEC) developed a basic "phase" schedule for warhead research, development, and production that lasted throughout the Cold War: Phase 1 was for concept definition studies by the two nuclear design laboratories (see below); Phase 2 was for joint feasibility studies; Phase 3 was development engineering with one of the two laboratories being assigned the development; Phase 4 was for production engineering; Phase 5 was for initial production; Phase 6 was for quantity production; and Phase 7 was for withdrawal from the stockpile and eventual disassembly.

The characteristics of atomic demolition munitions (ADM) are provided under their warhead descriptions:

ADM	W31
ADM-B	Mk 7
Medium ADM	W45
Special ADM	W54
T4	Mk 4
T-ADM	W30

U.S. nuclear weapons were developed at the Los Alamos National Laboratory in Los Alamos, New Mexico, and the Lawrence Livermore National Laboratory near Oakland, California.

Designation: Initially mark numbers (Mk) were used for U.S. nuclear weapons, but subsequently the letters B for bomb, TX for bomb prototype, and W for warhead (for missiles and artillery projectiles) were applied. Some nuclear warheads have had multiple prefixes. Several suffix schemes have been used for nuclear warheads—mod for modification, Y for yield, and simply a hyphenated suffix number, as B60-1 for B60 Mod 1. In this volume the designations listed are those most commonly used or are listed in primary documents (e.g., Standard Aircraft Characteristics sheets). While the prefix may vary, the numerical sequence remains the same throughout the text.

Little Boy

The Little Boy was the first atomic bomb used in combat. Development of the world's first nuclear weapons—Little Boy and Fat Man—was begun at Los Alamos, New Mexico, in early 1943.

Design: The steel gun barrel of the weapon was approximately six feet long and weighed about 1,000 pounds. The subcritical uranium projectile was fired through the barrel into a subcritical uranium target propelled by several pounds of cordite. Projectile velocity reached 1,000 feet/second. This action produced the supercritical mass necessary for a nuclear detonation.

The Little Boy used about ten times more fissile material than the contemporary Fat Man—approximately 141 pounds of uranium compared to 13.5 pounds of plutonium.

Designation: The term "gadget" was generally used during the war when referring to the atomic bombs. The terms Fat Man and Little Boy were reportedly assigned as code words for protection against possible espionage efforts to the implosion and gun-type bombs, respectively. It was hoped that enemy agents who heard those terms would believe that they referred to aircraft being modified to

Little Boy atomic bomb. (U.S. Air Force)

carry Prime Minister Winston Churchill (Fat Man) and President Roosevelt (Thin Man—an elongated bomb design that was discarded). The terms Big Boy, Round Man, and Big Fellow also were applied at various times to the implosion bomb.

Operational: The major components of the Little Boy dropped on Hiroshima were carried to the U.S. base complex on Tinian in the Mariana Islands by the heavy cruiser *Indianapolis* (CA 35), arriving on 26 July 1945; three C-54 cargo aircraft subsequently flew out the last of the U_{235}, and minor parts for the bomb and arrived at Tinian on 28 July.

The bomb was dropped on the Japanese city of Hiroshima on 6 August 1945, delivered by the B-29 *Enola Gay* of the 509th Composite Group based on Tinian. The bomb detonated about 1,900 feet above Hiroshima with an explosive force of 15 kilotons.[1] The city was instantly devastated.

The effects of the explosion killed an estimated 70,000 people in Hiroshima. Another 20,000 to possibly as many as 70,000 persons died by the end of the year because of burns, radiation, and disease caused by the bomb.

Production: There have been reports that in addition to the Hiroshima bomb, five Little Boy weapon assemblies were produced between August 1945 and February 1950. However, no further Little Boy bombs are known to have been produced because of the large quantities of uranium required.

Inventory:	1 Aug 1945–Jan 1951	
Yield:	15 KT	
Weight:	9,700 lb (4,400 kg)	
Dimensions:	length	10 ft (3.05 m)
	diameter	28 in (711 mm)
Weapons:	aircraft bomb	
Aircraft:	B-29	

Fat Man/Mk 3

The Fat Man was the second atomic bomb to be used in combat although it was the world's first nuclear weapon to be detonated. It was developed at Los Alamos. The designation Mk 3 was used for subsequent production weapons.

Design: The Fat Man's large diameter was required to accommodate the two layers of high explosive "lens" necessary to implode the plutonium core. The later Mk 6 and Mk 18 implosion weapons, for example, had cores surrounded with a 17-inch layer of high explosive charges. Accordingly, about one-half of the weight of the Fat Man was high explosives required to compress the nuclear charge.

The bomb had to be assembled shortly before use because of the limited life of the batteries required for the detonators; the bomb had to be partially dismantled to replace or recharge them.

The production Mk 3 weapons had improved fuzing and other features. However, "shelf life" still was a problem for these weapons; once installed the batteries—necessary for detonators—had a life of only nine days (during which time they had to be recharged twice). After nine days the bomb had to be dismantled to remove or replace the batteries. Also, the plutonium core of the weapon had to be removed after ten days or the thermal effects from the alpha decay could damage bomb components.

Fat Man atomic bomb. (U.S. Air Force)

1 Estimates of the Hiroshima and Nagasaki bomb yields vary considerably; the numbers used here are from the definitive analysis by John Malik of the Los Alamos National Laboratory in his *The Yields of the Hiroshima and Nagasaki Nuclear Explosions* (Los Alamos, N.M.: Los Alamos National Laboratory, September 1985).

The Fat Man used about 13 pounds of plutonium.

Operational: The second atomic bomb used in combat was a Fat Man dropped on the Japanese city of Nagasaki on 9 August 1945, released by the B-29 *Bockscar* of the 509th Composite Group based on Tinian. The plane's original target was the city of Kokura; however, clouds and smoke swathed that city, and flak bursts erupted around the plane. The pilot and weaponeer opted for the alternate target of Nagasaki.

The bomb detonated 1,650 feet above the city with a force of 21 kilotons. Hemmed in by hills around the impact point, the explosion caused less damage than the Hiroshima bomb. Still, in the instant of the explosion, an estimated 35,000 Japanese died and 60,000 were injured; additional deaths were caused by injuries and radiation, raising the death toll to perhaps 70,000 by the end of 1945.

B-29s flew three Fat Man bomb assemblies to Tinian in July 1945, with a C-54 subsequently flying out the critical components (including the plutonium core) for one bomb. The plutonium core for a second Fat Man bomb was scheduled to be dispatched from Los Alamos on 12–13 August, to be ready for use against another Japanese target, depending upon weather conditions, after 17 August. The dispatch of that weapon was halted when it appeared that Japan would accept the allied surrender terms.

Two Fat Man–type bombs were detonated in the 1946 Operation Crossroads tests at Bikini atoll in the Marshall Islands. Crossroads was originally intended to determine the effects of atomic bombs on naval ships, but the program rapidly expanded to include numerous scientific and military tests supported by the Manhattan Project as well as the military services. A vast array of U.S. Navy ships as well as surrendered German and Japanese warships were employed as targets, their decks loaded with vehicles, aircraft, ordnance, and test animals, as well as technical instrumentation.

In test Able on 1 July 1946, the B-29 *Dave's Dream* dropped a Mk 3 bomb that detonated at an altitude of 520 feet above the target ships. The abandoned battleship *Nevada* (BB 36) was the aim point, the center of a target array of 78 ships and craft. The bomb missed the aim point by 2,130 feet. The 50-year-old unmanned *Nevada* was burnt and battered by the 21-kiloton blast but not sunk; several other ships did go down as a result of the blast, and others were wrecked or damaged.

For test Baker a Mk 3 bomb was suspended 90 feet below the landing ship *LSM 60*. That bomb, of approximately 21 kilotons, was detonated on 25 July. A column of water nearly a half-mile across rose almost 6,000 feet into the air, tossing warships around like toys. Several additional target ships were sunk and damaged.

A planned third test, an underwater detonation 1,000 feet below the surface, was cancelled, ostensibly because scientists and military officers felt that there were sufficient data on the effects of nuclear weapons. However, there is evidence that the third detonation—test Charlie—was called off because of the limited number of atomic weapons available, and the fissionable material needed for additional bombs was in short supply.[2]

The Bikini tests demonstrated that ships were vulnerable to atomic attack but also that ships could survive such attacks if dispersed. Studies led to eventual shipboard protective features and formation tactics to mitigate the effects of nuclear weapons. The Bikini atoll subsequently was used for 21 additional nuclear weapon tests from 1954 to 1958.

(Possibly having a more lasting impact than the weapons tested there, Paris clothing designer Louis Réard introduced his own bombshell—the bikini swimsuit—following the first Bikini bomb tests.)

Production: By the summer of 1945 production rates for Fat Man bomb components and critical material were two bombs in July (one detonated at Alamogordo, New Mexico, and one over Nagasaki), one in August, and a planned production of three in September, with the production rate expected to increase to seven or more per month by December 1945. However, production and subassembly was halted in mid-August 1945 when Japan announced capitulation.

By 30 June 1946—ten months after the end of the war—there were nine nuclear components and assemblies for Fat Man–type weapons. These were now designated Mk 3, the original Fat Man and Little Boy designs having had the first two numerical places although not actual designations.

2 Memorandum from James V. Forrestal to President Truman, 30 August 1946, in "NSC Atomic Test—Misc." folder, Box 201, Harry S. Truman Library.

A Mk 4 atomic bomb being loaded into the forward bomb bay of a B-50 Superfortress bomber. The aircraft in jacked up forward to permit the five-foot-diameter weapon to be loaded into the bay. The B-50 was a variant of the B-29 and served in the bomber role in the Strategic Air Command until 20 October 1955, when the last B-50D was retired. The RB-50 reconnaissance variant was also retired that year. (U.S. Air Force)

Inventory:	1 Aug 1945–1950	
Yield:	21 KT (see Operational notes)	
Weight:	10,300 lb (4,681 kg)	
Dimensions:	length	10 ft 8 in (3.25 m)
	diameter	5 ft (1.52 m)
Weapons:	aircraft bomb	
Aircraft:	AJ/A-2[3]	
	B-29	
	B-36	
	B-50	

Mk 4

This was an improved Fat Man/Mk 3 weapon and the first U.S. nuclear weapon to be produced on an assembly-line basis and stockpiled in large numbers.

The yield of this weapon ranged between 1 and 31 kilotons, depending upon the nuclear capsule (core) that was inserted into the casing. When addressing the potential use of atomic bombs in the Korean War, the official Air Force history discusses the use of 30- and 40-kiloton weapons, presumably referring to the Mk 4.[4]

Design: The slightly heavier Mk 4 featured an improved ballistic shape and better fuzing, and new composite cores made of plutonium and uranium, which were more efficient uses of fissile materials, as well as levitated cores. These designs were successfully tested in Operation Sandstone in the spring of 1948. The weapon yield could be changed by using different nuclear cores (plutonium, uranium, or composite). Another improvement was that the nuclear core could be more easily inserted after the aircraft was airborne and withdrawn prior to landing.

Final ground assembly of the Mk 4 required at least eight hours plus two hours for checkout.

The design was also incorporated into the first atomic demolition munition (ADM) known as the T4, which was in the stockpile from 1957 to 1963. The operational concept for employing the T4 was to clandestinely parachute a five-man team with the weapon (in several components) into enemy territory. After several failures in exercises, the decision was made that future ADMs would be a deployed as a single package.

Operational: Designs for the Mk 4 cores were tested in the Eniwetok nuclear tests (Operation Sandstone) in April–May 1948. Before the end of 1948, the first production Mk 4 bombs were being delivered to AEC nuclear storage sites.

3 The latter designation indicates the redesignation of U.S. military aircraft effective 1 October 1962.
4 Robert F. Futrell, *The United States Air Force in Korea 1950–1953* (New York: Duell, Sloan and Pearce, 1961), pp. 655–56.

The Mk 4 was the first U.S. nuclear weapon to be deployed outside of the United States. Early in the Korean War, President Truman authorized the transfer of Mk 4 assemblies to Guam in the Western Pacific. Nine assemblies were sent in July and August 1950 without their nuclear cores. The tenth was involved in a nuclear weapon accident when the B-29 transporting it crashed at Fairfield-Suisan AFB (later Travis AFB), California, on 5 August 1950. The high-explosive component detonated, killing 19 crew members and rescue personnel.

Inventory:	Mar 1949–1953	
Yield:	variable 1, 3½, 8, 14, 21, 31 KT	
Weight:	10,800 (4,900 kg)	
Dimensions:	length	10 ft 8 in (3.25 m)
	diameter	5 ft (1.52 m)
Weapons:	aircraft bomb	
Aircraft:	AJ/A-2	
	B-29	
	B-36	
	B-50	

B5

The B5 bomb was the first U.S. lightweight nuclear weapon. With its smaller diameter and reduced weight, it could be carried by more types of aircraft, especially by Navy carrier-based planes. It was assigned for development to Los Alamos in May 1948. The warhead was also used as a missile warhead (see separate entry).

These bombs were supplied to Great Britain in the early 1960s for the Valiant bomber aircraft under Project "E."

A Mk 5 atomic bomb casing on display with the nose "doors" open for loading the plutonium capsule in the weapon. (National Atomic Museum)

Operational: The first production unit was manufactured in May 1952, and the B5 was in production until July 1953. The B5 was first tested as shot Easy of Operation Greenhouse on 21 April 1951, producing a yield of 47 KT. It was further tested in 1952, 1953, and 1955 and was used as the primary for the Operation Ivy thermonuclear test code-named Mike on 1 November 1952.

Withdrawal from the inventory began in 1957 and it was completely retired by January 1963.

Inventory:	1952–Jan 1963	
Yield:	variable 6, 16, 55, 60, 100, 120 KT	
Weight:	3,025–3,175 lb (1,373–1,440 kg)	
Dimensions:	length	10 ft 9 in–11 ft (3.28–3.35 m)
	diameter	3 ft 7¾ in (1.11 m)
Weapons:	aircraft bomb	
Aircraft:	AJ/A-2	B-47
	A3D/A-3	B-50
	B-29	B-66
	B-45	Valiant (British)

Mk 5

This variation of the B5 bomb was developed as a lightweight warhead for the ship/submarine-launched Regulus I, Rigel, and Triton land-attack missiles, and for the land-launched Matador cruise missile. The Regulus I warhead was retired in 1956 and was replaced by the W27.

Inventory:	1954–1958	
Yield:	variable 10 to 45 KT	
Weight:	2,400–2,650 lb (1,089–1,202 kg)	
Dimensions:	length	6 ft 4 in (1.93 m)
	diameter	3 ft 7¾ in (1.11 m)
Weapons:	Matador surface-to-surface missile 1955–1961	
	Regulus I surface-to-surface missile 1954–1958	
	Rigel surface-to-surface missile (cancelled)	
	Triton surface-to-surface missile (cancelled)	

Mk 6

This was the first U.S. nuclear warhead to be produced in large numbers—that is, in excess of 1,000 weapons. It was also the first bomb that could have the weapon's burst altitude selected while the aircraft was in flight.

A Mk 6 atomic bomb being loaded into the forward bomb bay of a B-50 Superfortress bomber. (U.S. Air Force)

The Mk 6 was assigned to Los Alamos, and the first warhead was produced in July 1951; production continued until April 1954 with an estimated 1,100 produced.

Design: This was a successor to the Fat Man and Mk 4 weapons; the outer dimensions of all of these weapons were approximately the same. The Mk 6, Mk 13, Mk 18, and Mk 20 bombs all used the same external casing, which was made of aluminum in place of the ⅜-inch thick mild armored shell of previous weapons, saving, in some models, more than 3,000 pounds.

The variable yields were achieved by using different insertable plutonium cores.

Operational: The Mk 6 was phased out of service in the late 1950s as lighter weight, smaller diameter fission and thermonuclear bombs became available; all were retired by January 1961.

Inventory:	1951–Jan 1961	
Yield:	variable 8, 26, 80, 154, 160 KT	
Weight:	7,600–8,500 lb (3,448–3,857 kg)	
Dimensions:	length	10 ft 8 in (3.25 m)
	diameter	5 ft 10 in (1.78 m)
Weapons:	aircraft bomb	
Aircraft:	A3D/A-3	B-47
	AJ/A-2	B-50
	B-29	B-52
	B-36	B-66

B7/Mk 7

The versatile Mk 7 was put to many uses by the Army, Navy, and Air Force and was produced in ten different models or variants. This was the first nuclear bomb developed to be carried externally by fighter/attack aircraft. In the initial stages the warhead was developed under the nickname Thor (the Norse god of thunder) by Los Alamos.

The last aircraft bombs were retired in 1967. These bombs were supplied to Great Britain for Canberra B(I).8 aircraft under Project "E" from 1960 to 1968.

The Betty was the first Western nuclear depth bomb.

The rocket-boosted BOAR (bombardment aircraft rocket) configuration was the first standoff weapon developed for naval aircraft.

Inventory:	1952–1967	
Yield:	B7 variable 8, 19, 22, 30, 61 KT	
Weight:	1,645–1,700 lb (746–771 kg)	
Dimensions:	length	15 ft 3 in (4.65 m)
	diameter	30½ in (778 mm)
Weapons:	B7 aircraft bomb July 1952–1957	
	atomic demolition munition (ADM/Project B)	
	Betty nuclear depth bomb (Mod 1) 1955–1963	
	BOAR aircraft rocket (Mod 1) 1956–1963	
	Corporal surface-to-surface missile 1955–1964	
	Honest John surface-to-surface missile 1954–1958	
Aircraft:	AD/A-1	F3H/F-3
	A3D/A-3	FJ/F-1
	AJ/A-2	F7U-3
	B-45	P5M/P-5
	B-57	P2V-3C
	F-84	S2F/S-2
	F-100	HSS-1/SH-34
	F-101	Canberra (British)
	F2H/F-2	

Mk 7 atomic bomb. (National Atomic Museum)

Mk 90 Betty nuclear depth bomb with the Mk 7 warhead.

B8/Mk 8 Elsie

This weapon was developed by Los Alamos specifically for naval use against hardened land targets—such as submarine "pens" (bunkers)—and buried structures and was thus the first U.S. nuclear weapon fuzed to detonate after striking the ground as opposed to an air burst. It was an improved gun-assembly Little Boy design.

The nickname Elsie was apparently derived from the designation LC for "light case" bomb. Three B8 "emergency bombs" were produced in March 1951. Production began at the end of 1951 and bombs were produced in limited numbers until May 1953. All the bombs were retired by May 1957 as the Mk 11 entered the stockpile.

The Regulus II ship/submarine-launched attack missile, at one point to have carried a Mk 8 warhead, was cancelled in 1959.

Mk 8 atomic bomb. (National Atomic Museum)

Inventory:	Dec 1951–May 1957	
Yield:	variable 20 to 30 KT	
Weight:	3,230–3,280 lb (1,466–1,488 kg)	
Dimensions:	length	9 ft 8 in–11 ft (2.95–3.35 m)
	diameter	14½ in (368 mm)
Weapons:	aircraft bomb	
	Regulus II surface-to-surface missile (cancelled)	
Aircraft:	AD/A-1	B-45
	AJ/A-2	F2H/F-2
	A3D/A-3	F3H/F-3
	A4D/A-4	F7U-3

Mk 9

This artillery-fired atomic projectile (AFAP) was developed by Los Alamos for the Army's "atomic cannon." The Mk 9 and Mk 19 were the largest nuclear artillery projectiles produced by the United States except for the naval 16-inch (406-mm) projectile with the Mk 23 warhead.

The nuclear components of the shell were nearly identical to the Mk 8 bomb. A full-scale nuclear test (shot Grable) of a Mk 9 being fired from an atomic cannon was conducted at the Nevada Proving Ground on 25 May 1953, witnessed by several prominent military and officials and members of Congress. The enormous 280-mm guns were deployed with U.S. Army units to West Germany, South Korea, and Okinawa.

The weapon was developed under the code name Button.

Inventory:	1952–1957	
Yield:	15 KT	
Weight:	800–850 lb (363–386 kg)	
Dimensions:	length	4 ft 6¾ in (1.37 m)
	diameter	11 in (280 mm)
Weapons:	280-mm artillery projectile T-124	

B10

This joint Air Force–Navy project was initiated by Los Alamos in February 1951 and cancelled in 1952. It was to have been a lightweight, gun-type, air burst weapon.

The warhead was considered for the Loon missile, the U.S. naval version of the German V-1 cruise missile. That weapon was to have provided an interim submarine-launched missile capability pending the availability of the Regulus missile.

Inventory:	cancelled
Yield:	variable 12 to 15 KT
Weight:	1,500–1,750 lb (681–794 kg)
Dimensions:	
Weapons:	Loon surface-to-surface missile (cancelled)
	aircraft bomb

B11/Mk 91

The B11 was an improved Mk 8 penetration bomb developed for naval use against hardened targets such as protected submarine pens. It was developed by Los Alamos and was produced as the Mk 91 Mod 0 bomb.

The weapon was intended to penetrate up to 22 feet of reinforced concrete, 90 feet of packed sand, or 120 feet of clay. It was fitted with a delayed action fuse that would detonate 1½ to 2 minutes after impact. It was designed external carriage at supersonic speeds (as well as internal, bomb bay carriage).

This weapon was a replacement for the Mk 8 bomb and, in turn, was succeeded by the B43 bomb. About 40 Mk 91 bombs were produced through 1957.

A W11 missile warhead program was assigned to a laboratory in October 1951 but was cancelled in May 1955.

Inventory:	Sep 1956–June 1960	
Yield:	variable 10 to 20 KT	
Weight:	3,210–3,500 lb (1,456–1,588 kg)	
Dimensions:	length	12 ft 2 in (3.71 m)
	diameter	14 in (355 mm)
Weapons:	aircraft bomb	
Aircraft:	AD/A-1	B-45
	AJ/A-2	F2H/F-2
	A3D/A-3	F3H/F-3
	A4D/A-4	

Mk 91 penetration bomb with B11 warhead. (National Atomic Museum)

B12 atomic bomb. (National Atomic Museum)

B12

This was a small, tactical bomb carried externally by supersonic fighter-type aircraft as well as subsonic attack aircraft. It was developed by Los Alamos.

The planned use of the B12 as a warhead in several missiles—including the BOAR and Talos—was cancelled in November 1955 in favor of the W30. It was thus produced only in the aircraft bomb configuration.

The B12 was developed under the code name Brok.

Inventory:	Mod 1 Dec 1954–Oct 1958	
	Mod 2 Apr 1956–June 1962	
Yield:	12 or 14 KT	
Weight:	1,200 lbs (545 kg)	
Dimensions:	length	12 ft 11 in (3.93 m)
	diameter	22 in (558 mm)
Weapons:	missile warhead	
	aircraft bomb	
Aircraft:	A3D/A-3	F9F-8B/AF-1J
	A4D/A-4	F3H/F-3
	AJ/A-2	FJ/F-1
	F-86	F7U-3

B13

This high-yield version of the Mk 6 was intended for Air Force and Navy use. The project began in August 1954, but was cancelled later that year in favor of thermonuclear designs that had been successfully tested earlier that year. It was planned for the Snark and Navaho strategic surface-to-surface missiles as well as an aircraft bomb.

The B13 was an implosion-fission weapon developed at Los Alamos.

Inventory:	cancelled 1954
Yield:	several hundred KT
Weight:	7,400 lb (3,358 kg)

Dimensions:	length	10 ft 8 in (3.25 m)
	diameter	5 ft 10 in (1.78 m)
Weapons:	aircraft bomb	
	missile warhead	

Mk 14

The first U.S. deliverable hydrogen bombs. A handful of TX-14s were provided to the military in early 1954 on an "emergency capability" basis. The huge, almost 15-ton bomb could only be carried in specially modified B-36s. Only five were produced with the designation EC 14 (EC indicating emergency capability), and they were soon replaced by the Mk 17 bomb.

The Mk 14 was developed by Los Alamos.

(The next thermonuclear weapon to enter the U.S. stockpile was the 22-ton TX-17 in late 1954 followed by the 2-ton TX-15 in March 1955.)

The weapon was developed with the code name Alarm Clock.

Operational: The warhead was tested in the Union shot of Operation Castle on 26 March 1954 with a yield of 6.9 MT.

Inventory:	Feb 1954–Oct 1954	
Yield:	MT range	
Weight:	28,954 lb (13,133.5 kg)	
Dimensions:	length	18 ft 6 in (5.64 m)
	diameter	5 ft 1½ in (1.56 m)
Weapons:	aircraft bomb	
Aircraft:	B-36	

B15

The B15 was the first lightweight hydrogen bomb and the first fusion weapon to be series produced in the United States. The design was assigned to Los Alamos in October 1953 and was in production from April 1955 until February 1957, with about 1,200 bombs being produced.

The weapon was nicknamed Zombie during its development.

This bomb was supplied to Great Britain for Valiant aircraft under Project "E."

Operational: The B15 was tested in Operation Castle in shot Nectar on 13 May 1954, producing a yield of 1.69 MT.

A B-47 Stratojet bomber jettisoned a B15 bomb early on 5 February 1958, off the Georgia coast after a collision with an F-86 Sabre fighter. Both aircraft were damaged but were able to land. The bomb was never recovered.

Inventory:	July 1955–1965	
Yield:	3.4 MT	
Weight:	6,850–7,600 lb (3,108–3,448 kg)	
Dimensions:	length	11 ft 4 in–11 ft 8 in (3.45–3.56 m)
	diameter	2 ft 10½ in (880 mm)
Weapons:	aircraft bomb	
Aircraft:	AJ/A-2	B-47
	A3D/A-3	B-52
	B-36	Valiant (British)

TX-16

Prior to the U.S. detonation of the first nuclear fusion or thermonuclear device on 31 October 1952—the Mike device—Los Alamos began planning a weaponized version of the liquid-fueled Mike device. This would be a standby effort in the (unlikely) event that the solid-fuel, thermonuclear devices under development failed to work or to achieve the expected high yields.

The weapon would require a supply of cryogenic fuel (liquid hydrogen) to be provided from tanks in the carrying aircraft. This was an extremely complex process with respect to storage, handling, and transfer (in flight) to the weapon. An estimated five casings for the weapon (to be designated EC 16) were fabricated.

The project was cancelled on 2 April 1954. It was given the name Jughead during development at Los Alamos.

Inventory:	cancelled	
Yield:	6–10 MT (untested estimate)	
Weight:	approx 60,000 lb (27,216 kg)	
Dimensions:	length	25 ft (7.62 m)
Weapons:	aircraft bomb	
Aircraft:	B-36	

Mk 17

The Mk 17 and the similar Mk 24 were large hydrogen bombs that could be only be carried by the B-36.

In May 1954 some EC17/24 bombs were stockpiled. These EC bombs were modernized into Mod 0 status by the end of the year.

In August 1957 all Mk 17s were retired in favor of the B36 bomb. About 200 Mk 17s were produced.

Operational: The design was tested on 4 May 1954 in shot Yankee during Operation Castle. The detonation produced a yield of 13.5 MT.

A Mk 17 was involved in a nuclear weapons accident on 22 May 1957, when it was accidently released from a

Mk 17 hydrogen bomb. (National Atomic Museum)

B-36 bomber near Kirtland AFB, New Mexico. The unarmed bomb did not contain its nuclear capsule but the high explosive did detonate upon impact.

Inventory:	May 1954–Aug 1957	
Yield:	approx 13.5 MT	
Weight:	42,000 lb (19,056 kg)	
Dimensions:	length	24 ft 8 in (7.51 m)
	diameter	5 ft 10 in (1.78 m)
Weapons:	aircraft bomb	
Aircraft:	B-36	

B18

The B18—referred to as the Super Oralloy—used components from the B6 and B13 bombs and was originally conceived of as a high-yield backup weapon if the thermonuclear programs were delayed or failed. The bomb design was assigned to Los Alamos in August 1952 and was produced from July 1953 until February 1955.

This was the largest pure fission (i.e., not boosted) weapon in the U.S. stockpile and used as much as 125 to 200 kilograms of uranium per bomb with probably fewer than 100 weapons entering the stockpile. They were soon withdrawn as several thermonuclear types began deployment achieving higher yields and using less uranium.

Operational: The bomb was tested in the King shot during Operation Ivy on 15 November 1952. It was dropped from a B-36 aircraft and produced a yield of 500 KT.

Inventory:	1953–Apr 1956	
Yield:	500 KT	
Weight:	8,600 lb (3,902 kg)	
Dimensions:	length	10 ft 8 in (3.25 m)
	diameter	5 ft (1.52 m)
Weapons:	aircraft bomb	
Aircraft:	B-36	
	B-47	

Mk 19

Developed for the Army's 280-mm atomic cannon, this gun-type artillery projectile was assigned to Los Alamos in April 1953. It was basically a reduced-weight Mk 9 AFAP with an increase in range. It replaced the Mk 9 in overseas locations.

The first U.S. atomic artillery projectile was the Mk 19 for the Army's 280-mm atomic cannon (left). The U.S. Navy, not to be excluded from nuclear gunfire, sponsored development of the Mk 23 for the 16-inch (406-mm) guns of the *Iowa* (BB 61) class battleships (below). For most of the time that the Mk 23 was in inventory there were no battleships in commission to fire the projectile. (National Atomic Museum)

The Mk 9 and Mk 19 were the largest nuclear projectiles produced by the United States except for the naval 16-inch (406-mm) projectile with the Mk 23 warhead.

Inventory:	1956–1962	
Yield:	variable 15 to 20 KT	
Weight:	600 lb (272 kg)	
Dimensions:	length	4 ft 6 in (1.37 m)
	diameter	11 in (280 mm)
Weapons:	280-mm artillery projectile T-315	

B20

Intended as an Air Force–Navy weapon, development of the B20 began at Los Alamos in October 1953. It was cancelled in August 1954 in favor of the B15. The bomb was to be an improved, higher yield version of the Mk 13, in the lineage of the Mk 6 and earlier Fat Man designs.

Inventory:	cancelled	
Yield:	50–70 KT	
Weight:	6,400 lb (2,904 kg)	
Dimensions:	length	10 ft 8 in (3.25 m)
	diameter	5 ft (1.52 m)
Weapons:	aircraft bomb	

B21

This was the largest nuclear bomb developed in the United States in terms of its physical dimensions. The bomb was a redesign of the Mk 17 configured to fit into the weapons bay of the B-47 bomber.

Laboratory assignment was to Los Alamos in May 1954. The warhead was produced from December 1955 to July 1956.

Stockpile life was extremely short and the weapons were converted to B36Y1 configuration between June and November 1957. Approximately 275 of these bombs were produced.

A missile warhead configuration (W21) was proposed for the Navaho missile, but that effort was cancelled.

Inventory:	Dec 1955–1957	
Yield:	4–5 MT	
Weight:	15,000–17,700 lb (6,806–8,031 kg)	
Dimensions:	length	12 ft 6 in (3.81 m)
	diameter	4 ft 8 in (1.42 m)
Weapons:	missile warhead (cancelled)	
	aircraft bomb	
Aircraft:	B-36	
	B-47	
	B-52	

B22

Laboratory assignment was to Livermore—that laboratory's first nuclear weapon—in August 1953. After design problems were revealed, the warhead was cancelled in April 1954 in favor of the Los Alamos–developed B21.

Inventory:	cancelled	
Yield:	1 MT	
Weight:	18,000 lb (8,167 kg)	
Dimensions:	length	
	diameter	4 ft 3 in (1.30 m)
Weapons:	aircraft bomb	

Mk 23

This was the only nuclear naval projectile developed by the United States. (The only other known nuclear projectile for naval guns was the 6-inch/152-mm round produced by the Soviet Union for use in *Sverdlov*-class cruisers.)

Laboratory assignment of the Mk 23 was to Los Alamos in August 1953. Production began in December 1956. Each ship was to carry nine nuclear rounds. Based on available documents, no nuclear rounds were ever placed aboard the ships.

Its nickname was Katie.

Operational: Only the four battleships of the *Iowa* (BB 61) class could fire this projectile; the last two ships of the class were laid up in reserve in early 1958 (after Korean War service). Thus, ships were available to use the weapon for about one year. (One ship was briefly reactivated in 1968–1969, and all four were returned to active service in the mid-1980s, with all again deactivated in 1991–1992 and subsequently discarded; the nuclear round was not available during those periods.)

The maximum range of the 16-inch guns with conventional munitions was 27 nautical miles.

Inventory:	Dec 1956–Oct 1962	
Yield:	15–20 KT	
Weight:	1,500–1,900 lb (681–862 kg)	
Dimensions:	length	5 ft 4 in (1.63 m)
	diameter	16 in (408 mm)
Weapons:	16-inch (406-mm) artillery projectile	

B24

This large strategic bomb, developed by Los Alamos, was similar to the B17.

In May 1954 some EC17/24 bombs were placed in the stockpile. These EC bombs were modernized to Mod 0 status by the end of the year. By October 1956 all B24 bombs had been discarded in favor of the B36 bomb.

A total of 105 B24s were produced.

Operational: The design was tested in shot Yankee during Operation Castle on 4 May 1954. The detonation produced a yield of 13.5 MT.

Inventory:	1954–1956	
Yield:	10–15 MT	
Weight:	42,000 lb (19,056 kg)	
Dimensions:	length	24 ft 8 in (7.52 m)
	diameter	5 ft 10 in (1.55 m)
Weapons:	aircraft bomb	
Aircraft:	B-36	

W25

The Genie was one of two air-to-air nuclear weapons developed by the United States, the other being the Falcon with the W54 warhead.

Laboratory assignment was to Los Alamos in November 1954 and production engineering began in December 1956. A few EC warheads were produced in November/December 1956, but the large production run did not begin until May 1957. Over the next three years—through May 1960—some 5,500 weapons were produced.

The W25 was considered for a lightweight ADM but that project was cancelled in February 1956. It was also considered for a short-range Honest John (Honest John Junior) battlefield rocket, but that project was also cancelled in February 1956.

Operational: On 19 July 1957, in shot John of Operation Plumbob, a Genie was fired from an F-89J aircraft. The rocket traveled 4,635 yards in 4.5 seconds and detonated at an altitude of 15,000 feet to produce a yield of about 2 KT.

At peak deployment in 1958 there were 42 air defense fighter squadrons deploying Genie rockets:

11 F-89J squadrons (260 aircraft) with 112 Genies per squadron

F-101B squadrons (400 aircraft) with 148 Genies per squadron

14 F-106A squadrons (250 aircraft) with 107 Genies per squadron

In addition, the Genie was carried by Canadian CF-101F strategic interceptors under the operational control of the joint North American Air Defense Command (NORAD). The last Canadian Genies were withdrawn in 1984.

Inventory:	1957–1984	
Yield:	1.5 KT	
Weight:	219 lb (99.36 kg)	
Dimensions:	length	2 ft 2½ in (673 mm)
	diameter	17 in (440 mm)
Weapons:	Genie air-to-air rocket	
	missile warhead (cancelled)	
Aircraft:	F-89J	
	F-101B/F	
	F-106A/B	
	F-15	

B26

The B26 was to have been a thermonuclear weapon similar in size to the B21. It was assigned to Los Alamos in May 1954 but was cancelled in favor of the B21.

Inventory:	cancelled	
Yield:		
Weight:	15,000–17,700 lb	
Dimensions:	length	12 ft 6 in (3.81 m)
	diameter	1 ft 8 in (221 mm)
Weapons:	aircraft bomb	

B27

Laboratory assignment of the B27 bomb was to Livermore in September 1955. Production began in November 1958 and was completed in March 1960.

The bomb version of this warhead was used only by the Navy. The bomb had a unique, spherical shape that was rigidly attached to two 275-gallon expendable fuel tanks carried in the tunnel-like weapons bay of the A3J/A-5 Vigilante attack aircraft. These tanks were emptied en route to the target, and in the attack the bomb-tank package was ejected rearward, along rails, with the empty fuel tanks serving to stabilize the weapon as the package sped toward the target. (In addition, four wing pylons could each carry a B28 or B43 nuclear weapon, or conventional bombs of up to a total 2,000 pounds.)

B27 hydrogen bomb. (National Atomic Museum)

Inventory:	Nov 1958–Oct 1964	
Yield:	2 MT	
Weight:	3,150–3,300 lb (1,429–1,497 kg)	
Dimensions:	length	10 ft 5 in–11 ft 10 in (3.18–3.61 m)
	diameter	2 ft 6 in (762 mm)
Weapons:	aircraft bomb	
Aircraft:	A3D/A-3 AJ/A-2 A3J/A-5	

W27

The W27 was a replacement for the Mk 5 warhead in the Navy's Regulus I land-attack cruise missile. Development was assigned to Livermore.

The design was also considered for warheads for the Matador, Snark, Rascal, Redstone, and Triton missiles, and the weapons pod of the B-58 Hustler bomber.

Inventory:	W27	Sep 1957–June 1963
	Mod 1	Apr 1963–Oct 1965
Yield:	2 MT	
Weight:	2,800 lb (1,270 kg)	
Dimensions:	length	6 ft 3 in (1.90 m)
	diameter	2 ft 6 in (762 mm)
Weapons:	Regulus I surface-to-surface missile	

B28

The B28 was produced in greater numbers than any other U.S. aircraft bomb—probably more than 4,500 bombs for Air Force, Navy, and Marine Corps, plus a small number provided to Great Britain for Valiant bombers under Project "E." Including both bomb and missile warheads, the B/W28-series was produced in more variants than any other U.S. nuclear weapon except for the B61.

The weapon was assigned for development to Los Alamos in August 1954 and production engineering began in January 1957. Quantity production began in August 1958 and continued until May 1966.

The Mk 28 bomb was produced in five variants: IN: free-fall internal carriage; EX: free-fall external carriage; RE: parachute-retarded external carriage; RI: parachute-retarded internal carriage; and FI: fully fuzed, parachute-retarded internal carriage.

The warheads began to be disassembled in September 1975. Periodically throughout the 1970s and 1980s, as new bombs entered the stockpile, B28s were retired and dismantled with the last disassembly completed on 15 April 1992.

Design: The original B28 was intended primarily for the B-52, which could carry four weapons in internal bomb bays. A spherical version was developed for the Navy's A3J Vigilante attack aircraft (see B27 entry).

Beyond yield variations, there were different bomb configurations for internal and external carriage, and for free-fall and parachute retarded configurations.

Some variants were modified under the DOE Stockpile Improvement Program to incorporate new electrical equipment, safety features, and high explosive material more resistant to detonation in a fire or crash. The B28RE was equipped with a CAT B PAL and the B28-0 and -1 were fitted with a CAT D PAL.

Operational: Prototype B28s were tested in Operation Redwing in the South Pacific in May–July 1956, specifically in shots Erie, Seminole, Flathead, and Dakota.

On 17 January 1966, an Air Force B-52G Stratofortress bomber and an Air Force KC-135 tanker collided near Palomares, Spain, at an altitude of 30,500 feet. Four B28 bombs tumbled from the stricken bomber, three falling to

B28 hydrogen bomb. (National Atomic Museum)

to the ground and one into the Mediterranean Sea. Both planes crashed, and three of the seven crewmen of the B-52G and all four men in the KC-135 were killed.

The Navy initiated a major search/recovery effort for the bomb that fell into the sea, recovering the weapon on 7 April 1966 from a depth of 2,500 feet.

Inventory:	Sep 1958–1991	
Yield:	Y1 1.1 MT or 1.45 MT	
	Y2 1 MT	
	Y3 70 KT	
	Y4 (unknown)	
	Y5 1.45 MT	
Weight:	1,980–2,320 lb (898–1,053 kg)	
Dimensions:	length	8 ft–14 ft 2 in (2.44–4.32 m)
	diameter	1 ft 8 in–1 ft 10 in (508 mm–558 mm)
Weapons:	aircraft bomb	
	A3D/A-3	F-101
	A4D/A-4	F-104
	A3J/A-5[5]	F-105
	A2F/A-6	F-111
	A-7	FB-111
	B-47	P2V/P-2
	B-52	S2F/S-2
	B-66	Valiant (British)
	F-4[5]	F-100

W28

The W28 was the missile warhead variant of the B28 bomb. It was proposed as a replacement for the Regulus missile's W27 warhead with the smaller yield being compensated for by the smaller circular error probability (CEP); that project was cancelled as was the planned submarine-launched Triton missile that could have carried this warhead (as well as the W27).

Development of the W28 began in 1955 at Los Alamos.

Approximately 600 warheads were produced for the Hound Dog and 100 for the Mace.

Inventory:	Mods 1, 2	Hound Dog 1960–1975
	Mods 1–3	Mace Aug 1959–1969
Yield:	Y1 1.1 MT	
	Y2 350 KT	
	Y3 70 KT	
Weight:	1,500–1,725 lb (681–783 kg)	
Dimensions:	length	5 ft (1.52 m)
	diameter	1 ft 8 in (508 mm)
Weapons:	Hound Dog air-to-surface missile	
	Mace surface-to-surface missile	

B29

The B29 thermonuclear bomb was assigned to Los Alamos for development in September 1954 but was cancelled in August 1955 in favor of the B15/B39.

Inventory:	cancelled
Yield:	
Weight:	
Dimensions:	
Weapons:	aircraft bomb

W29

The W29 was the missile warhead configuration of the B29 bomb; it was cancelled in 1955 in favor of the W39 warhead.

Inventory:	cancelled
Yield:	
Weight:	
Dimensions:	
Weapons:	Navaho strategic cruise missile (cancelled)
	Redstone surface-to-surface missile

W30

This was one of two nuclear warheads used by the Navy for surface-to-air missiles, the other being the W45-3 warhead for the Terrier-BTN. The warhead was also planned for the land-based BOMARC air-defense missile, but the W40 warhead was instead provided for that weapon. The W30 warhead was also planned for the cancelled AIM-68 Quetzalcoatl ("Big Q") air-to-air missile.

An estimated 600 W30 warheads were produced between February 1959 and January 1965. Laboratory assignment was to Los Alamos in April 1955.

5 Only U.S. Air Force F-4 Phantoms carried nuclear weapons; U.S. Navy and Marine Corps F4H/F-4 aircraft could not; Navy A-7 Corsairs could carry nuclear weapons, but not Air Force A-7D variants.

Operational: The Talos was fitted in seven U.S. Navy cruisers including the nuclear-propelled *Long Beach* (CGN 9).

Inventory:	Mod 1	T-ADM 1959–1966
	Mod 2	Talos Feb 1959–Mar 1979
Yield:	Y1	approx 0.5 KT
	Y2	5 KT
Weight:	438–490 lb	(199–222 kg)
Dimensions:	length	4 ft (1.22 m)
	diameter	1 ft 10 in (558 mm)
Weapons:	T-ADM	
	Talos surface-to-air missile	

W31

This was one of the most widely used tactical nuclear warheads.

Laboratory assignment was to Los Alamos in December 1954. Quantity production began in October 1958 and was completed in December 1961. The Honest John was originally deployed with the Mk 7 warhead. The W31 was considered for the Crossbow missile, but that program was cancelled on 24 January 1957.

Note the broad range of yields available with this warhead.

An estimated 1,750 warheads were produced for the Honest John, 3,000 for the Nike-Hercules, and 300 for the ADM—a total of approximately 5,050.

The last W31 warheads were disassembled in September 1989.

Inventory:	Mod 1–3	ADM 1960–1965
	Mod 1–3	Honest John Oct 1959–1984
	Mod 1, 2, 4	Nike-Hercules Oct 1958–1984
Yield:	1, 2, 12, 20, 40 KT	
Weight:	900–945 lb	(408–429 kg)
Dimensions:	length	3 ft 4 in (990 mm)
	diameter	2 ft 4 in–2 ft 6 in (711 mm–762 mm)
Weapons:	ADM	
	Honest John surface-to-surface missile	
	Nike-Hercules surface-to air missile	

Mk 32

The W32 was a Los Alamos designed candidate warhead for 240-mm artillery. It was cancelled in May 1955 in favor of the W48 artillery projectile.

Inventory:	cancelled
Yield:	sub-KT
Weight:	
Dimensions:	
Weapons:	240-mm artillery projectile

W33

The W33 was the warhead for the M422 8-inch AFAP. It was a gun assembly oralloy (uranium) fission weapon with three separate internal projectile assemblies, two for the lower yields and one for the high yield. Each warhead contained approximately 60 kilograms of uranium.

Laboratory assignment was to Los Alamos in December 1954. Quantity production began in January 1957 and was complete in January 1965. An estimated 1,200 rounds were produced.

The projectile had an XM83 mechanical combination lock PAL device.

In 1984–1985 a portion of the inventory was removed and 440 projectiles were dismantled. On 6 October 1992, the Department of Energy announced that all remaining W33 warheads had been dismantled.

Plans to adopt the warhead to naval 8-inch guns in heavy cruisers were cancelled.

Operational: The warhead was in service with U.S. Army and Marine Corps artillery units while several NATO armies practiced for the potential use of nuclear rounds. At the peak period, there were approximately 900 M422 rounds stored in Europe.

Inventory:	1957–1991	
Yield:	5–10 and up to 40 KT	
Weight:	243 lb (110.25 kg)	
Dimensions:	length	3 ft 1 in (940 mm)
	diameter	8 in (203 mm)
Weapons:	8-inch artillery projectile	

W34

The W34 warhead was employed in several Navy antisubmarine weapons. On 6 May 1954, the Navy's Bureau of Ordnance requested consideration of a small-diameter

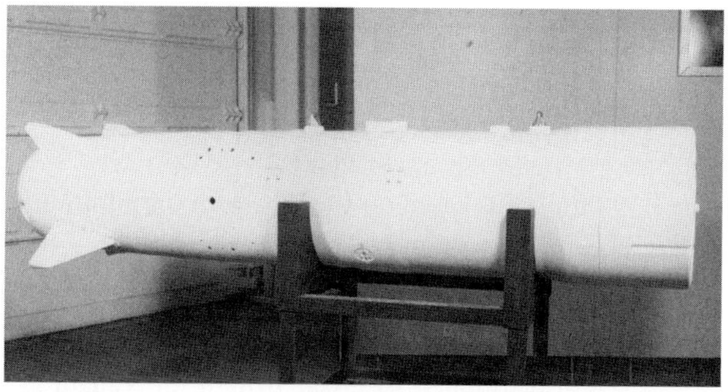

Mk 105 Hotpoint bomb with W34 warhead. (National Atomic Museum)

Mk 101 Lulu nuclear depth bomb with W34 warhead. (National Atomic Museum)

atomic depth bomb. Operation Teapot, conducted in Nevada from February to May 1955, confirmed the practicality of small, boosted-implosion devices. Further studies in 1955–1957 investigating a strike bomb and torpedo were incorporated into what eventually became the W34 warhead.

The warhead was assigned to Los Alamos in April 1955 and entered production engineering in January 1957. It was produced between June 1958 and September 1962.

An estimated 3,200 warheads were built. The weapons used a Los Alamos fission design nicknamed Python.

Operational: The Lulu could be carried by all contemporary Navy ASW aircraft. It was replaced in the depth bomb role by the B57.

The ASTOR (anti-submarine torpedo) was the only nuclear torpedo developed in the West and was carried in U.S. torpedo-attack (SS/SSN) and ballistic missile (SSBN) submarines. It was replaced by the Mk 48 conventional torpedo.

The Hotpoint was primarily a "laydown" bomb that struck the ground before detonating; it could be fuzed for air burst or underwater burst. It weighed 1,500 pounds.

Inventory:	Lulu	1958–1971
	Hotpoint	1958–1965
	ASTOR	1960–1977
Yield:	10 to 15 KT	
Weight:	311–320 lb (141.1–145.2 kg)	
Dimensions:	length	2 ft 8 in (813 mm)
	diameter	1 ft 5 in (432 mm)
Weapons:	Mk 45 ASTOR torpedo	
	Mk 101 Lulu depth bomb (Y4)	
	Mk 105 Hotpoint bomb (Y1)	
Aircraft:	AD/A-1	P5M/P-5
	P2V/P-2	S2F/S-2
	P3V/P-3	HSS-2/SH-3

Mk 35

In January 1952 the Department of Defense requested a study to investigate a warhead for the Atlas ballistic missile. Also considered that year for ICBMs were the W7 and W13. The W35 eventually became a candidate warhead for four types of strategic missiles. It was assigned to Los Alamos in July 1956 but was cancelled in August 1958 in favor of the W49.

Inventory:	cancelled	
Yield:		
Weight:		
Dimensions:		
Weapons:	Atlas	ICBM
	Jupiter	IRBM
	Thor	IRBM
	Titan I	ICBM

B36

The B36 was a large thermonuclear bomb, successor to the B21 and with the same external case as the B21.

The project was assigned to Los Alamos in June 1954.

More than 900 B36 bombs were produced. The bombs were rapidly retired in late 1961 to early 1962, causing the overall stockpile megatonnage to drop by almost half.

Inventory:	1956–1961	
Yield:	Y1 19 MT	
	Y2 6 MT	
Weight:	17,500 lb (7,940 kg)	
Dimensions:	length	12 ft 6 in (3.81 m)
	diameter	4 ft 10½ in (1.49 m)
Weapons:	aircraft bomb	
Aircraft:	B-36	
	B-47	
	B-52	

B36 hydrogen bomb. (National Atomic Museum)

W37

Laboratory development of this warhead began in April 1954 at Los Alamos. It was cancelled in October 1956, becoming the W31Y2.

Inventory:	cancelled
Yield:	
Weight:	
Dimensions:	
Weapons:	Nike-Hercules surface-to-air missile

W38

The W38 warhead was deployed on 153 intercontinental ballistic missiles: 27 Atlas E, 72 Atlas F, and 54 Titan I. Initially the warhead was also planned for the Jupiter and Thor IRBMs, but on 13 May 1959, those missiles were deleted from the W38 program. (The IRBMs were armed with the W49 warhead.)

The W38s were produced between May 1961 and January 1963. All were retired between January and May 1965.

The warhead was developed by the Livermore laboratory.

Inventory:	Atlas E	Oct 1961–Mar 1965
	Atlas F	Sep 1962–Apr 1965
	Titan I	Apr 1962–Apr 1965
Yield:	4.5 MT	
Weight:	3,080 lb (1,398 kg)	
Dimensions:	length	6 ft 10½ in (2.1 m)
	diameter	2 ft 8 in (813 mm)
Weapons:	Atlas E, F	ICBM
	Titan II	ICBM

B39

The B39 bomb was based on the B15 with the addition of a parachute to retard descent speed for low-altitude release and a new, boosted primary. The W39Y1 was fitted in a fuel-bomb pod for use by the B-58 Hustler. The B39 bomb was also supplied to Great Britain for Valiant aircraft under the Project "E" program.

Development began at Livermore in 1955.

This weapon was also used as a missile warhead (see W39).

Operational: The B39 was tested at full yield during Operation Redwing (shot Cherokee) in 1956 and Operation Hardtack I (Teak and Orange) in 1958.

Inventory:	Feb 1957–1966	
Yield:	3.8 MT	
Weight:	Mod 0	6,650 lb (3,017 kg)
	Mod 1	6,750 lb (3,063 kg)
Dimensions:	length	11 ft 4 in (3.45 m)
	diameter	2 ft 11 in (890 mm)
Weapons:	aircraft bomb	
Aircraft:	A3D/A-3	B-58
	B-36	Valiant (British)
	B-52	

W39

The W39 was a missile warhead based on the B39/15 bomb. It was also considered for the Navaho missile, which was cancelled in July 1957, and for the B-58 bomber weapon pod.

Laboratory assignment was to Los Alamos in January 1956. The warhead was produced from April 1958 to July 1958. About 30 Snark and 60 Redstone warheads were produced.

The weapon was also used as an aircraft bomb (see B39).

Inventory:	1958–1965	
Yield:	3–4 MT	
Weight:	6,230–6,400 lb (2,827–2,904 kg)	
Dimensions:	length	8 ft 10 in (2.69 m)
	diameter	2 ft 11 in (890 mm)
Weapons:	Redstone surface-to-surface missile	
	Snark strategic cruise missile	

W40

In January 1956 feasibility studies were initiated for a nuclear warhead for the Lacrosse and BOMARC missiles. In May 1956 it was proposed that the primary components of the B28 bomb be used for those missiles, and Los Alamos was given the design responsibility.

Approximately 350 warheads for the BOMARC and 400 warheads for the Lacrosse were produced between September 1959 and May 1962.

The warhead was also considered for the Navy Corvus air-to-surface missile, which was cancelled on 11 August 1960.

Inventory:	Mod 1	BOMARC 1959–1972
	Mods 1, 2	Lacrosse 1959–1964
Yield:	10 KT	
Weight:	350–380 lb	(158.8–174.7 kg)
Dimensions:	length	31⅔ in (804 mm)
	diameter	1 ft 6 in (457 mm)
Weapons:	BOMARC surface-to-air missile	
	Lacrosse surface-to-surface missile	

B41

The B41, with its 25-MT yield, was the only three-stage weapon to ever enter the U.S. stockpile. It was the largest aircraft bomb in the U.S. arsenal from 1962 to 1975.

The B41 was assigned for development to Livermore on 19 February 1957. It entered production engineering in October 1958 and was produced between September 1960 and June 1962. An estimated 500 bombs were produced.

A warhead version designated W41 was assigned to Livermore in November 1956 but was cancelled in July 1957.

Inventory:	1960–1975	
Yield:	25 MT	
Weight:	10,670 lb (4,840 kg)	
Dimensions:	length	12 ft 2 in (3.71 m)
	diameter	4 ft 2 in (1.27 m)
Weapons:	aircraft bomb	
Aircraft:	B-47	
	B-52	

W42

This small nuclear warhead was initiated for use in a variety of surface-to-air missiles. In March 1956 the Department of Defense requested the Atomic Energy Commission to study the feasibility of a warhead for an Air Force air-to-air missile. A further request was made on 4 April 1958 for the GAR-9, which would become the Falcon missile. In September 1956 a request was initiated for a nuclear warhead for the Army Hawk missile. In May 1957 a request was initiated for a nuclear warhead for the Navy Sparrow missile. In January 1960 a warhead study was requested for the Navy Eagle missile.

The Sparrow/W42 was cancelled 13 January 1958, the Hawk/W42 on 20 December 1960, and the Eagle/W42 on 22 June 1961. None of those missiles was ever fielded with a nuclear warhead. The Falcon was provided with the W54 warhead.

Laboratory assignment was to Livermore in May 1957.

Inventory:	cancelled	
Yield:	sub-KT	
Weight:	75–92 lb (34–41.7 kg)	
Dimensions:	length	1 ft 6½ in (470 mm)
	diameter	1 ft 2 in (356 mm)
Weapons:	Eagle air-to-air missile (cancelled)	
	Falcon air-to-air missile	
	Hawk surface-to-air missile (cancelled)	
	Sparrow air-to-air missile (cancelled)	

B43 nuclear bomb. (National Atomic Museum)

B43

The B43 was a versatile nuclear bomb capable of being delivered by a variety of strategic and tactical aircraft. Bombs were provided to Great Britain for use in Canberra and Valiant aircraft under Project "E."

The B43 had five yields (Y1–Y5) with several modifications with fuzing options including laydown, retarded,

The W44 nuclear warhead of an ASROC anti-submarine rocket detonates during a 1962 weapons test. The ASROC was fired from the amidships Mk 116 "pepper box" launcher of the destroyer *Agerholm* (DD 826). Nuclear/conventional-capable ASROCs were provided to 262 U.S. Navy cruisers, destroyers, and escort ships, for many their principal ASW weapon. (U.S. Navy)

and freefall ground burst, and retarded or freefall air burst. In November 1960 the B43 was made available to the Air Force's Tactical Air Command, becoming the most powerful weapon assigned to TAC up to that time (for use with F-series aircraft). Only Air Force F-4 Phantoms were configured to carry nuclear weapons.

Laboratory assignment was to Los Alamos in October 1956 and production engineering began in February 1959. Quantity production of about 1,000 weapons began in April 1961 and was completed in October 1965.

The last of these bombs were disassembled in April 1991.

The B43 was replaced in strategic forces by the B83 and in tactical forces by the B61-3 and B61-4.

Design: The B43 was produced with two different nose configurations: The Mod 0 had a steel spike to penetrate hard targets and hold the bomb in place for a delayed surface burst ("laydown"). The Mod 1 nose contained a radar fuzing device to set air bursts. The B43 could be fitted with rigid fuel tank attachments for use with the Navy A3J/A-5 Vigilante attack aircraft (see B27 and B28 descriptions).

Inventory:	Apr 1961–1991	
Yield:	70 KT, 500 KT, 1 MT	
Weight:	2,060–2,125 lb (935–964 kg)	
Dimensions:	length	12 ft 6 in–13 ft 8 in (3.81 m–4.16 m)
	diameter	1 ft 6 in (457 mm)

Weapons:	aircraft bomb	
Aircraft:	AD/A-1	F-4
	A3D/A-3	F-100
	A4D/A-4	F-101
	A2F/A-6	F-104
	A3J/A-5	F-105
	A-7	F-111
	B-47	FJ/F-1
	B-52	FB-111
	B-57	Canberra (British)
	B-58	Valiant (British)

W44

The W44 was a small nuclear weapon for use as the warhead in the nuclear depth bomb (Mk 17) fitted to the ASROC anti-submarine rocket on board surface warships. The depth bomb consisted of a nose cap, fuze section, tailcone, and fin extension tips, which were collectively referred to as the Adaption Kit (AK) Mk 28 with the W44-0 warhead. ASROC could also carry a conventional Mk 44 or Mk 46 homing torpedo.

On 1 November 1954 the Department of Defense requested the Atomic Energy Committee to collaborate with the Navy's Bureau of Ordnance in a feasibility study of a rocket-propelled depth bomb. Laboratory assignment was to Los Alamos in November 1956 and production engineering began in August 1959. The warhead was produced between May 1961 and March 1968.

Several proposals to upgrade ASROC never reached

fruition and the nuclear ASROC was retired with only limited replacement by the vertical-launch ASROC (VLA), a nonnuclear weapon. The last ASROC nuclear warhead was dismantled on 27 August 1991.

Some 35,000 ASROC missiles were produced for U.S. and foreign use, only a fraction of which were provided with nuclear warheads.

Operational: A full-scale proof test of the ASROC/W44 weapon was conducted on 11 May 1962, during Operation Dominic (shot Swordfish) about 370 n.miles southwest of San Diego, California. The destroyer *Agerholm* (DD 826) fired an ASROC at a target raft 4,000 yards away with the warhead detonating underwater.

U.S. Navy cruisers, destroyers, and frigates were fitted with nuclear-capable ASROC systems.

Inventory:	May 1961–1989	
Yield:	5 KT	
Weight:	170 lb (77.1 kg)	
Dimensions:	length	2 ft 1½ in (640 mm)
	diameter	1 ft 1¾ in (350 mm)
Weapons:	ASROC anti-submarine rocket	

W45

This relatively small warhead was employed in several tactical missiles as well as the medium ADM (MADM).

Laboratory assignment was to Livermore in November 1956, and production engineering began in October 1959. Quantity production began in January 1962 and was completed in June 1966.

The MADM was deployed with U.S. Army units in western Europe and with Marine Corps units. It weighed just over 400 pounds.

The Terrier-BTN (beam-riding Terrier nuclear) was the nuclear version of that weapon.

The last W45 warheads were dismantled in late 1988.

Operational: The Terrier was fitted in U.S. Navy cruisers and destroyers. Three aircraft carriers and one converted destroyer were also fitted with Terrier missiles but apparently were not fitted with the BTN version.

The Air Force employed the Bullpup-B; the nuclear version was not carried by Navy or Marine Corps aircraft.

Inventory:	Mod 1	Bullpup-B 1962–1976
	Mods 2, 3	LittleJohn Mar 1962–1965
	Mods 2–4	MADM 1965–1986
	Mod 1	Terrier-BTN Oct 1962–1987
Yield:	0.5, 1, 5, 8, 10, 15 KT	
	MADM variable 1 to 15 KT	
Weight:	150 lb (68 kg)	
Dimensions:	length	2 ft 3 in (685 mm)
	diameter	11¾ in (298 mm)

Components of the medium atomic demolition munition based on the W45 nuclear warhead. (Department of Defense)

Weapons:	Bullpup-B air-to-surface missile
	LittleJohn surface-to-surface missile
	medium atomic demolition munition
	Terrier surface-to-air missile

TX-46

This warhead was envisioned as a replacement of the B21 aircraft bomb and as a warhead for the Army's Redstone missile.

This Air Force–Navy project was begun at Los Alamos in July 1957 but was cancelled on 16 October 1958.

Inventory:	cancelled
Yield:	10 MT
Weight:	
Dimensions:	
Weapons:	aircraft bomb
	Redstone surface-to-surface missile

Mk 47

This was the first warhead developed for a U.S. Navy submarine-launched ballistic missile (SLBM).

Laboratory assignment was to Livermore in August 1957 and production engineering began in November 1958. The warhead was manufactured between June 1960 and July 1964 with an estimated 350 produced. The early versions had an EC designation.

Operational: The Polaris A-1 missile was deployed in five U.S. submarines (16 missiles each) from 1960 to 1965, and the Polaris A-2 in 13 submarines (16 missiles each) from 1962 to 1974.

On 6 May 1962, the USS *Ethan Allen* (SSBN 608) launched a Polaris A-1 missile almost 1,200 nautical miles in the Pacific with a nuclear detonation. This was the first and, to date, only full systems test of a U.S. strategic missile from an operational platform or launch facility. The test had the code name Operation Frigate Bird.

Inventory:	Mod 1	May 1960–1965
	Mod 2	1962–1974
Yield:	Y1 600 KT	
	Y2 1.2 MT	
Weight:	Y1 717–720 lb (325.3–326.7 kg)	
	Y2 733 lb (332.6 kg)	
Dimensions:	length	3 ft 10½ in (1.18 m)
	diameter	1 ft 6 in (457 mm)
Weapons:	Mod 1	Polaris A-1 SLBM
	Mod 2	Polaris A-2 SLBM

W48

The W48 was the warhead for the M454 155-mm AFAP.

Safety and arming features included a mechanical combination PAL device. It had a variable time (VT)/mechanical fuze.

Laboratory assignment was to Livermore in July 1957, and production engineering was assigned in January 1961. Production began in October 1963 and was completed in March 1968 with an estimated 1,000 nuclear rounds produced.

Dismantling began in April 1992 and was completed in February 1996.

The W48 was planned to be replaced by the W82 projectile starting in 1990, but that program was cancelled.

Operational: The W48 was deployed with the U.S. Army and Marine Corps, with nuclear rounds being deployed with artillery units in West Germany, Guam, Okinawa, and South Korea.

Inventory:	1963–1991	
Yield:	0.1 KT	
Weight:	128 lb (58.0 kg)	
Dimensions:	length	2 ft 10 in (863 mm)
	diameter	6.1 in (155 mm)
Weapons:	155-mm artillery projectile	

Mk 49

The Mk 49 was essentially a repackaged W28 with a new arming and fuzing system for ICBM/IRBM application. The Mk 49 Mod 2 was fitted in the Atlas D intercontinental missiles as well as both of the intermediate-range ballistic missiles that were deployed by the United States.

Development began in 1957 at Los Alamos.

Operational: The first Atlas D ICBM was placed on alert at Vandenburg AFB, California, on 31 October 1959. At full strength there were 30 Atlas D missiles deployed from 1961 to 1963.

The Jupiter IRBMs were deployed in Italy and Turkey under U.S. Air Force control. Thirty missiles were deployed in Italy; the basing of Jupiter missiles in Turkey was a catalyst for the Cuban missile crisis of 1962. The first missile in Turkey became operational on 22 October 1962—during the crisis.

Sixty Thor IRBMs were based in Britain in four equal-size wings operated by the Royal Air Force. Their W49 warheads were the responsibility of four detachments of the U.S. Air Force 99th Aviation Depot Squadron at Lakenheath, with the missiles located at Feltwell, Hemswell, North Luffenham, and Waddington.

Inventory:	Atlas D	Nov 1959–1964
	Jupiter	1961–1963
	Thor	Sep 1958–Aug 1963
Yield:	1.4 MT	
Weight:	1,640–1,680 lb (744–762 kg)	
Dimensions:	length	4 ft 6 in–4 ft 10 in (1.38 m–1.47 m)
	diameter	1 ft 8 in (508 mm)
Weapons:	Atlas D ICBM	
	Jupiter IRBM	
	Thor IRBM	

W50

Laboratory assignment for the W50 warhead was to Los Alamos in August 1958 and production engineering in June 1961. Quantity production began in March 1963 and was completed in December 1965.

The W50 warhead was planned for the Navy's Hopi air-to-surface missile, which was cancelled on 15 December 1958, and for the Army's Nike-Zeus ABM missile, which was cancelled in January 1961.

Operational: The Pershing Ia was deployed in West Germany with the U.S Army and the West German Air Force. The Mod 1 was deployed in 1963 and retired in 1978 and replaced by a Mod 2 warhead. Three warheads with differing yields were available (M28, M141, and M142).

The Pershing Ia system was replaced in U.S. Army units by the Pershing II/W85 beginning in December 1983. The W50 warheads on German missiles were returned to the United States after June 1990 and some 160 warheads were disassembled between February 1990 and August 1991.

Inventory:	Mods 1–3	Pershing Ia 1964–1990
Yield:	variable 60, 200, 440 KT	
Weight:	410 lb (186 kg)	
Dimensions:	length	3 ft 8 in (1.11 m)
	diameter	1 ft 3¾ in (390 mm)
Weapons:	Nike-Zeus anti-ballistic missile (cancelled)	
	Pershing Ia surface-to-surface missile	
	Hopi air-to-surface missile (cancelled)	

Mk 51

In February 1958 the Pentagon requested the AEC to conduct a feasibility study for a warhead for the Air Force's Falcon air-to-air missile. It recommended that the same warhead be used for the Falcon and the Army's Davy Crockett short-range weapon; Livermore was assigned to develop the weapon in June 1958.

The warhead was redesignated XW-54 in January 1959 and transferred to Los Alamos.

Inventory:	cancelled
Yield:	sub-KT
Weight:	
Dimensions:	
Weapons:	Falcon air-to-air missile
	Davy Crockett battlefield rocket

Mk 52

Laboratory assignment was to Los Alamos in May 1960, in the midst of the U.S.-Soviet nuclear testing moratorium. The warhead was produced between May 1962 and April 1966 with some 300 estimated to have been manufactured.

Inventory:	1962–1978	
Yield:	200 KT	
Weight:	950 lb (431 kg)	
Dimensions:	length	4 ft 8¾ in (1.44 m)
	diameter	2 ft in (610 mm)
Weapons:	Sergeant surface-to-surface missile	

B53

This was one of the largest nuclear bombs deployed by the United States in quantity (also see Mk 17, Mk 24, Mk 41). The B53 was a high-yield thermonuclear, strategic bomb; it was carried internally by B-47 and B-52 strategic bombers, and in the fuel-bomb pod carried by the B-58 Hustler (Y1).

In March 1958 the Strategic Air Command (SAC) requested the development of a new weapon with full-fuzing options to succeed the B41 bomb. Laboratory assignment was to Los Alamos on 23 December 1958, and production engineering occurred in July 1960. Quantity production of the Y1 Mod began in August 1962; quantity production of the Y2 Mod began in June 1964 and was completed in June 1965.

The B53 was an all-oralloy weapon without plutonium. B53 fuzing options had to be selected on the ground to provide air burst, contact burst, laydown, free-fall, or retarded delivery. It had a frangible aluminum honeycomb "false nose" for impact absorption and shock mitigation during laydown deliveries.

Operational: For an unknown period of time the B53 was in the "inactive reserve" before being finally disassembled sometime after 2003. At the beginning of 1987 it was returned to operational status until it was succeeded by the B61-11 earth penetrator warhead and B83 bombs.

Inventory:	1962–approx 2004	
Yield:	9 MT	
Weight:	8,850 lb (4,015 kg)	
Dimensions:	length	12 ft 4 in (3.76 m)
	diameter	4 ft 2 in (1.27 m)
Weapons:	aircraft bomb	
Aircraft:	B-47	
	B-52	
	B-58	

W53

The W53 was the largest warhead carried by a U.S. strategic missile. It was a high-yield thermonuclear warhead carried by the Mk 6 reentry vehicle of the Titan II ICBM. The W53 warhead—like the B53 bomb—was an all-oralloy system without plutonium.

Laboratory assignment was to Los Alamos in October 1960, with production engineering beginning in May 1961. Production began in December 1962 and was completed in December 1963, with approximately 65 warheads produced.

Operational: The Titan II missile was retired beginning in October 1982, with one missile withdrawn per month thereafter, the last being removed from alert status on 5 May 1987. For much of the Cold War the Titan II and the Minuteman series comprised the U.S. ICBM force.

Inventory:	1962–May 1987	
Yield:	9 MT	
Weight:	6,200 lb (2,813 kg)	
Dimensions:	length	8 ft 7 in (2.61 m)
	diameter	3 ft 1 in (940 mm)
Weapons:	Titan II ICBM	

W53 hydrogen warhead. (National Atomic Museum)

W54

The W54 was a low-yield, Los Alamos–designed warhead used in three quite different weapon systems and deployed with Army, Air Force, Navy (SEALs), and Marine Corps units.[6] Some NATO armies were also trained to use the special atomic demolition munition (SADM).

The W51 was redesignated W54 on 15 January 1959 and transferred from Livermore to Los Alamos.

The Falcon air-to-air missile was one of only two U.S. air-to-air weapons with a nuclear warhead, the other being the Genie unguided rocket (W25). It was widely deployed at Air Force bases within the continental United States as well as in Alaska, Canada, Hawaii, Okinawa, the Philippines, and Spain.

The Mod 2 was used as the warhead for the Army's Davy Crockett tactical-battlefield weapon system with the nuclear projectile being designated M338. The Davy Crockett was deployed to Guam, Hawaii, Okinawa, South Korea, and West Germany

The Mods 1 and 2 were used as the warhead for SADM. It was produced between August 1964 and June 1966. The SADM was a man-portable (163-pound) demolition carried in a single backpack. The SADM had the designations M129 and M159. The last SADM warheads were dismantled in August 1991.

Inventory:	Mod 2	Davy Crockett
		1961–1971
	Mod 0	Falcon
		1961–1972
	Mods 1, 2	SADM
		1964–1988
Yield:	Davy Crockett .01, .02 KT	

6 SEAL = sea-air-land team, i.e., Navy special forces.

	Falcon	.25 KT
	SADM	variable .01 to 1 KT
Weight:	50–60 lb (22.7–27.2 kg)	
Dimensions:	length	1 ft 4 in–1 ft 6 in (406 mm–457 mm)
	diameter	10¾ in (273 mm)
Weapons:	Davy Crockett battlefield rocket Falcon air-to-air missile special atomic demolition munition (SADM)	

W55

The W55 was a low-yield warhead for the SUBROC (submarine rocket), a nuclear-only weapon fired from standard 21-inch (533-mm) torpedo tubes in U.S. attack submarines beginning with the USS *Permit* (SSN 594), completed in 1962.

Laboratory assignment was to Livermore in March 1959 and production engineering began in November 1961. The W55 was manufactured in two production runs, the first between June 1964 and March 1968, and the second between May 1972 and April 1974. An estimated 285 warheads were produced.

Final dismantlement began in January 1992 and was completed by the end of 1996.

There were plans to replace the SUBROC with the Sea Lance missile (earlier called the ASW standoff weapon) in the early and mid-1980s, but that weapon was cancelled.

Inventory:	Jan 1965–1988	
Yield:	variable 1 to 5 KT	
Weight:	460 lb (208.71 kg)	
Dimensions:	length	3 ft 3 in (1.0 m)
	diameter	1 ft 1 in (330 mm)
Weapons:	SUBROC anti-submarine rocket	

W56

The W56 was a megaton-yield thermonuclear warhead carried by the Minuteman ICBM's Mk 11 reentry vehicle. It was fitted in some 650 Minuteman IB and, subsequently, to 500 Minuteman II missiles.

Laboratory assignment was to Livermore in December 1960, and production engineering began in September 1961. Quantity production of the Mod 1 began in March 1963, of the Mod 2 in August 1964, of the Mod 3 in February 1967, and of the Mod 4 in May 1967. Production of all Mods was completed in May 1969 with some 950 warheads estimated to have been manufactured.

Inventory:	Minuteman IB	1963–1969
	Minuteman II	1966–1991
Yield:	1.2 MT	
Weight:	680 lb (308.53 kg)	
Dimensions:	length	3 ft 11 in (1.19 m)
	diameter	1 ft 5½ in (445 mm)
Weapons:	Minuteman ICBM	

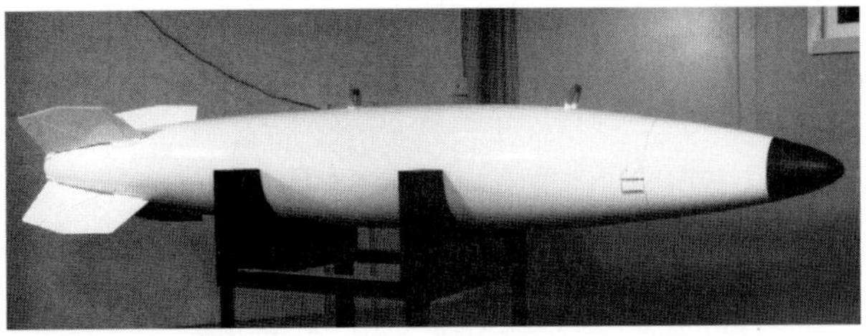

B57 atomic bomb.

B57

The B57 was a lightweight, multipurpose, air-launched, nuclear bomb designed for special tactical strikes and ASW operations. Produced in large numbers, it was used by the U.S. Air Force, Navy, and Marine Corps, with ASW aircraft of three NATO air arms certified to use the B57 nuclear depth bomb—the Dutch P-3 Orion, British Nimrod, and Italian Atlantic. The RAF Tornado was also certified to deliver the B57.

Only Air Force F-4 Phantoms carried nuclear weapons; the Navy and Marine Corps variants of those aircraft were not nuclear armed. The Air Force A-7D Corsairs were not nuclear capable.

The QH-50C DASH was an unmanned helicopter, the only U.S. drone aircraft ever fitted to carry a nuclear weapon. It was flown from U.S. Navy cruisers, destroyers, and frigates; it could carry either Mk 44 or Mk 46 ASW homing torpedoes or the B57 nuclear depth bomb.

The B57 had selectable delivery options including free-fall air burst, retarded-fall air burst, and retarded-fall underwater burst (i.e., depth bomb).

On 22 December 1959, the Military Liaison Committee approved the characteristics for what would become the B57. The weapon was assigned to Los Alamos and quantity production began January 1963. Six Mods were produced over the next four years with production completed in May 1967. An estimated 3,100 B57s were produced. Several B57 mods incorporated permissive action links (PAL).

The dismantling of B57s began in 1975. In 1991 all remaining B57 bombs were removed from aircraft carriers and stored ashore. The majority of the disassemblies took place from 1990 to March 1995. It was replaced by the B61 and a planned B90 strike/depth bomb, which was subsequently cancelled.

Inventory:	Jan 1963–1992	
Yield:	variable 5 to 20 KT (4 yield options)	
Weight:	490–510 lb (222–231 kg)	
Dimensions:	length	9 ft 10 in (3.0 m)
	diameter	1 ft 2¾ in (375 mm)
Weapons:	aircraft bomb	
Aircraft:	A3D/A-3	F-105
	A4D/A-4	F-111
	A2F/A-6	FB-111
	A3J/A-5	P2V/P-2
	A-7	P3V/P-3
	B-52	S2F/S-2
	DSN/ QH-50C	S-3
	F-4	HSS-2/SH-3
	F-16	Atlantic (Italian)
	F/A-18	Nimrod (British)
	F-100	P-3 Orion (Netherlands)
	F-104	Tornado (British)

W58

This was the only multiple reentry vehicle (MRV) warhead to be deployed on a U.S. missile. The three-RV warhead was carried by the Polaris Mk 2 reentry body, which was 54 inches long and weighed 309 pounds. The three RVs were "shotgunned" onto a single target.

Development began in August 1960 at Livermore. An estimated 1,450 warheads were produced.

Operational: The Polaris A-3 missile was initially fitted in 23 U.S. nuclear-propelled missile submarines (16 missiles each); an additional 18 submarines originally armed with A-1 and A-2 variants were later backfitted with the A-3 missile.

At the peak force, for a four year period from 1967 to 1970, the A-3 missile was fitted in 28 U.S. Polaris submarines. Four British nuclear-propelled submarines were fitted with the Polaris A-3 carrying British-developed warheads. All Polaris submarines carried 16 missiles.

Inventory:	Mar 1964–1981	
Yield:	200 KT	
Weight:	257 lb (116.6 kg)	
Dimensions:	length	3 ft 4 in (1.02 m)
	diameter	1 ft 3½ in (396 mm)
Weapons:	Polaris A-3 SLBM	

W59

Los Alamos was given laboratory assignment for this warhead in December 1960, to be developed for land-launched and air-launched strategic missiles.

Quantity production of the W59 took place between June 1962 and July 1963. About 175 warheads were produced for use on 150 Minuteman IA missiles. The Minuteman IA had the Mk 5 single-warhead RV (the IB model used the Mk 11 with the W56 warhead.)

The W59 warhead was planned for the Skybolt air-launched ballistic missile. The Skybolt, to be carried by U.S. B-52H strategic bombers and British V-series bombers, was cancelled and that warhead program was terminated on 12 March 1963. The Mk 7 reentry vehicle was to be carried in the Skybolt missile. The Skybolt warhead was to weigh 565 pounds, have a length of 4 feet, and diameter of 1 foot 11¾ inches, with the Mk 7 assembly weighing 812 pounds.

Operational: The first Minuteman IA missile went on alert in October 1962, during the Cuban missile crisis. At peak deployment there were 150 Minuteman IA missiles operational.

Inventory:	Oct 1962–Jan 1969	
Yield:	1 MT	
Weight:	550 lb (250 kg)	
Dimensions:	length	3 ft 11¾ in (1.19 m)
	diameter	1 ft 4 in (406 mm)
Weapons:	Minuteman IA ICBM	
	Skybolt air-launched strategic missile (cancelled)	

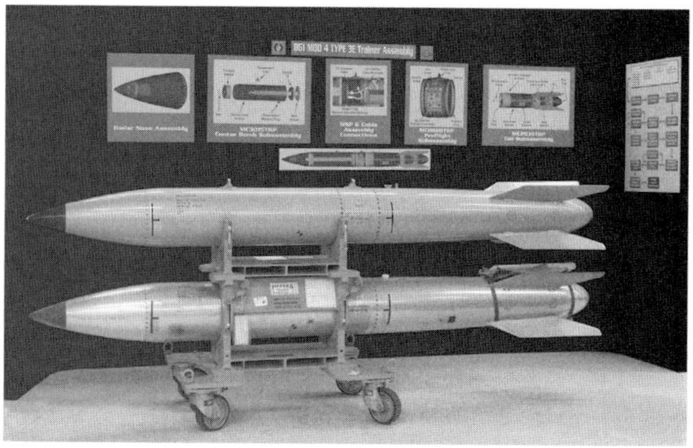

B61 atomic bombs. (National Atomic Museum)

W60

This was the warhead planned for U.S. Navy's Typhon surface-to-air missile. Development began in October 1961; the program was cancelled in 1964. The Typhon air-defense system was to have been fitted in a new class of guided missile frigates (DLG/DLGN). Typhon technology was applied to the subsequent advanced surface missile system (ASMS) and Aegis system.

Inventory:	cancelled
Yield:	
Weight:	
Dimensions:	
Weapons:	Typhon surface-to-air missile (cancelled)

B61

The B61 is a lightweight, multipurpose, thermonuclear bomb produced or proposed in 12 modifications. It is carried by a large number of U.S. Air Force, Navy, and Marine Corps aircraft, as well as aircraft flown by Belgium, Great Britain, Greece, Italy, the Netherlands, Turkey, and West Germany.

The bomb is both a tactical and a strategic weapon. Approximately 3,000 bombs were produced: 900 B61-0/1/6/7; 520 B61-3; 680 B61-4; 200 B61-10; 265 B61-2; and 435 B-61-5. Variations of the B61 design were used for the ALCM/SCLM W80 warhead and for the Pershing II W85 warhead.

The first B61-11 bombs were introduced at the end of 1996; these were B61-7s configured for earth penetration to strike deeply buried targets, replacing the B53 in that role. The strengthened B61-11 casing was designed to penetrate 15 to 25 feet before detonation. All except the B61-1 had PAL features. Mods 8 and 9 were safety upgrades that were not produced as new bombs.

The B61-7/11 bombs are being upgraded.

A W61 variant was cancelled in October 1992.

Los Alamos developed the B61.

Operational: In 2008 the Air Force held B61 bombs for both strategic (Mods 7/11) and tactical (Mods 3/4) use. After the reductions announced at the end of 2007 the numbers were reduced significantly with an estimated 500 Mods B61-3/4, 420 B61-7, and 35 B61-11 in operational and reserve status. Until 2008 seven bases in six European countries—Belgium (1), England (1), Germany (1), Italy (2), The Netherlands (1), and Turkey (1)—hosted an estimated 350 B61-3/4 gravity bombs for delivery by various U.S. and NATO aircraft. In 2008 nuclear weapons were removed from RAF Lakenheath, England. The U.S. Air Force 4th Fighter Wing at Seymour Johnson AFB, North Carolina, also has a nuclear strike mission in support of overseas contingencies. Additional tactical bombs are in reserve status stored at Kirtland AFB, New Mexico, and Nellis AFB, Nevada.

All B61s have been removed from bases in Greece and from Ramstein Air Base in Germany.

Inventory:	1968–current	
Yield:	variable sub-KT to 360 KT	
	B61-3	0.3, 1.5, 60, 170 KT
	B61-4	0.3, 1.5, 10, 45 KT
	B61-7	variable to 360 KT
	B61-10	0.3, 5, 10, 80 KT
	B61-11	variable to 360 KT
Weight:	Mods 0/1/2/5	715 lb (324.41 kg)

	Phase 3	Phase 4	Phase 5	Phase 6	Completed
B61-0	Jan 1963	May 1965	Oct 1966	Jan 1967	June 1969
B61-1	Jan 1963	May 1965	—	Feb 1969	Apr 1971
B61-2	Aug 1971	June 1972	Mar 1975	June 1975	Jan 1977
B61-3	Apr 1972	Dec 1976	May 1979	Oct 1979	
B61-4	Apr 1972	Dec 1976	May 1979	Aug 1979	
B61-5	June 1975	June 1975	June 1977	Sep 1977	Sep 1979
B61-6/8	Apr 1984	Nov 1987	cancelled in Feb 1992		
B61-7	May 1979	Mar 1982	June 1985	Sep 1985	
B61-9	(none)	Jan 1989	cancelled in Sep 1991		
B61-10	(none)	Apr 1989	May 1990	June 1990	

	Mods 3/4/9/10	
		755 lb (342.55 kg)
	Mods 6/7/8	
		763 lb (346.19 kg)
	Mod 11	1,250 lb (567.4 kg)
Dimensions:	length	11 ft 10 in (3.6 m)
	diameter	1 ft 1⅓ in (338 mm)
Weapons:	aircraft bomb	
Aircraft:	A4D/A-4M	F-104G/S
	KA-6D	F-105
	A-6E	F-111
	A-7E	FB-111A
	F-4D/E	F-117
	F-15E	B-1B
	F-16	B-2A
	F/A-18	B-52G/H
	F-100	Tornado (NATO)

Mk 12 reentry vehicle with a W62 nuclear warhead.

W62

This warhead was developed for Minuteman III and Poseidon C-3 missiles, but was cancelled for the latter weapon. These MIRV warheads are carried by the Mk 12 reentry body.

Laboratory assignment was to Livermore in June 1964 and production engineering began in March 1967. The first production unit was completed in March 1970; quantity production began the same month and was completed in June 1976. Beginning in October 2006, W87 warheads from retired MX/Peacekeeper ICBMs began to replace W62 warheads on the remaining wing of Minuteman IIIs that carried the warhead.

Approximately 1,800 warheads were produced.

Inventory:	1970–current	
Yield:	170 KT	
Weight:	253 lb (114.79 kg)	
Dimensions:	length	3 ft 3⅓ in (1.0 m)
	diameter	1 ft 7¾ in (500 mm)
Weapons:	Minuteman III ICBM	
	Poseidon C-3 SLBM	

W63

This was an enhanced radiation (ER) warhead assigned to Livermore for development engineering in July 1964. The W63 was cancelled in November 1966 in favor of the W70 warhead.

Inventory:	cancelled
Yield:	
Weight:	
Dimensions:	
Weapons:	Lance battlefield missile

W64

This warhead for the Lance missile was assigned to Los Alamos for development engineering in July 1964 but was cancelled in September 1964 in favor of the W63 warhead. The W64 was also to be an ER weapon.

Inventory:	cancelled
Yield:	
Weight:	
Dimensions:	
Weapons:	Lance battlefield missile

Mk 65

This ER warhead for the Sprint anti-ballistic missile (ABM) system was assigned to Livermore in October 1965 with the designation W65. The W65 was cancelled in January 1968 in favor of the W66.

Inventory:	cancelled
Yield:	
Weight:	
Dimensions:	
Weapons:	Sprint anti-ballistic missile

W66

This ER warhead was intended for the cancelled Sprint ABM system. Development began in 1968 with warheads becoming available for the missile in 1974; the system was never fully activated and was dismantled in 1976.

Sprint warheads were maintained in a reserve status until they were dismantled from August 1981 to August 1985.

Inventory:	cancelled	
Yield:	1 KT	
Weight:	150 lb (68 kg)	
Dimensions:	length	2 ft 11 in (890 mm)
	diameter	1 ft 6 in (457 mm)
Weapons:	Sprint anti-ballistic missile	

W67

This was an alternative warhead for the Minuteman III and Poseidon C-3 strategic missiles. Development began in June 1966 at Los Alamos but was cancelled in 1967 in favor of the W62 for Minuteman III and W68 for Poseidon.

Inventory:	cancelled
Yield:	150 KT
Weight:	
Dimensions:	
Weapons:	Minuteman III ICBM
	Poseidon C-3 SLBM

W68

The W68 was the Poseidon C-3 warhead, which was carried in the Mk 3 MIRV reentry vehicle. With an estimated 5,250 warheads produced, it has been fabricated in larger numbers than any other U.S. nuclear weapon. The Poseidon carried the largest number of warheads of any U.S. ballistic missile, with 8 to 14 RVs being fitted (in even numbers); it was also the first U.S. missile to become operational with a MIRV warhead.

Laboratory assignment was to Livermore in December 1966, and the warhead entered production engineering in May 1968. The first production unit was delivered in May 1970, and quantity production began in December 1970.

Production was completed in June 1975 with an estimated 5,250 warheads being built for the loadout of 31 strategic missile submarines carrying a total of 496 C-3 missiles (i.e., 16 missiles per submarine).

Warhead dismantling was completed in May 1995.

Operational: Thirty-one Polaris missile submarines were modified to carry the Poseidon C-3 SLBM (16 missiles per submarine).

Inventory:	1970–1991
Yield:	40–50 KT
Weight:	367 lb (166.5 kg)
Dimensions:	length
	diameter
Weapons:	Poseidon C-3 SLBM

W69

The W69 warhead was developed for the short-range attack missile (SRAM) carried by strategic bombers.

Laboratory assignment was to Los Alamos in January 1967, and the warhead entered production engineering in January 1968. The first production unit was completed in October 1971 with quantity production from February 1972 to August 1976.

Some 1,250 warheads were produced. Concerns over W69 safety led to its retirement in the early 1990s; the warheads were dismantled in 1996–1998.

Inventory:	1972–1990	
Yield:	170 to 200 KT	
Weight:	275 lb (127.8 kg)	
Dimensions:	length	2 ft 6 in (762 mm)
	diameter	1 ft 3 in (381 mm)
Weapons:	SRAM-A air-to-surface missile	

W70

The W70 was an ER warhead developed for the U.S. Army's Lance short-range tactical ballistic missile. The warhead came in several variants with different yields with an ER version (Mod 3). The two yields for the W70-3 consisted of about 40 percent fission and 60 percent fusion.

Early development work on the W70 went on at Livermore from November 1966 until January 1968; see W63 and W64 entries.

Work resumed on the W70 and laboratory assignment was formally made in April 1969 and production engineering began in December 1970. The first production warheads for the W70-0/1/2 began in June 1973 and quantity production was completed in July 1977.

Laboratory assignment was reinitiated for the W70-3 ER version in April 1976. The first production warhead was completed in May 1981 with completion of production in February 1983.

Some 380 warheads of the original production run were modified to an ER configuration. Those warheads were stored at Seneca Army Depot in Romulus, New York, and Sierra Army Depot in Herlong, California, and were never deployed to Europe because of political controversy over ER warheads. The warheads were dismantled from 1992 to 1996.

Safety and arming features included a PAL built into the warhead section.

Inventory:	1973–1991	
Yield:	W70-0/1/2	sub-100 KT
	W70-3	sub-KT, 1+ KT
Weight:	465 lb (210 kg)	
Dimensions:	length	3 ft 5 in (1.04 m)
	diameter	1 ft 6 in (457 mm)
Weapons:	Lance battlefield missile	

W71

The W71 was the warhead for the Spartan ABM missile, part of the Safeguard ABM system. After a very brief service life the warheads were kept in an inactive reserve status for almost 20 years until they were dismantled.

Development engineering began at Livermore in March 1968 and production engineering started in January 1972. The first production warheads were delivered in July 1974.

The 43 W71 warheads were disassembled between August 1994 and July 1995.

Operational: A full-scale underground test (Cannikan) of the W71 was conducted at Amchitka, Alaska, on 6 November 1971.

Inventory:	1975	
Yield:	5 MT	
Weight:	2,850 lb (1,293 kg)	
Dimensions:	length	8 ft 5 in (2.57 m)
	diameter	3 ft 6 in (1.07 m)
Weapons:	Spartan anti-ballistic missile	

W72

This was the nuclear warhead for Air Force variant of the Walleye guided missile. The W72 was similar to the W54. Development began at Los Alamos in 1969, and 300 warheads were produced from September 1970 to April 1972.

The Navy's Walleye as well as some Air Force variants had conventional warheads.

Inventory:	1970–Sep 1980
Yield:	.06 KT
Weight:	
Dimensions:	
Weapons:	Walleye air-to-surface missile

W73

Development of this warhead began in May 1969, for the Navy's Condor missile. It was a Los Alamos design derived from the B61 bomb. The missile and the W73 warhead were cancelled in September 1970.

Inventory:	cancelled
Yield:	
Weight:	
Dimensions:	
Weapons:	Condor air-to-surface missile (cancelled)

W74

Laboratory assignment for the W74 was to Los Alamos in March 1970. It was intended to replace the M454 (W48) projectile. Congress cancelled funding for the warhead in June 1973, which led to development of the XM785 (W82) projectile in its place.

Inventory:	cancelled
Yield:	
Weight:	
Dimensions:	
Weapons:	6.1-inch (155-mm) artillery projectile

W75

The W75 was a Livermore design for an 8-inch artillery projectile. Laboratory assignment occurred in June 1971. Congress cancelled funding for the weapon in June 1973, which led to development of the M753 (W79) projectile in its place.

Inventory:	cancelled
Yield:	
Weight:	
Dimensions:	
Weapons:	8-inch (203-mm) artillery projectile

W76

The current U.S. strategic submarine force of 14 submarines of the *Ohio* class (SSBN 726) all carry the Trident D-5 missile with either the Mk 4 RV with W76 warheads or the Mk 5 RV with W88 warheads. Originally the missiles were deployed with up to eight warheads per missile. Beginning about 2003, reductions have taken place with six or fewer warheads per missile in the 2008 force; by 2012 the number is to drop to four. The exact inventory mix is not known, but most Trident D-5 missiles have the Mk 4/W76 payload.

The Navy is modernizing the Mk 4/W76 warhead, providing upgrades and replacing the weapon's air-burst fuze with a ground-burst fuze. The first "new" warhead—designated Mk 4A/W76-1—was delivered in late 2007. About 800 of these RVs will be upgraded by 2012.

The W76 was also the warhead for the Mk 4 MIRV of the Trident C-4 missile, all of which have now been discarded.

The W76 laboratory assignment was to Los Alamos in May 1973 and production engineering began in November 1975. The first production unit was completed in June 1978; quantity production began in November 1978 and ended in July 1987. An estimated 3,200 warheads were produced.

Operational: The Trident C-4 missile was initially fitted in 8 U.S. nuclear-propelled submarines (24 missiles each) becoming operational in October 1979. The D-5 missile is fitted in 14 submarines (24 missiles), of which 4 were refitted C-4 submarines. (Four *Ohio*-class submarines have been converted from ballistic to cruise/guided missile submarines, the SSBN/SSGN 726–729.)

The four British *Vanguard*-class SSBNs are fitted with the Trident D-5 SLBM carrying a British-developed warhead that has similar characteristics to the W76.

Inventory:	1979–current
Yield:	100 KT
Weight:	212 lb (96.19 kg) total RV weight
Dimensions:	length
	diameter
Weapons:	Trident C-4 SLBM
	Trident D-5 SLBM

B77

The SAC identified the need for a new high-yield bomb in 1966 that led to warhead studies initiated that year. It was intended to replace the B28 and B43.

Laboratory assignment for the B77 was to Livermore in April 1974. The cancellation of the B-1 bomber in 1977 placed the B77 bomb in jeopardy; the Office of Management and Budget in December 1977 proposed cancelling the B77 for budgetary reasons, against the objections of the Secretary of Defense. In January 1978 President Jimmy Carter approved the OMB proposal in favor of modifying the B43. That approach was not successful and in January 1979, the program was reinstated as the B83, a lower cost version of the B77.

Inventory:	cancelled
Yield:	
Weight:	
Dimensions:	
Weapons:	aircraft bomb

W78

The W78 warhead is fitted in the Mk 12A MIRV originally deployed on 300 of 500 Minuteman III missiles. Beginning on 12 July 2007, the Minuteman III force was reduced by 50 missiles, reducing the W78 inventory to 250 missiles.

Laboratory assignment was to Los Alamos in July 1974, and production engineering began in March 1977. The first production unit was completed in August 1978,

Three Mk 12A reentry vehicles each with a W78 nuclear warhead. (U.S. Air Force)

and quantity production began in September 1978. An estimated 1,000 warheads were produced by October 1982.

Inventory:	1979–current	
Yield:	335 KT	
Weight:	approx 385 lb (175 kg)	
Dimensions:	length	5 ft 8 in (1.72 m)
	diameter	1 ft 9¼ in (540 mm)
Weapons:	Minuteman III ICBM	

W79

The W79 was an ER warhead for the M753 AFAP for 8-inch (203-mm) artillery. It was to replace the M422A1 (W33) projectile for use with the U.S. M110, M110A1, M110A2, and M115 howitzers.

Research and development for the W79 began almost immediately after cancellation of the W75 program in June 1973. Laboratory assignment was to Livermore in January 1975, and in March 1977 production engineering began.

On 6 June 1977, a story in *The Washington Post* identified production funds for an enhanced radiation version of the Lance missile warhead. On 24 June 1977, a follow-on story identified ER programs for 8-inch and 155-mm projectiles. On 30 September 1977 production engineering was suspended with President Carter, who deferred production until the views of European allies were heard. Subsequently, in October 1978 he directed that ER efforts be resumed.

The first production unit was completed in July 1981, and quantity production began in September 1981. Production of an ER and non-ER versions were completed in August 1986, with an estimated 550 warheads produced. The ER versions were stored at Seneca Army Depot in Romulus, New York, and were never deployed outside of the United States. The warheads subsequently were dismantled in 1996–1997.

The M753 projectile included a rocket-assist motor that doubled the range of the projectile from 11 to 18 miles. It had a CEP of 120 meters at 11 miles versus 210 meters for the M422A1 (W33) projectile. The M753 projectile was ballistically similar to a conventional 8-inch high-explosive round, thus eliminating the need for a spotting round as was the case in the M422 (W33).

Inventory:	Nov 1981–1991	
Yield:	variable up to 1.1 KT	
Weight:	200 lb (90.74 kg)	
Dimensions:	length	3 ft 8 in (1.12 m)
	diameter	8 in (203 mm)
Weapons:	8-inch (203 mm) artillery projectile	

W80

The W80 was a common warhead used in the advanced cruise missile (ACM), air-launched cruise missile (ALCM), and Tomahawk land-attack missile (TLAM). The warhead is a modification of the B61 bomb.

Laboratory assignment was to Los Alamos in June 1976. The W80-1 entered production in January 1982 and the W80-0 in March 1982. The first production W80-1 unit was completed in January 1981, and quantity production began in February 1982. An estimated 1,850 warheads were produced.

The first production W80-0 unit was completed in December 1983, and quantity production began in March 1984. An estimated 350 warheads were produced before production ended in early 1990.

The Mod 0 uses low neutron/gram/second (ngs) plutonium to produce low intrinsic radiation for personnel protection on board submarines. Safety features for both variants include insensitive high explosive and PAL.

Programs to upgrade the W80-0 and -1 to become the W80-2 and -3, respectively, have been proposed.

Approximately 300 Tomahawk land-attack missile/nuclear (TLAM-N) missiles are stored at the strategic weapons facilities at Bangor, Washington, and Kings Bay, Georgia; none is deployed on attack submarines (SSN), and they cannot be quickly put aboard those craft because of certification, equipment, and training considerations. The TLAM-N missiles were taken out of service in 1992.

Operational: The ACM was withdrawn from service in 2007 and the number of ALCMs has been reduced. The Air Force has said that there will be 528 ALCMs in service in 2012.

Inventory:	1982–current	
Yield:	variable 5 to 150 KT	
Weight:	290 lb (131.58 kg)	
Dimensions:	length	2 ft 7⅓ in (0.80 m)
	diameter	11¾ in (300 mm)
Weapons:	Mod 0 Tomahawk land-attack missile	
	Mod 1 ALCM/ACM air-to-surface missiles	

W81

This weapon was the planned replacement for the ship-launched Terrier-BTN missile. It was approved in 1979 but cancelled in 1985. Thus, the Standard-series missiles that replaced the nuclear-capable Terrier and Talos missiles have only conventional warheads.

The W81 was a modification of the B61 "primary" already in the stockpile and tested. Laboratory assignment was to Los Alamos in October 1977. It had an enhanced radiation capability; safety features included insensitive high explosives and PAL.

Inventory:	cancelled
Yield:	approx 1 KT
Weight:	
Dimensions:	
Weapons:	Standard-ER surface-to-air missile

W82

The W82 was a warhead for the XM785 artillery-fired atomic projectile (AFAP) for 155-mm howitzers. These included the U.S. M114A2, M109, M109A, and M198 howitzers, and the non-U.S. NATO M109G, FH70, and SP70 systems. The W82 was to replace the W48 M454 AFAP.

Five U.S. Army studies from 1963 to 1966 consistently supported the need for new AFAP. The report Nuclear Cannon Artillery Projectiles (NUCAP) of November 1973 and NUCAP II dated June 1975 also supported a new AFAP.

The warhead was assigned to Livermore in February 1978, and the initial rounds were produced by September 1984. Funding cuts, technical problems, and congressional restrictions delayed the program. In February 1985 a new operational date of December 1989 was established. Production engineering began in May 1986, and production began in February 1990 but was terminated before the W82 was deployed, having been cancelled by President George H. W. Bush in May 1992.

The warhead was built in components, allowing easy conversion from fission yield to an ER yield that would have used large amounts of tritium. Safety and arming features include PAL, a nonviolent disable system in the M617 storage container, and improved radar fuzing for more accurate height of burst. An 18½-mile range was the goal for all howitzer rounds. It had a CEP of 125 meters at a range of 8½ miles versus 180 meters for the M454 (W48).

Inventory:	cancelled	
Yield:	less than 2 KT	
Weight:	95 lb (43.1 kg)	
Dimensions:	length	2 ft 10¼ in (874 mm)
	diameter	6.1 in (155 mm)
Weapons:	155-mm artillery projectile	

B83

The B83 is a megaton-yield strategic bomb with improved low-level delivery features. It was designed to be effective against hardened Soviet ICBM silos and launch complexes, command installations, and nuclear storage sites. It can survive high-speed impacts against hard, irregular targets such as railroad yards, missile sites, command structures, factories, and steel-reinforced concrete runways.

The B83 replaced the B28, B53, and B43 bombs. It is the lineal descendant of the aborted B77 bomb. In January 1978, President Carter approved cancellation of the B77 in favor of modifying the B43. That approach was not successful, and in January 1979 the program was reinstated as the B83, a lower cost version of the B77. Some of the changes included incorporating a simpler retardation and electrical system, limiting required compatibility to only the B-52 and FB-111, and deleting the requirement for the bomb to be carried by Air Force and Navy tactical aircraft. In October 1981, the B-1B bomber was added and, later, the B-2A.

The basic weapon design was developed in the B77 program. It was assigned to Livermore in January 1979, and the warhead reached production engineering in September 1980. The first production unit was achieved in June 1983, and quantity production began in September 1983. The last B83 was assembled in March 1990.

The B83 delivery modes include low-level, high-speed flight profiles. The B83 incorporates the full range of fuzing options, including free-fall air burst and free-fall ground burst, and parachute-retarded air burst; its primary delivery mode is parachute-retarded delayed ground burst (laydown). It is capable of "laydown" delivery at velocities up to Mach 1.2 (from an altitude of 150 feet or higher) and free-fall delivery at velocities up to Mach 2.0 (up to 50,000 feet).

Safety features include insensitive high explosives and PAL.

Approximately 600 B83s were produced.

Operational: Much of the B83 nuclear testing was completed during the accelerated high-yield test program

Components of a B83 hydrogen bomb, including the packed Kevlar-nylon parachute (left), air filter pack, and test and control equipment. "DTU" indicates drop test unit. (Sandia National Laboratories)

in 1975–1976 and during a windup phase of the B77 program in 1978.

Inventory:	1984–current	
Yield:	variable up to 1 MT	
Weight:	2,465 lb (1,118 kg)	
Dimensions:	length	12 ft 1 in (3.688 m)
	diameter	1 ft 6 in (457 mm)
Weapons:	aircraft bomb	
Aircraft:	B-1B	
	B-2A	
	B-52G/H	
	FB-111	

W84

The W84 warhead was developed for the Gryphon ground-launched cruise missile (GLCM), the land-based derivative of the Tomahawk missile.

Laboratory assignment was to Livermore in September 1978 with the warhead derived from the B61 Mod 3/4. Production engineering began in December 1980, and the first production unit was completed in June 1983. Quantity production began in September 1983 and continued until February 1988.

An estimated 550 warheads were planned for the GLCM system. By the time of the Intermediate-range Nuclear Forces (INF) treaty of 1987 some 400 had been produced. Gryphons were forward deployed in Britain, West Germany, and Sicily beginning in January 1984; they were returned to the United States starting in 1988. After the treaty was implemented, the warheads were not disassembled but were kept in the inactive reserve for the next two decades.

The safety and arming features include insensitive high explosives and PAL.

Inventory:	1983–1991	
Yield:	variable 0.2 to 150 KT	
Weight:	388 lb (176.0 kg)	
Dimensions:	length	2 ft 8 in (864 mm)
	diameter	1 ft 1 in (330 mm)
Weapons:	Gryphon surface-to-surface missile	

W85

The W85 warhead for the Pershing II missile was configured for air and surface bursts. It was an adaptation of the B61 Mod 3/4 aerial bombs.

Laboratory assignment of the W85 was to Los Alamos in May 1979, and the warhead entered production engineering in September 1980. The first production unit was completed in February 1983; quantity production began in May 1983 and was completed in July 1986.

Approximately 200 warheads were produced for the 120 Pershing II missiles deployed by the U.S. Army in West Germany. After the INF treaty banned the Pershing II, the warheads were removed and converted into B61 Mod 10 bombs, with some redeployed to western Europe.

Other warheads considered for the Pershing II were adaptations of the W70 Mod 1 and Mod 2, the W80, and the W86.

The W85 safety features included insensitive high explosives and PAL.

Inventory:	1983–1991	
Yield:	variable 0.3 to 80 KT	
Weight:	880 lb (400 kg)	
Dimensions:	length	3 ft 11½ in (1.20 m)
	diameter	1 ft ⅛ in (305 mm)
Weapons:	Pershing II battlefield missile	

W86

The W86 warhead was configured as an earth penetrator for the Pershing II missile.

Development began at Los Alamos in May 1979; it was cancelled in September 1980 when engineering production began on the W85.

Inventory:	cancelled
Yield:	
Weight:	
Dimensions:	
Weapons:	Pershing II battlefield missile

W87

The W87 was an advanced ICBM warhead originally developed for the Peacekeeper (originally MX) missile and subsequently selected for the Small ICBM (commonly referred to as "Midgetman"). The latter weapon was cancelled in 1991. An estimated 200 W87 warheads are being fitted in Minuteman III ICBMs, with additional warheads being held in reserve.

The principal emphasis in developing the W87 was reducing the use of special nuclear material. The Phase 1 development study was published in November 1976. The Mk 21—known as the advanced ballistic reentry vehicle (ABRV)—was selected from several candidate systems. The principal competitor was the General Electric Mark 12A, which was used on a portion of the original Minuteman III force.

Laboratory assignment of the W87 was to Livermore in February 1982. The warhead entered engineering development in October 1983. The first production unit was completed in March 1986; quantity production began in August 1986 and was completed in December 1988.

Approximately 560 warheads were produced for the 50 Peacekeeper ICBMs. (The original Peacekeeper program provided for the deployment of 200 missiles.)

The W87-1 variant planned for the Small ICBM entered was assigned to Livermore in November 1987. That warhead was to have an increased yield by adding additional oralloy. The W87-1 was terminated with the cancellation of the SICBM by President George H. W. Bush in September 1991.

Mk 21 advanced ballistic reentry vehicles each with a W87 nuclear warhead. (U.S. Air Force)

The now-defunct START II treaty banned MIRVed ICBMs and the W87 warheads were removed from the Peacekeeper ICBMs.

The W87 has most of the modern arming and safety features, the latter including insensitive high explosives, fire-resistant "pit," and enhanced nuclear detonation safety.

Inventory:	1986–current	
Yield:	300 KT	
Weight:	450 lb (204 kg) Mk 21 RV weight	
Dimensions:	length	5 ft 9 in (1.75 m)
	diameter	1 ft 9 in (533 mm)
Weapons:	Peacekeeper ICBM	
	Midgetman ICBM (cancelled)	

W88

The W76 and W88 warheads arm the U.S. Navy's current force of 14 Trident missile submarines of the *Ohio* class. These missiles with the Mk 5 RV originally carried eight W88 warheads but are being reduced to comply with U.S.-Russian arms agreements (also see W76 entry).

The Los Alamos–developed warhead was assigned to engineering development in March 1984 and to production engineering in April 1986. The first W88 warheads were produced in September 1988 and quantity production began in June 1989. Some 400 warheads were produced before the Department of Energy halted production of all nuclear warheads in mid-1990. Each Trident D-5 can carry up to 8 MIRV W88 warheads.

The Navy originally proposed that the entire Trident submarine force would be armed with Mk 5/W88 warheads. Navy planning in the late 1980s envisioned an eventual Trident force of some 24 submarines, all to be armed with the Mk 5/W88 warhead—a total of some 4,600 W88 warheads. President George H. W. Bush, during his State of the Union message on 28 January 1992, announced that the United States would cease production of the W88 warhead. However, the Department of Energy reestablished small-scale warhead pit production at Los Alamos. The laboratory produced its first certifiable W88 pit in 2003; the Los Alamos goal is to manufacture ten W88 pits per year by 2007.

With the post–Cold War cutback in nuclear forces, the decision was made to complete ten submarines (24 missiles each) with the D-5 and then refit four earlier submarines, originally completed with the Trident C-4 missile. (Four *Ohio*-class SSBNs with C-4 missiles have been removed from the strategic role and converted to cruise missile submarines—SSGN.)

Inventory:	1990–current	
Yield:	455 KT	
Weight:	500–600 lb (227–272 kg)	
Dimensions:	length	
	diameter	
Weapons:	Trident II D-5 SLBM	

W89

The W89 was being developed for the SRAM II strategic attack missile, which was cancelled in 1992. The missile was to have become operational in 1994.

Similarly, it was also to arm the Sea Lance ASW weapon, the replacement for both the ASROC/W34 and SUBROC/W55 weapons. The Sea Lance was cancelled in 1990.

Laboratory assignment to Livermore occurred in January 1988, and production engineering in June 1990.

The fuzing mode was to be air burst with backup contact fuzing.

Inventory:	cancelled	
Yield:	200 KT	
Weight:	324 lb (147.27 kg)	
Dimensions:	length	3 ft 4⅞ in (1.036 m)
	diameter	1 ft 1⅓ in (337 mm)
Weapons:	Sea Lance ASW missile (cancelled)	
	SRAM II air-to-surface missile (cancelled)	
Aircraft:	B-1B	
	B-2A	
	B-52G/H	

B90

The B90 bomb was planned to replace the B57 strike/depth bomb. Development was assigned to Los Alamos in June 1988. The warhead was cancelled in March 1991.

Approximately 1,500 B90 bombs were planned for U.S. Navy and for British, French, Dutch, and Italian ASW aircraft; it also would have been compatible with the French Atlantique. The U.S. Navy P-7 was the long-range air anti-submarine warfare capable aircraft (LRAACA), intended as a replacement for the P-3 Orion in the maritime patrol/ASW roles. The P-7 program was cancelled on 20 July 1990.

Inventory:	cancelled	
Yield:	low KT	
Weight:	780 lb (354 kg)	
Dimensions:	length	9 ft 10 in (2.99 m)
	diameter	1 ft 1⅓ in (338 mm)
Weapons:	strike/depth bomb	
Aircraft:	A-6E	S-3
	F/A-18	SH-3D/H
	P-3C	Atlantic (French)
	P-7	Nimrod (British)

W91

The W91 was to have been the warhead for the SRAM-T (Tactical) missile and a bomb. The SRAM II and the SRAM-T programs were terminated on 28 September 1991 by President George H. W. Bush (see W90).

Inventory:	cancelled
Yield:	variable 10 or 100 KT
Weight:	310 lb (140.6 kg)
Dimensions:	
Weapons:	aircraft bomb
	SRAM-T air-to-surface missile
Aircraft:	B-1B
	B-2A
	B-52G/H

CHAPTER THREE

Strategic Aircraft

Aircraft were the first method of delivering an atomic bomb. Strategic aircraft are generally considered those that can reach an opponent's homeland. For the United States in the nuclear age, these were the B-29 Superfortresses, which destroyed Hiroshima and Nagasaki in August 1945, and the numerous aircraft that followed the "Superfort" into service that could strike the Soviet Union and (after 1949) mainland China.

Despite the attendant publicity and awe that followed the Hiroshima and Nagasaki attacks and the subsequent explosion of two atomic bombs in tests at the Bikini atoll in 1946 (Operation Crossroads), the U.S. nuclear striking force was quite small in that period.[1] Into the early 1950s the U.S. nuclear force consisted of a relatively few B-29s and improved B-50 Superfortress bombers, with a limited number of atomic bombs, relatively few trained air crews, and even fewer specialists needed to assemble the bombs.

The U.S. strategic bombing campaigns planned in the late 1940s called for a massive bomber assault against the Soviet Union using the few available atomic bombs. Most of the air assaults were to be carried out with conventional bombs and would attempt to destroy Soviet urban and industrial centers.[2]

Manned aircraft were the only means available to the United States for delivering atomic weapons at intercontinental ranges until the Atlas D intercontinental ballistic missile (ICBM) became operational in late 1959. Although a large number of ICBMs and other strategic missiles were operational during the Cold War, manned aircraft remained a major U.S. nuclear delivery vehicle. The B-29s of the AAF were followed by a succession of land-based piston and then turbojet bombers. The U.S. Navy, from 1949, employed carrier-launched P2V-3C Neptune and subsequent heavy attack aircraft for the delivery of nuclear weapons. Beginning in 1952 with the Mk 7 atomic bomb, Navy fighter and light-attack aircraft could deliver nuclear weapons. The Navy's carrier-based fighter and light-attack aircraft could strike the Soviet and Chinese homelands when launched from aircraft carriers in the Norwegian and Mediterranean seas, and from western Pacific areas. The latter aircraft, the AD Skyraider and A4D Skyhawk are listed in this chapter because, in their nuclear configuration, they were intended primarily for attack of the Soviet Union and China.

Also described in this chapter is the P6M Seamaster. Although designated as a "patrol" aircraft, it was designed as a long-range aerial minelayer and, secondarily, as a nuclear strike aircraft. In those roles, the P6M was to have been supported in forward areas by replenishment submarines that could refuel and possibly rearm the aircraft while in the water.

In the aftermath of the Cold War, only the USAF B-52 and B-2 manned bombers retain a nuclear strike capability in ALCM air-to-surface missiles and B61 and B83 gravity bombs. (The only tactical nuclear weapon left in the U.S. Air Force service was the B61 bomb with Air Force fighter-bomber squadrons.)

Absent from these entries is the Northrop B-49 Flying Wing, the radical-design turbojet aircraft that first flew in 1947. Although in the film version of H. G. Wells's classic

---

1 Of the two atomic bombs detonated in the 1946 tests at Bikini atoll, one was an air burst from a B-29 (test Able) and one a shallow underwater burst, the second bomb having been suspended beneath a landing craft (test Baker).

2 See Gian P. Gentile, "Planning for Preventing War, 1945–1950," *Joint Forces Quarterly* [Department of Defense] (Spring 2000), pp. 68–74; David A. Rosenberg, "The Origins of Overkill," *International Security* (Spring 1983), pp. 3–71; David A. Rosenberg and Steven T. Ross, *America's Plans for War against the Soviet Union, 1945–1950* (New York: Garland Publishers, 1989); and Walton S. Moody, *Building a Strategic Air Force* (Washington, D.C.: Air Force History and Museum Program, 1995).

The War of the Worlds (1953) a B-49 drops an atomic bomb in an (unsuccessful) effort to destroy the Martian spacecraft, in fact the aircraft's bomb bay could not accommodate an atomic weapon.

One other "aircraft" should be mentioned—the X-20 Dyna-soar project. The X-20 was a combination of missile and manned aircraft, with the Boeing and Martin firms being named the project's prime contractors on 17 June 1958. This was to have been an intercontinental strategic bomber concept derived from a German design for a trans-Atlantic ballistic missile.

Conceived in mid-1940—before Germany was at war with the United States—the A-10 rocket was to be the first stage for a weapon with a range in excess of 3,000 miles; the second stage would be a V-2 rocket or the planned A-9, a lighter version of the winged variant of the V-2 (designated A-4b). Launch weight for the two-stage weapon was 191,000 pounds, including 36,300 pounds for the second stage with a payload of just less than one ton of high explosives. Although the weapon would have poor accuracy—probably a circular error probability (CEP) of several miles—the weapon would have been used to attack American morale rather than specific targets.

The U.S. X-20 program was officially begun in 1957 and was to have been a "rocket sled," a hypersonic reentry glider that would be launched by rocket boosters and orbit the Earth by skipping in and out of the atmosphere. The glider, weighing 10,000 pounds or more, was envisioned as a potential bombing vehicle. A modified Titan missile booster was to be the launch vehicle. Neither the German nor the U.S. Air Force strategic glider program progressed beyond the paper stage.

Aircraft Designations: From 1945 to 1962 the U.S. Army Air Forces and (from 1947) the U.S. Air Force and the U.S. Navy had separate aircraft designation systems. A unified system effective on 1 October 1962, merged the two systems into one based on that of the USAF.

The basic AAF/USAF system dates to 1924 with a prefix letter for type followed by the sequential number of the aircraft within the type. Thus, the B-29 Superfortress was the 29th bomber type developed since 1924. Other aircraft types that would have a nuclear capability was F for fighter, which replaced the AAF letter P for pursuit in 1948 (previously the F prefix indicated reconnaissance aircraft).

From 1922 to 1962 the Navy had its own designation scheme that indicated the aircraft mission, sequence of that aircraft type produced by the manufacturer, manufacturer (hyphen) model, and modification. Thus AD-2N indicated the first series of attack aircraft built by Douglas, the second model (2), and modified for night operation (N); the second Douglas attack aircraft was A2D, the third A3D, and so on. That scheme became unwieldy as the number of manufacturers of naval aircraft increased. For example, the letter F was used for Grumman (as F9F) because G was already assigned to Gallaudet; Y was used for Consolidated (as PBY) with C used by Curtiss and, later, Cessna and Culver as well; A was used for Brewster (as F2A) because B was previously assigned to Boeing and, later Beech and Budd Manufacturing. Also, the same aircraft flown by different services had different designations. The B-29 had the Navy designation P2B, the North American B-25 was used by the Navy and Marine Corps as the PBJ, and the McDonnell Phantom entered service as the F4H in the Navy and F-110 in the Air Force.

Under the unified scheme of 1962 all existing and new naval aircraft were redesignated. The Navy-flown AD Skyraider became the first plane in the new attack series, the A-1; the Navy's F4H and Air Force F-110 became the F-4 in the new system. There was no P-1 or S-1 as the new system picked up the Navy's P2V Neptune and S2F Tracker as the P-2 and S-2, respectively. The improved P3V Orion was the obvious candidate for P-3 and the P5M Marlin, the Navy's last combat flying boat, for P-5. The designation P-4 was used, albeit briefly, for the drone versions of the Privateer (the P4Y-2K, formerly PB4Y-2). The designations P-4 and P-6 are sometimes cited as having been reserved for the P4M Mercator and the P6M Seamaster. But the last of the combination piston-turbojet Mercators were gone by 1962 and the turbojet Seamaster flying boat had been cancelled in 1959. The next patrol aircraft was to be the P-7 (which was the cancelled long-range air anti-submarine warfare capable aircraft). The P-8 is the U.S. Navy's next maritime patrol/ASW aircraft; it will replace the P-3 Orion.

NUCLEAR-POWERED AIRCRAFT

During the Cold War the U.S. Air Force proposed a nuclear-powered strategic strike aircraft, designated WS-125A. The aircraft would have been very large—on the order of 600,000 pounds gross takeoff weight—and was envisioned to cruise the periphery of the Soviet Union, prepared to launch nuclear-tipped missiles when conflict was initiated.

After some 15 years of research and development, with a concomitant expenditure of funds, the project was cancelled in the late 1950s. Reportedly, the Kennedy administration official who would finally cancel the project—on which $1 billion had been spent by 1960—was amazed at how long the project had survived. It was, he said, upon discovering that the nuclear aircraft was still in the budget, like "coming downstairs in the morning and finding a dead walrus in your living room."

At last, however, the aircraft was cancelled in the spring of 1961. Although the issue of a nuclear-propelled aircraft was raised from time to time in congressional hearings, by the Navy as well as by the Air Force.

ATTACK AIRCRAFT

AIRCRAFT[a]			SERVICE[b]	IOC	TYPE
A—7		Corsair	N	1966	carrier
	A-12	Avenger	N	cancelled	carrier
AD	A-1	Skyraider	N	1946	carrier
A2D		Skyshark	N	cancelled	carrier
A3D	A-3	Skywarrior	N	1956	carrier
A4D	A-4	Skyhawk	N/MC	1956	carrier
A2F	A-6	Intruder	N/MC	1963	carrier
AJ	A-2	Savage	N	1949	carrier
A2J		Savage	N	cancelled	carrier
A3J	A-5	Vigilante	N	1961	carrier

a Second column indicates 1962 redesignation.
b With nuclear capability; N = Navy; MC = Marine Corps.

A-1 Skyraider (see AD)

A-2 Savage (see AJ)

A-3 Skywarrior (see A3D)

A-4 Skyhawk (see A4D)

A-5 Vigilante (see A3J)

A-6 Intruder (see A2F)

A-7 Corsair

The Corsair was a carrier-based, light attack aircraft in first-line U.S. Navy service from 1966 until 1991. It saw extensive combat in the Vietnam War with the last two A-7E squadrons flying in the Gulf War (1991). The U.S. Air Force—urged on by Congress—procured the A-7D variant. Only the Navy aircraft were nuclear capable.

The aircraft was developed as a replacement for the Navy's A4D/A-4 Skyhawk light attack aircraft with the Navy choosing the LTV proposal in a 1963 design competition. The A-7 was optimized for daylight strike and close air support operations. (It was succeeded in carrier air wings by the F/A-18 Hornet.) The A-7 was not flown by Marine attack squadrons (the Marine Corps having instead opted for the AV-8 Harrier vertical/short takeoff and landing aircraft).

LTV A-7E Corsair. (U.S. Navy)

Design: The A-7 had a simplistic design that was derived directly from the highly successful F8U/F-8 Crusader fighter. However, the A-7 had a fixed, shoulder-position wing (the fighter aircraft's wing pivoted upward for a better angle of attack for carrier landings). The A-7 was a single-engine, single-seat aircraft, with highly simplified maintenance requirements for its time.

Weapons and drop tanks (as well as buddy refueling stores) could be carried on six wing pylons; two "cheek" pylons held Sidewinder air-to-air missiles. Two 20-mm cannon were fitted in early aircraft; a single 20-mm multi-barrel Gatling gun was fitted in the interim A-7C and the definitive A-7D and A-7E aircraft.

Sixty U.S. Navy TA-7C aircraft were converted to the training role with dual controls (converted from A-7A and A-7C aircraft). A proposed RA-7E with reconnaissance pods was dropped in favor of the F-14 carrying a reconnaissance pod. Other proposals included the KA-7F tanker and A-7G developed to Swiss requirements; none was pursued. A proposed twin-engine A-7 variant lost out in the concept stage to the F/A-18 as the new Navy attack aircraft.

From 1965 to 1983 a total of 1,567 Corsairs were delivered by LTV—1,013 A-7s for the U.S. Navy; 459 A-7D and 30 two-seat A-7K variants for the U.S. Air Force; and 65 A-7H and TA-7H variants for Greece. (Two A-7D aircraft were upgraded to YA-7F prototypes for an improved close air support aircraft, first flying in late 1989. No mass conversion or production followed.)

The A-7A entered service with Navy squadrons in October 1966; the A-7E became operational in April 1970.

Many former U.S. Navy aircraft were modified for foreign use.

The following table lists Navy attack aircraft only.

Aircraft	Built	First flight
A-7A	199	1965
A-7B	196	1968
A-7C	67	
A-7E	551	1968

Operational:	Navy Oct 1966–Apr 1991	
Design:	LTV	
Aircraft data:	A-7A	
Crew:	1	
Engines:	1 Pratt & Whitney TF30-P-6 turbofan; 11,350 lbst	
Weights:	empty	15,497 lb
	max takeoff	26,922 lb
Dimensions:	length	40 ft 1 in
	wingspan	38 ft 9 in
	wing area	375 ft^2
	height	16 ft
Speed:	maximum	684 mph at sea level
	cruise	546 mph
Ceiling:	41,000 ft	
Range:	radius	2,880 miles
Radar:	AN/APQ-116 search	
Guns:	2 20-mm Mk-12 (250 rounds per barrel)	
Ordnance:	1 nuclear bomb B28 or 2 nuclear bombs B43, B57, B61	
or	15,000 lb conventional bombs, mines, rockets, missiles	

A-12 Avenger

The A-12 Avenger was intended to replace the A-6E Intruder as the "medium" all-weather strike aircraft in Navy carrier air wings. It was also to replace the A-6E in five Marine attack squadrons, but the Marine Corps early in the program decided to forego the A-12 in favor of flying the two-seat F/A-18D in the all-weather attack role. The A-12 would have possessed a nuclear strike capability.

The Navy and the Department of Defense cancelled the A-12 program on 7 January 1991, citing "the inability of the contractors to design, develop, fabricate, assemble, and test A-12 aircraft within the contract schedule and to deliver an aircraft that meets contract requirements." Also cited were major cost overruns, and misrepresentations by the Naval Air Systems Command; the latter resulted in several officers being censored.[3] Subsequently, General Dynamics and McDonnell Douglas sued the government for improper termination of their contract. On 9 December 1994, the U.S. Court of Federal Claims decided that "testimony and other evidence at trial showed that the A-12 contract was not terminated because of contractor default. The contract was terminated because the Office of the Secretary of Defense withdrew support and funding from the A-12. Prior to that the Navy did not believe that the contractors' performance justified termination for default."

General Dynamics–McDonnell Douglas A-12 Avenger. (General Dynamics)

The design and cost of the A-12 were highly classified until January 1990, when a "slip" in congressional testimony led to revelations that the 620-plane program was to have a unit cost of $96.2 million, making it the most expensive Navy aircraft yet built. Peak A-12 production for the run of 620 aircraft was to be 36 units in fiscal year 1994. (When the Marine Corps also planned to buy the A-12, the procurement goal was 858 aircraft.)

The Air Force had been forced by the Defense Department to join the A-12 program to develop a deep-strike aircraft to succeed the F-111 and F-15E Strike Eagle in that role. Production of the Air Force version—called the advanced tactical aircraft (ATA)—was deferred in 1990 to

3 The Commander, Naval Air Systems Command was retired and the program's manager and executive officer were transferred, the two latter officers also receiving administrative letters of censure.

beyond 1997. The Air Force had considered procuring up to 400 ATA variants of the aircraft.

The A-12 program began in 1984 with the first of six A-12 prototypes originally scheduled to fly in June 1990; at the time that the program was cancelled in January 1991, the first flight was expected to occur in March 1992.

Design: The A-12 was to be a low-observable, or stealth, aircraft with a flying-wing configuration sans significant tail-fin structures. The official artist's drawings of the A-12—heavily retouched—showed a classic delta flying-wing configuration with the "fuselage" extending slightly rearward from the trailing edge; the wing has a leading-edge sweep of about 48°. There were two large, trapezoidal inlets for the engines just aft and below the leading edges of the wing; no afterburners were fitted. The tandem-seat cockpit was close to the nose of the aircraft, which was sharply pointed and "blended" into the leading edge of the wing. The aircraft was to have been fitted with a digital fly-by-wire control system. The wings folded for carrier stowage.

Operational:	cancelled	
Design:	General Dynamics and McDonnell Douglas	
Aircraft data:	A-12	
Crew:	2 (pilot, navigator-bombardier)	
Engines:	2 General Electric F412-GE-400 turbofan	
Weights:	empty	37,327 lb
	max takeoff	70,146 lb
Dimensions:	length	38 ft
	wingspan	70 ft 3 in
	wing area	1,317 ft^2
	height	11 ft 7 in
Speed:	maximum	621 mph
	cruise	
Ceiling:	45,000 ft	
Range:	900 miles with 5,500 lb bombs, missiles	
	645 miles with 10,000 lb bombs, missiles	
Radar:	AN/APQ-183 multifunction	
Guns:	none	
Ordnance:	1 nuclear bomb	
or	12,000 lb conventional weapons	

AD Skyraider (later A-1)

The Douglas Skyraider represented the ultimate development in piston-engine attack aircraft and saw extensive first-line service as a conventional attack aircraft with the U.S. Navy and Marine Corps. Nuclear-capable Skyraiders were in Navy service, operating from aircraft carriers, from 1952 to 1965. It was the world's first single-engine aircraft to carry a nuclear weapon.

The aircraft was also flown in the tactical fighter-attack role (without nuclear capability) by the U.S. Air Force as well as several other nations.

The Skyraider was adaptable to 28 configurations—possibly more than any aircraft in history—including airborne early warning, electronic countermeasures, night all-weather attack, transport, casualty evacuation, photographic, anti-submarine, target-tug, drone control, and nuclear attack.

Design: Skyraider pilot and author Rosario "Zip" Rausa wrote that the aircraft's "most distinctive characteristic . . . was its straightforward, uncomplicated design. Its unadorned style gave it an aura of simplicity and strength."[4] It was a low-wing, wide-fuselage aircraft with a large radial-piston engine; weapons were carried on fuselage and wing pylons. Designed from the outset for carrier operation, the plane's wings folded for carrier stowage.

The aircraft had a remarkable lift capacity with the AD-7 having an external payload of 8,000 pounds on 15 wing and fuselage stations in addition to four 20-mm fixed cannon; in most aircraft the centerline position could carry 3,600 pounds and the two inboard wing positions stores up to 3,000 pounds. In May 1953 a standard AD-4B flew

Douglas AD-6 (A-1H) Skyraider. (U.S. Navy)

4 Capt. Rosario Rausa, USNR (Ret.), *Skyraider: The Douglas A-1 "Flying Dump Truck"* (Baltimore, Md.: Nautical & Aviation, 1982), p. 3.

at a gross weight of 26,739 pounds—a world record for a single-engine piston aircraft, albeit not reflective of a combat loadout; the aircraft took off with an empty weight of 11,791 pounds and carried 10,500 pounds of bombs plus 4,441 pounds for fuel and pilot.

Designation: Originally designated XBT2D-1; changed to AD-series on 11 March 1946; and changed to A-1 series on 1 October 1962.

The aircraft was originally known as the Dauntless II, for the highly successful Dauntless SBD of wartime fame; changed in 1946. The AD was popularly referred to as the "Spad," after the popular World War I fighter.

Operational: The first flight of XBT2D-1 was on 18 March 1945. The AD-1 initial operational capability (IOC) was 6 December 1946. The Skyraider was in U.S. Navy service until 1968 and was flown by 44 Navy attack squadrons and 2 fighter squadrons; the Marine Corps assigned the plane to 13 attack squadrons from 1951 until 1958.

Specialized Skyraiders were flown by the Navy until April 1968, the last being the EA-1F electronics variant. The USAF flew 150 A-1E Skyraiders from 1964 to November 1972, most in Vietnam for close air and rescue support; the surviving USAF aircraft were turned over to the South Vietnamese Air Force. No Marine or USAF aircraft were nuclear capable. Nuclear delivery training flights of up to 13 hours were flown by Skyraiders.

The U.S. Navy and Marine Corps flew the Skyraider extensively in the Korean War; the Navy and USAF as well as the South Vietnamese Air Force employed Skyraiders extensively in the Vietnam conflict. (Britain, France, and Cambodia also flew the Skyraider.)

Production: A total of 3,180 ADs were produced from 1945 to February 1957, with the AD-4 being the principal production model. Douglas produced 165 AD-4B aircraft specifically to carry nuclear weapons (originally the Mk 7 and Mk 8 bombs); another 29 basic AD-4 aircraft were similarly converted. The subsequent (and similar) AD-6 and AD-7 models also had a nuclear capability; they were retained in naval service until about 1965.

The nuclear-capable aircraft were:

Aircraft		Built	First flight
AD-4		29	1949
AD-4B	A-1D	165	
AD-6	A-1H	713	1956
AD-7	A-1J	72	

Operational:	Navy Dec 1946–Apr 1968	
Design:	Douglas	
Aircraft data:	AD-6/A-1H	
Crew:	1	
Engines:	1 Wright R-3350-26WD piston; 2,700 hp	
Weights:	empty	11,902 lb
	max takeoff	25,000 lb
Dimensions:	length	39 ft 2 in
	wingspan	50 ft
	wing area	400 ft^2
	height	15 ft 8 in
Speed:	maximum	348 mph at 15,800 ft
		319 mph at sea level
	cruise	196 mph
Ceiling:	combat	31,300 ft
Range:	838 miles with 1 2,000-lb bomb	
	2,275 miles with 1,660-lb bomb + 2 300-gal drop tanks	
Radar:	none	
Guns:	4 20-mm M-3 cannon (200 rounds per barrel)	
Ordnance:	1 nuclear bomb B7, B8, B11, B12, Mk 34 Hotpoint, Mk 101 Lulu, B43, B57, B91, BOAR	
or	6,500 lb conventional bombs, rockets, torpedoes, mines	

A2D Skyshark

The A2D Skyshark was a turboprop derivative of the celebrated AD/A-1 Skyraider, which, if produced for operational service, would have been nuclear capable. The aircraft was to be capable of level, dive-bombing, and torpedo attacks.

Douglas XA2D-1 Skyshark. (U.S. Navy)

Development of a gas-turbine propelled attack aircraft was begun in June 1945 with the Navy choosing a turboprop design over a pure jet because of the limitations of existing turbojet engines for carrier operation. Excessive vibration and other problems associated with the T40 engine ultimately led to cancellation of the aircraft. (The aircraft's first flight slipped more than one year because of engine difficulties.)

Design: The A2D had a conventional aircraft design with the exhausts from the T40 engine located on both sides of the fuselage, flush with the sides, aft of the cockpit. The T40 had three-blade, contra-rotating propellers.

The radar was fitted in the large fiberglass propeller spinner/radome.

Ordnance was carried under the fuselage and on ten wing stations. Three stations could carry a 2,000-pound bomb, 1,700-pound aerial torpedo, or 300-gallon drop tank; eight stations could each carry a single 5-inch rocket. Later variants were to have up to 22 of the smaller weapon stations and a model with a more powerful T40-A-8 engine and wings swept back some 35° to increase speed; a further improvement in propellers was to have additionally enhanced performance.

Like the original AD Skyraider, several specialized electronic variants of the A2D were envisioned.

The aircraft was to be fully carrier capable with folding wings.

Designation: The A2D was derived from a Douglas design that had the original Navy designation XBT3D-1.

Operational: The first flight of XA2D-1 was on 26 May 1950.

Production: Following the 2 XA2D-1 prototypes, production orders for 341 aircraft were placed in mid-1950. However, the crash of the first XA2D-1 on 14 December 1950, and subsequent engine difficulties led the Navy to begin termination of the aircraft in February 1952. In the end, only 10 production A2D-1 aircraft were completed, the last being delivered in 1955. (Four of the 10 production aircraft were never flown.)

Operational:	cancelled	
Design:	Douglas	
Aircraft data:	A2D-1	
Crew:	1	
Engines:	1 Allison XT40-A-2 turboshaft; 5,035 shp	
Weights:	empty	5,044 lb
	max takeoff	27,835 lb
Dimensions:	length	40 ft 9 in
	wingspan	50 ft
	wing area	401 ft^2
	height	15 ft 4 in
Speed:	maximum	495 mph at 28,200 ft
		450 mph at sea level
Ceiling:	combat	43,300 ft
Range:	1,550 miles with 2 1,000-lb bombs, 8 5-inch rockets, 2 300-gal drop tanks	
Radar:	AN/APG-30 fire control	
	AN/APS-19A search/intercept	
Guns:	4 20-mm T31 cannon (140 rounds per barrel)	
Ordnance:	1 nuclear bomb (provisional)	
or	8,700 lb conventional bombs, torpedoes, rockets, mines	

A3D Skywarrior (later A-3)

The Douglas A3D Skywarrior was developed as a carrier-based nuclear strike aircraft and was the largest aircraft to regularly operate from aircraft carriers. It survived in operational service more than a decade after its planned successor—the A3J/A-5 Vigilante—was retired. Beyond serving aboard U.S. carriers in the nuclear strike role from 1956 until 1971, the A3D operated in the electronic intelligence (ELINT), photo-reconnaissance, and tanker roles. A few aircraft were flown in the navigator-bombardier trainer and the VIP transport configuration. Only the "straight" attack aircraft could carry weapons.

Development of the A3D began in late 1945 with a Navy requirement for a carrier aircraft with a gross takeoff weight of up to 100,000 pounds to carry 8,000 pounds of conventional bombs. Subsequently, the load requirement was increased to 10,000 pounds, the weight of an early atomic bomb. Douglas designer Ed Heinemann set a goal of 78,450 pounds for his design, which became the A3D, for which the Navy awarded a prototype contract on 31 March 1949. It would be able to operate from Navy carriers of the *Essex* (CV 9) size and larger.

The Navy briefly considered using the A3D for task force air defense carrying air-to-air missiles and stationed as an aerial picket 90 to 200 miles out from the task force; the concept was not pursued.

The U.S. Air Force flew the similar B-66 Destroyer in the bomber, electronic, reconnaissance, weather, and research roles (see separate entry).

Design: The A3D was the largest aircraft intended for

The Douglas Skywarrior was the largest aircraft to regularly operate from aircraft carriers. The 35-ton "whale"—developed for the nuclear strike role—also served as a conventional bombing aircraft (in Vietnam) and flew in tanker, photo-reconnaissance, electronic countermeasures, training, and VIP transport configurations. This A3D-1 (A-3A) was assigned to Heavy Attack Squadron (VAH) 3. (U.S. Navy)

regular operation from U.S. aircraft carriers. The aircraft had swept-back (36°), shoulder-mounted wings with twin turbojet engines in pods under the wings. The three-place cockpit was located relatively far forward, with most of the fuselage devoted to fuel and a large weapons bay. Twin 20-mm cannon were fitted in a remote-controlled tail turret. The wings and tail fin folded for carrier stowage.

(The KA-3B tanker aircraft had a normal gross weight up to 78,000 pounds including 34,178 pounds of fuel, of which about two-thirds could be transferred to other aircraft. This was the heaviest aircraft to operate from carriers.)

Designation: Changed from A3D to A-3 on 1 October 1962. The Skywarrior was known in the fleet as the "Whale" because of its size.

Operational: The first flight was on 28 October 1952. IOC for the A3D-1 was 31 March 1956, with Heavy Attack Squadron (VAH) 1. The first operational deployment was from 15 January to 22 July 1957, to the Mediterranean with VAH-1 embarked in the USS *Forrestal* (CVA 59). Peak deployment was reached in 1961 with five squadrons based at NAS Sanford, Florida (Heavy Attack Wing 1) and six squadrons based at NAS Whidbey Island, Washington (Heavy Attack Wing 2), with up to 12 aircraft per squadron.

A few Skywarriors carried out conventional bombing missions in the Vietnam War; however, various specialized aircraft were employed extensively in that conflict, both from shore bases and aircraft carriers. ELINT variants flew from shore bases in the 1991 Gulf War.

The last operational aircraft was an EA-3B variant, retired in September 1991 (from 1987 until 1991 the aircraft were restricted to operations from land bases).

Production: The Navy took delivery of 282 Douglas-built Skywarriors including 66 specialized aircraft that did not have a nuclear-strike capability—24 for electronic surveillance (A3D-2Q/EA-3B), 30 for photo reconnaissance (A3D-2P/RA-3B), and 12 trainers (A3D-2T/TA-3B). Many "straight" attack aircraft were later modified to electronic, photo-reconnaissance, and tanker configurations. The aircraft built as attack variants are listed below. The last A3D was delivered in January 1961.

Aircraft	Built	First flight
XA3D-1	2	1952
YA3D-1 YA-3A	1	1953
A3D-1 A-3A	49	1953
A3D-2 A-3B	164	1956

Operational:	Mar 1956–1991	
Design:	Douglas	
Aircraft data:	A3D-2/A-3B	
Crew:	3 (pilot, navigator-bombardier, engineer)	
Engines:	2 Pratt & Whitney J57-P10 turbojets; 10,500 lbst each	
Weights:	empty	37,077 lb
	max takeoff	73,000 lb
Dimensions:	length	74 ft 9 in
	wingspan	72 ft 6 in
	wing area	779 ft^2
	height	22 ft 9 in
Speed:	maximum	590 mph at 35,000 ft 643 mph at sea level
	cruise	528 mph
Ceiling:	combat	41,500 ft

Range:	2,760 miles with 6,650 lb weapons (high altitude attack)
	2,500 miles with 3,900 lb weapons (low altitude attack)
Radar:	AN/ASB-1A bomb director
Guns:	2 20-mm M-3 cannon (500 rounds per barrel)
Ordnance:	1 nuclear bomb Mk 4, Mk 5, Mk 6, B7, B15, B18, B27, Mk 34 Hotpoint, B39
or	12,000 lb conventional bombs, mines

A4D Skyhawk (later A-4)

The Douglas A4D was a carrier-based attack aircraft developed specifically to deliver a single nuclear weapon on targets in the Soviet Union under daylight, clear-weather flight conditions. The U.S. Navy envisioned flights of A4Ds being led to their target area by larger A3D/A-3 Skywarrior attack aircraft that provided navigation. Subsequently the A4D evolved into a highly versatile ground-attack aircraft for the U.S. Navy and Marine Corps; it was also flown in the fighter role from U.S. ASW aircraft carriers.

The Skyhawk was employed extensively by the U.S. Navy–Marine Corps in adversary air combat training (with the nickname "Mongoose") and as a two-seat trainer (TA-4 series); a few aircraft were fitted with special electronics (EA-4) and as tactical air control aircraft (OA-4). Several foreign air forces have also flown the Skyhawk.

Design: The basic Skyhawk was a single-seat, single-engine aircraft with a delta wing (which did not fold for carrier stowage). Ordnance on early aircraft was carried on three pylons; the A4D-5 and later aircraft had five with a capacity of 9,155 pounds. A pair of 20-mm cannon were fitted in the wing roots. A fixed in-flight refueling probe was introduced during A4D-2 production; Skyhawks also served as tankers with "buddy store" external drogue pods and fuel tanks.

Douglas A-4F Skyhawk. (U.S. Navy)

All operational U.S. Skyhawks had a nuclear weapons capability, including the two-seat trainers, although the latter were not wired for such weapons.

The A4D-2N and later aircraft had improved avionics including terrain clearance radar and the low-altitude bombing system (LABS). The A4D-3 was planned as an all-weather attack aircraft.

Skyhawks through the A4D-2N had the J65 engine; subsequent aircraft had the J52. Proposals to produce a Skyhawk with the Rolls-Royce Spey RB 168-20 engine (12,000 lbst) were not pursued.

Designation: Originally the A4D series, the aircraft was changed to A-4 on 1 October 1962. The aircraft's diminutive size led to the early nicknames "Scooter" and "Heinemann's Hot Rod," the latter honoring its brilliant designer, Ed Heinemann.

Operational: The first flight of XA4D-1 was on 22 June 1954; Navy IOC was in September 1956 and Marine Corps IOC was in September 1957. The A4D-2 became operational in September 1957, the A4D-2N in February 1960, A4D-5 in January 1963, and the A-4M in April 1971.

Skyhawks served in active U.S. Navy attack squadrons until 1975 and in reserve units until 1978; U.S. Marine Corps attack squadrons flew the aircraft until 1992; Skyhawks served in Navy utility roles for another few years. TA-4J variants served in Navy training squadrons until 1998 and in a utility role until 2003.

The aircraft was flown extensively by the U.S. Navy and Marine Corps in the Vietnam War. The Israeli Air Force made extensive use of A4Ds in Middle East combat while the Argentine Navy and Air Force effectively flew Skyhawks in strikes against British ships during the 1982 conflict in the Falklands. The surviving A-4K models of the Kuwaiti Air Force flew combat strikes from Saudi Arabia after the Iraqi invasion of Kuwait in August 1990.

Several countries continue to fly the Skyhawk, both new models and refurbished U.S aircraft.

Production: A total of 2,960 Skyhawks were built for U.S. and foreign use, of which 555 were two-seat training aircraft. The last Skyhawk built was an A-4M, delivered on 27 February 1979, giving a production run of 24½ years—the longest production run of any U.S. combat aircraft.

The following list contains only aircraft built for U.S. Navy–Marine Corps use; another 277 Skyhawks were built specifically for Australia, Israel, Kuwait, and New Zealand. (Australia operated their Skyhawks from the aircraft carrier *Melbourne.*) Hundreds of ex-U.S. aircraft were refurbished for foreign use.

The A4D-3, A4D-4, and A4D-6 models were cancelled.

Aircraft		Built	First flight
XA4D-1		1	1954
A4D-1	A-4A	165	
A4D-2	A-4B	542	1956
A4D-2N	A-4C	638	1959
A4D-5	A-4E	499	1961
	TA-4E	2	1965
	A-4F	147	1966
	TA-4F	238	
	TA-4J	293	
	A-4M	158	1970

Operational:	Navy/Marine attack role Sep 1956–June 1992	
Design:	Douglas	
Aircraft data:	A4D-2N/A-4C	
Crew:	1 (2 in TA-4 variants)	
Engines:	1 Wright J65-W-16A turbojet; 7,700 lbst	
Weights:	empty	9,146 lb
	max takeoff	22,500 lb
Dimensions:	length	39 ft 5 in
	wingspan	27 ft 6 in
	wing area	260 ft^2
	height	15 ft
Speed:	maximum	647 mph at 4,000 ft
		591 mph at 35,000 ft
	cruise	496 mph
Ceiling:	41,000 ft	
Range:	1,565 miles with 1 Mk 28 bomb + 2 300-gal drop tanks	
	550 miles with 2 1,000-lb bombs + 6 500-lb bombs	
Radar:	none	
Guns:	2 20-mm cannon (100 rounds per barrel)	
Ordnance:	1 nuclear bomb B7, B8, B12, B28, Mk 34 Hotpoint, B43, B57, B61, B91 (Elsie), BOAR	
or	9,000 lb conventional bombs, rockets, missiles	

Grumman A2F-1 (A-6A) Intruder. (U.S. Navy)

A2F Intruder (later A-6)

A highly versatile attack aircraft, the Intruder was the most capable U.S. carrier-based aircraft with respect to weapons payload and all-weather, day/night attack capability.

The concept for the Intruder came from experience in the Korean War when tactical aircraft were hard pressed to operate over communist supply routes at night and in rain. In 1956 the Navy requested proposals from the aviation industry for an attack aircraft able to operate at night in all weather conditions without recourse to external navigation aids. The Grumman response was selected, becoming the A2F-1. (The USAF acquired the B-57 Canberra to meet a similar requirement.)

The Intruder was to be replaced by the A-12 Avenger; however, with cancellation of that aircraft the A-6 remained in carrier service until late 1996. Its numerical replacement in U.S. Navy and Marine Corps squadrons was the F/A-18 Hornet.

Design: The Intruder was a two-seat, twin-turbojet aircraft. All weapons were carried on five attachment points, each with a 6,200-pound capacity with a maximum mission payload of 18,000 pounds of ordnance and drop tanks. The definitive A-6E could carry up to three nuclear weapons as well as a variety of laser-guided bombs and guided missiles. The planned loadout for a nuclear strike was generally one nuclear bomb and four 300-gallon drop tanks. No guns were fitted.

The A-6A aircraft's digital integrated attack and navigation equipment (DIANE) integrated several subsystems that together computed navigation and other data to provide information for precise navigation and weapons delivery. An Intruder could be catapulted from a carrier deck at night, fly a complex strike mission, and return aboard the carrier without resorting to visual contact with the real world outside the cockpit until the pilot acquired the mirror landing aid on his final approach. The mission and complexity of avionics required a two-man crew, the pilot and the navigator-bombardier being seated side by side under a large canopy.

The A-6E variant had the target recognition attack multisensor (TRAM), forward-looking infrared (FLIR), a combination laser designator/range finder, and a laser-designation receiver. It could carry HARM, Harpoon, Maverick, Sidewinder, Skipper, and SLAM missiles in addition to "iron" and laser-guided bombs and mines.

An improved A-6F, with new avionics and the AN/APQ-173 radar, the General Electric F404 engines, and composite-material wings, was being developed with an IOC planned for the late 1980s; that project was cancelled, as was a proposed new A-6G aircraft. As late as 1990 the Navy requested cost data from Grumman for 500 A-6G variants—300 new aircraft and 200 remanufactured A-6E aircraft—because of delays in the A-12 program.

KA-6D tankers—all converted from earlier aircraft—had their avionics deleted from the after fuselage to provide space for the refueling hose and reel; up to five 500-gallon drop tanks could be carried to permit the transfer of over 21,000 pounds of fuel immediately after takeoff or 15,000 pounds at a distance of 288 n.miles from the carrier. The first A-6A tanker conversion flew in April 1970, designated NA-6A as a land-based tanker test aircraft.

The EA-6A Intruder and EA-6B Prowler electronic warfare variants were developed from the basic Intruder.

All Intruder variants had a prominent fixed refueling probe forward of the cockpit; wings folded for carrier stowage.

Designation: Originally A2F, changed to A-6 on 1 October 1962.

Operational: The first flight of A2F-1 was on 19 April 1960. A2F-1 Navy IOC was in February 1963 and Marine IOC was in October 1964. KA-6D IOC was in September 1970, and A-6E IOC was in September 1971.

The Intruder was used extensively in the Vietnam conflict, with most U.S. carriers being assigned one squadron (10 A-6 and 4 KA-6D aircraft). In the 1980s the number of Intruders in some wings were doubled (up to 25 A-6Es) to enhance their strike capability.

Intruders were also used in combat over Grenada and Lebanon (1983) and in strikes against Libya and Libyan naval forces (1986), and in the Gulf War (1991).

The last Marine A-6E squadron was disbanded in 1993 and the last Navy A-6E squadron gave up the Intruder in December 1996, with the last aircraft being retired in March 1997. (The EA-6B variant continues in Navy–Marine Corps service.)

There were no foreign users of the aircraft.

Production: Grumman produced 684 attack variants, the last in January 1992. In addition, 21 EA-6A and 170 EA-6B aircraft were built. The attack variants are listed below; the A-6B and A-6C were small numbers of A-6A conversions; 240 A-6A aircraft were upgraded to the A-6E variant.

Aircraft		Built	First flight
A2F-1	A-6A	474	1960
A-6E		205	1970
A-6F		5	1987

Operational:	Feb 1963–Dec 1996	
Design:	Grumman	
Aircraft data:	A-6E	
Crew:	2 (pilot, navigator-bombardier)	
Engines:	2 Pratt & Whitney J52-P-8A/B turbojet; 9,300 lbst each	
Weights:	empty	25,980 lb
	max takeoff	58,600 lb
Dimensions:	length	54 ft 9 in
	wingspan	53 ft
	wing area	528.9 ft^2
	height	16 ft 2 in
Speed:	maximum	652 mph at sea level
	cruise	482 mph
Ceiling:	44,600 ft	
Range:	2,770 miles with 1 Mk 43 bomb + 4 300-gal drop tanks	
Radar:	AN/APQ-148	
Armament:	3 nuclear bombs B28, B43, B57, B61	
or	18,000 lb conventional bombs, mines, rockets, missiles	

AJ Savage (later A-2)

This North American was the first U.S. naval aircraft designed from the outset to carry a nuclear weapon. (Twelve P2V-3C Neptunes were earlier modified to carry nuclear weapons.) Savages also served in the photo-reconnaissance and tanker roles.

The AJ-1 was ordered on 24 June 1946, to carry conventional bombs because technical data on nuclear weapons, including dimensions, were not then available to the Navy. Subsequently the design was modified to enable a Fat Man/Nagasaki-type atomic bomb to be carried in the bomb bay, and the aircraft was delivered with a nuclear strike capability.

North American AJ-1 (A-2A) Savage. (*U.S. Navy*)

Design: The Savage was an all-weather, carrier-based aircraft with folding wings and tail fin to facilitate shipboard handling. Weapons were carried in an internal bomb bay. It had an unusual configuration with twin, nacelle-mounted piston engines and a turbojet engine mounted in the after fuselage for over-target burst speed. Wingtip fuel tanks for 300 gallons each were normally fitted.

No guns were fitted; the aircraft relied on speed to escape from hostile fighters.

The drogue refueling package could be removed from the tanker versions and they could be refitted to carry nuclear bombs.

Designation: Originally AJ; changed to A-2 on 1 October 1962.

Operational: The prototype XAJ-1 flew on 3 July 1948. The AJ-1 entered service in September 1949, although there were no nuclear bombs available for the planes until late 1950.

Although carrier capable, the Savages were unpopular aboard aircraft carriers in the early 1950s; some 30 other aircraft had to be taken off the ship and flown ashore to make room for a detachment of four Savages, and their readiness was relatively low—averaging just 47.3 percent aboard ship during the first six months of 1952. Subsequently, the development of angled flight decks eased the problems of operating the large aircraft.

The first overseas deployment of the AJ came in February 1951, when six AJ-1 and three P2V-3C aircraft from Composite Squadron 5 flew across the Atlantic to Port Lyautey (now Kenitra), Morocco. Beginning in March 1951, Savages periodically operated from the large aircraft carriers *Midway* (CVA 41) and *Franklin D. Roosevelt* (CVA 42), which were then operating in the Mediterranean. (The Neptunes operated only from shore bases overseas.) Subsequently, detachments of three to seven Savages were embarked in most forward-deployed carriers, the number depending upon the size of the ship.

Savages were operated by Navy attack squadrons until 1959, when they were relegated to photo-reconnaissance and tanker duties; the last AJ-2P Savage was retired in 1960.

Production: A total of 143 AJ-series aircraft were produced through 1954, including 30 aircraft built to a photo configuration (AJ-2P), which could not carry bombs. The "straight" attack variants are listed below.

Aircraft		Built	First flight
XAJ-1		3	1948
AJ-1	A-2A	55	1949
AJ-2	A-2B	55	1952

Operational:	Sep 1949–Jan 1960	
Design:	North American Aviation	
Aircraft data:	AJ-2	
Crew:	3 (pilot, navigator-bombardier, crewman)	
Engines:	2 Pratt & Whitney R-2800-44W piston; 2,300 hp each	
	1 Allison J33-A-10 turbojet; 4,600 lbst	
Weights:	empty	30,776 lb
	max takeoff	54,000 lb
Dimensions:	length	64 ft 1 in
	wingspan	71 ft 5 in (75 ft with tiptanks)
	wing area	836 ft^2
	height	21 ft 4 in
Speed:	maximum	443 mph at 32,000 ft
		355 mph at sea level
	cruise	269 mph
Ceiling:	33,000 ft	
Range:	1,700 miles with 1 B15 bomb + 2 300-gal tiptanks	
	2,470 miles with 1 Mk 5 bomb + 2 300-gal tiptanks + 1 500-gal bomb bay tank	
Radar:	AN/ASB-1 bomb director	
Guns:	none	
Ordnance:	1 nuclear bomb Mk 3, 4, Mk 5, Mk 6, B7, B15	
or	12,000 lb conventional bombs, mines	

A2J Savage

Intended as an improved version of the AJ Savage carrier-based attack plane, the A2J had twin turboshaft engines to provide higher performance. The A2J program was initiated in December 1947 when the Navy requested that North American Aviation submit a proposal for a turboshaft aircraft based on the XAJ-1 design. The turboshaft engine had fuel economies over a straight turbojet engine and increased speed over reciprocating engines. Adopting the basic AJ design would permit a one-third reduction in time and one-half reduction in cost compared to a completely new aircraft design.

Following delays caused by the turboshaft engines, the A2J program was cancelled with the all-jet A3D Skywarrior selected to provide the Navy's carrier-based (heavy) nuclear strike capability. The A2J program was cancelled on 8 June 1953. The A2J turboshaft engines were the same as in the A2D Skyshark attack aircraft, also cancelled.

Design: The A2J design was similar to that of the AJ with elongated engine nacelles and the provision of a remote-controlled tail turret. There was an internal bomb bay.

A defensive armament was provided in the aircraft's design (the AJ was unarmed). Initially two remote-controlled turrets were added for relatively large-caliber guns (20-mm) along with associated radar; this change added some 10,000 pounds to the aircraft's weight, and the armament was reduced to only the tail turret.

The aircraft was intended to operate from aircraft carriers of the *Midway* and larger classes (at the time, the takeoff weight was estimated at 72,000 pounds); it was later reduced to permit operation from modernized *Essex*-class carriers (as could the AJ). The wings and tail fin folded for shipboard stowage.

Operational: The first flight of XA2J-1 was on 4 January 1952; the second prototype did not fly.

North American XA2J-1 Savage. (U.S. Navy)

Production: North American received a contract on 28 September 1948 to produce two XA2J-1 prototype aircraft. Their delivery dates were originally scheduled for November 1949 and April 1950; the first aircraft was delivered almost two years late.

Operational:	cancelled	
Design:	North American Aviation	
Aircraft data:	A2J-1	
Crew:	3	
Engines:	2 Allison XT-40-A-6 turboshaft; 1,225 lbst each	
Weights:	empty	32,169 lb
	max takeoff	58,000 lb
Dimensions:	length	70 ft 2 in
	wingspan	71 ft 6 in
	wing area	850 ft^2
	height	22 ft 6 in
Speed:	maximum	489 mph at 30,000 ft
		446 mph at sea level
	cruise	400 mph
Ceiling:	37,600 ft	
Range:	2,180 miles with 8,000 lb bombs	
	1,760 miles with 10,500 lb bombs	
Radar:	AN/APS-31 search	
Guns:	2 20-mm cannon (500 rounds per barrel)	
Ordnance:	1 nuclear bomb	
or	8,000 lb conventional bombs	

A3J Vigilante (later A-5)

The North American A3J Vigilante was developed as a supersonic (Mach 2), carrier-based nuclear strike aircraft. The aircraft was a response, in part, to the increased capabilities of U.S. Air Force strategic bombers. However, soon after entry into service all Vigilantes were converted for combination reconnaissance-strike roles, although they were subsequently employed only as a recce aircraft. Significantly, the A3J/A-5 was retired more than a decade before the aircraft it was intended to replace, the A3D/A-3 Skywarrior.

The aircraft flew in the heavy attack role from 1961 to 1964, at which time A3J-2/A-5B aircraft in production were completed to the RA-5C reconnaissance-attack configuration.

A high-performance interceptor variant of the aircraft for land-based operation by the Air Force was considered from 1959 onward; that proposal included a liquid-fuel

The North American-built Vigilante was intended as successor to the A3D Skywarrior in the nuclear strike role. Instead, the "Vigie" gained success as a high-speed reconnaissance aircraft. This A3J-3-P (RA-5C) is assigned to Reconnaissance Heavy Attack Squadron (RVAH) 9.On these aircraft the large weapons "tunnel" was fitted with reconnaissance equipment, and nuclear bombs could be carried on four wing pylons. (U.S. Navy)

rocket installed in the weapons bay to boost high-altitude performance; later a variant with three turbojet engines was proposed. These air-defense concepts were not pursued.

Design: The Vigilante was a sleek-looking, swept-wing aircraft, with a midfuselage wing mounted above distinctive square air intakes for the aircraft's twin turbojet engines. The engines were housed in large nacelles that extended to the tail of the aircraft.

The aircraft had a unique, tunnel-like linear weapons bay in the fuselage (between the engine nacelles). The nuclear bomb was fitted in a spherical container that was rigidly linked to a pair of 275-gallon fuel tanks; all three were ejected from the rear of the aircraft to release the nuclear weapon. Weapons (including nuclear) and fuel tanks also could be carried on two or—in later aircraft—four wing pylons, each of which had a 2,500-pound capacity. Pods containing chaff and strobe flashers as well as fuel tanks (or weapons) could be carried on wing pylons of the RA-5C.

The RA-5C carried sensors in an elongated pod ("canoe") fitted to the bottom of the fuselage as well as in the fuselage, all correlated and oriented through the aircraft's inertial navigation system.

The Vigilante's wings and tail fin folded for carrier stowage. No guns were fitted.

Designation: Changed from A3J to A-5 on 1 October 1962.

Operational: The first flight of YA3J-1 was on 31 August 1958; squadron delivery of the A3J-1 began in June 1961; the RA-5C became operational in June 1963.

The aircraft's large size and high landing speed, and the need for a sophisticated shipboard intelligence center, restricted the aircraft to the larger U.S. carriers of the *Forrestal* and later classes. (Significantly, the heavier A3D/A-3 Skywarrior could operate from smaller carriers.) The RA-5C was flown extensively from carriers in the reconnaissance role during the Vietnam War.

The last RA-5C was retired from the fleet in November 1979.

The Vigilante established several performance records, perhaps the most impressive being a world altitude mark for aircraft carrying a 2,204-pound payload, reaching 91,450.8 feet on 13 December 1960.

Production: Only 156 Vigilantes were produced, a relatively small number for a significant U.S. naval aircraft. The last aircraft was delivered in November 1970.

In addition to new-build reconnaissance aircraft, starting in 1964 North American converted 43 A-5A aircraft to the RA-5C configuration.

Aircraft		Built	First flight
YA3J-1		2	1958
A3J-1	A-5A	47	1958
A3J-2	A-5B	6	1962
A3J-3P	RA-5C	101	1962

Operational:	June 1961–Nov 1979	
Design:	North American Aviation	
Aircraft data:	A3J-1	
Crew:	2 (pilot, navigator–systems operator)	
Engines:	2 General Electric J79-GE-8 turbojets; 17,000 lbst each with afterburner	
Weights:	empty	32,714 lb
	max takeoff	56,293 lb

Dimensions:	length	76 ft 6 in
	wingspan	53 ft
	wing area	700 ft^2
	height	19 ft 5 in
Speed:	maximum	1,319 mph at 40,000 ft
		805 mph at sea level
	cruise	560 mph
Ceiling:	41,400 ft	
Range:	2,010 miles with 1 Mk 28 bomb	
	2,670 miles with 1 Mk 28 bomb + 2 400-gal drop tanks	
Radar:	AN/ASB-12 bomb/nav	
Guns:	none	
Ordnance:	1 nuclear bomb B27, B28, B43, B57	
or	2 B43 nuclear bombs	
or	6,000 lb conventional bombs	

BOMBER AIRCRAFT

Aircraft		Service[a]	IOC	Type
Next Generation Bomber		AF	2018 (est.)	land
B-1	Lancer	AF	1985	land
B-2	Spirit	AF	1993	land
B-29	Superfortress	AF	1943	land
B-36	Peacemaker	AF	1948	land
B-46	—	AF	cancelled	land
B-47	Stratojet	AF	1951	land
B-50	Superfortress	AF	1948	land
B-52	Stratofortress	AF	1955	land
B-54	—	AF	cancelled	land
B-55	—	AF	cancelled	land
B-56	Stratojet	AF	cancelled	land
B-58	Hustler	AF	1960	land
B-59	—	AF	cancelled	land
B-60	—	AF	cancelled	land
B-68	—	AF	cancelled	land
B-70	Valkyrie	AF	cancelled	land
FB-111	—	AF	1971	land

a AF = Air Force

Next Generation Bomber

The U.S. Air Force in 2007 initiated the Next Generation Bomber program in response to Department of Defense and congressional mandates to produce a new long-range strike aircraft to be operational by 2018. This direction has led the Air Force to accelerate by about 20 years plans to acquire a new bomber aircraft.

The new aircraft would use existing or soon-to-mature technologies. The Air Force expects the 2018 bomber to serve as an "interim fix" to bridge a bomber capability gap to permit retirement of a portion of the current strategic bomber force and position itself for the development of a follow-on bomber with more advanced technologies such as hypersonic (Mach 5-plus) drive engines.

The Next Generation Bomber program will provide for the acquisition of approximately 100 aircraft capable of high-subsonic flight that can deliver both nuclear and conventional munitions. These will be low-observable (stealth) aircraft. A portion of them may be unmanned aircraft (i.e., unmanned aerial vehicles).

When this volume went to press, the Boeing and Lockheed Martin firms had already teamed and were working on preliminary bomber designs. Northrop Grumman, the only other viable competitor, may also join the competition given its experience with the B-2 Spirit aircraft and ongoing work on the Navy's unmanned strike aircraft.

Historical: The 2006 Quadrennial Defense Review (QDR) began laying the groundwork for the development of the Next Generation Bomber to either replace or augment the current force of approximately 180 strategic bombers—B-1B, B-2A, and B-52H types. Prior to the 2006 QDR, the Air Force had indicated that its current bomber force would suffice until 2037, when advanced technologies, such as hypersonic cruise, would potentially reach maturity and could be incorporated into follow-on bomber aircraft.

The Office of the Secretary of Defense, responding to the Air Force's desire to retire 38 B-52H bombers and concerned about the Air Force's ability to execute future long-range bombing requirements, accelerated Air Force plans for fielding a new aircraft by almost 20 years. The Department of Defense's 2002 Nuclear Posture Review (NPR) stated a requirement for 66 "operational" B-52Hs with an additional 28 B-52Hs in "attrition reserve." All aircraft receive standard maintenance and scheduled upgrades.

Subsequently, the Air Force was allowed to further reduce the B-52H force with Congress "strongly" discouraging the secretary of the Air Force from taking action to reduce the B-52H inventory below 76 aircraft prior to the Next Generation Bomber reaching IOC.

According to a Congressional Research Service report of 2008, "There is currently no consensus, within [the Department of Defense] or among military analysts, on the capabilities needed in the next generation bomber, or even whether an 'interim' bomber is needed."

Design: Rebecca Grant of the IRIS organization, a respected defense analyst and author, in December 2007 listed the capabilities and program milestones that were solidifying with respect to the Next Generation Bomber:

- Total buy ~100 aircraft
- Combat radius 2,300 miles minimum (unrefueled)
- Payload 28,000 pounds
- Speed high subsonic
- Two engines
- Very low observable–improved stealth technology
- Manned cockpit
- Nuclear capable
- Technology reaching maturity by 2009
- Demonstration flight by 2011

B-1 Lancer

The Rockwell B-1 is a long-range strategic bomber designed to penetrate advanced Soviet air defenses. Following the demise of the Soviet Union, beginning in 1993 the B-1 has been shifted from a strategic/nuclear to a conventional-only strike role.

B-1 development began in 1965 as the advanced manned strategic aircraft (AMSA). The aircraft had a gestation period of 20 years. When it became operational, it was found to lack the ability to penetrate Soviet air defenses without major upgrades.[5]

In the 1970s the U.S. Air Force had proposed 240 B-1s to replace the B-52 strategic bomber force. The cost of the B-1 coupled with rampant inflation of the late 1970s created an extremely expensive aircraft. The first prototype of the Rockwell B-1A flew in December 1974 with a production decision planned for 1976. Production was in fact approved on 2 December 1976, but the following month the Carter administration (1977–1981) entered office, and that June the entire B-1 program was cancelled. The offered alternative—because some strategic program was needed—was development of the air-launched cruise missile (ALCM) for use with the B-52 force. Beyond major cost overruns and higher-than-specified weight, the B-1B defensive electronics system (ALQ-161A), which was necessary for penetrating sophisticated anti-air defenses, continually failed to perform as intended.

The manned strategic bomber again became an issue in the 1980 presidential campaign. In August 1980, to help rationalize his decision to cancel the B-1, President Carter announced that the United States was in fact developing a new bomber in secret, the advanced technology bomber (ATB, later B-2) stealth aircraft. At that time, the announcement was considered to be a political ploy because it was too early for stealth technology to be considered for a production program.

The Reagan administration (1981–1989) gave immediate support to the B-1 program and in 1981 gained congressional funding for a 100-plane force in addition to continuing the ATB/B-2 development program.

While the B-1B is considered a "penetrating" bomber, it is unable to carry ALCM standoff missiles. It has a theoretical capability for maritime missions—minelaying and carrying Harpoon anti-ship missiles. With respect to conventional bombing, the B-1B as built could carry only the Mk 84 2,000-pound unguided bomb; upgrades have been undertaken to improve the aircraft's conventional payload, which is accommodated in three bomb bays.

Design: The aircraft has variable-geometry (swing) wings mounted at midfuselage. The wing pivot is located relatively close to the fuselage, which yields a large change in wing aspect ratio from its minimum sweep angle of 15° to the maximum sweep angle of 67.5°. In addition to a low-sweep angle, takeoff and landing performance are aided by leading and trailing edge lift devices. The four turbofan engines are housed in paired nacelles under the wings. They are fed through fixed-geometry intakes that limit maximum speed to approximately Mach 1.25.

The B-1B radar cross section is estimated to be $\frac{1}{7}$th that of the FB-111A and $\frac{1}{100}$th that of the B-52. This is achieved through fuselage shaping and the use of radar-absorbent materials.

The internal weapons bay can accommodate a maximum of 133,800 pounds.

Designation: Following the 1962 redesignation of U.S. military aircraft, the B-series was restarted with B-1. The B-70/RS-70 was the last bomber in the B-series followed by the SR-71 strategic reconnaissance aircraft in that same numerical series.

The B-1's nickname is "BONE" for its designation "B-One."

Operational: The first flight of B-1A was on 23 December 1974; first flight of the B-1B was on 18 October 1984. First squadron delivery of B-1B was in June 1985.

5 During the 13-year interval between delivery of the previous U.S. strategic bomber, the FB-111, and the first operational B-1B, the Soviet strategic air arm received an estimated 260 Backfire supersonic bombers plus a small number of Bear-H strategic missile-carrying aircraft.

The sleek-looking Rockwell B-1B Lancer was developed as a Cold War nuclear strike aircraft. However, its electronic systems for penetrating Soviet air defenses were not successful. With the end of the Cold War the surviving B-1s have been modified for the delivery of only conventional weapons, a role that they performed in the Balkans, Iraq, and Afghanistan conflicts. (U.S. Air Force)

A B-1B bird-strike accident occurred on 28 September 1987 with loss of the aircraft; thus the force only reached 99 aircraft with four USAF bomb wings flying the B-1B. In November 1988 two additional accidents reduced the B-1B force to 97 aircraft.

In 1999 there were 63 operational aircraft plus 2 aircraft allocated to test.

The B-1B was the only type of aircraft operational in the Air Force in 1991 that did not participate in the Gulf War. The combat debut of the B-1B occurred in the December 1998 strikes against Iraq. The aircraft was used in the subsequent Afghan and Iraqi invasions.

The aircraft holds almost 50 world records for speed, payload, range, and time to climb in its class. On 19 March 2008, a B-1B successfully flew at supersonic speeds while using synthetic fuel. The Air Force had been testing several aircraft using a 50/50 blend of petroleum and fuel produced from coal, but the B-1B was the first Air Force plane to fly at supersonic speeds on the blended fuel. The Air Force goal at the time was to have all aircraft using synthetic fuel blends by 2011.

Production: Four B-1A prototype aircraft were built. The Reagan administration requested a 100-bomber program.

A "buy-down" of another 50 or more bombers was proposed as a hedge against further delays in the ATB/B-2 program, a more economically sound posture for the aerospace industry, and as a response to the strong Soviet bomber program. However, Secretary of Defense Caspar Weinberger resisted these efforts, halting the program at 100 aircraft. The last of the 100-production B-1B aircraft was delivered in April 1988.

Operational:	June 1985–current	
Design:	Rockwell International	
Aircraft data:	B-1B (2008 data)	
Crew:	4 (pilot, copilot, 2 weapon systems officers)	
Engines:	4 General Electric F-101-GE-102 turbofan; 30,780 lbst each	
Weights:	empty	190,000 lb
	maximum	477,000 lb
Dimensions:	length	146 ft
	wingspan	136 ft 8½ in at 15° sweep 78 ft 2½ in at 67.5° sweep
	wing area	1,950 ft^2
	height	33 ft 7¼ in
Speed:	maximum	850 mph
	max low-level cruise	630 mph
Ceiling:	30,000+ ft	
Range:	radius: 2,070 miles unrefueled with full combat payload	
Radar:	AN/APQ-164 forward-looking phased array	
Guns:	none	

The Rockwell B-2A Spirit was the last of a long line of nuclear-capable strike aircraft developed during the Cold War. Developed under a veil of secrecy, the aircraft's low-observable (stealth) features are of a later generation than those of the F-117A Nighthawk (see Chapter 4). However, improvements in radar systems and their operation are a challenge to stealth aircraft. (U.S. Air Force)

Ordnance:	16 nuclear bombs B61-7 or B83 as built; nuclear capability subsequently deleted
or	75,000 lb conventional bombs, missiles, mines

B-2 Spirit

The Northrop B-2 represents the "ultimate" U.S. strategic bomber developed in the Cold War era. Incorporating stealth technology, the B-2 has a flying wing configuration reminiscent of the World War II–era Northrop XB-35 (piston) and XB-49 (turbojet) flying wing bombers, neither of which entered operational service.

The Air Force and Northrop initiated the B-2 program as a highly classified, "black" effort. Northrop began studies of low-observable aircraft design in 1975. The firm lost to Lockheed for development of the F-117 stealth attack aircraft but continued efforts to develop an "invisible" aircraft, leading to the flying wing bomber design. On 22 August 1980, Secretary of Defense Harold Brown announced that the United States was developing an advanced technology bomber (ATB) that would employ a new technology that "alters the military balance significantly." The disclosure came in response to criticism of the Carter administration for not supporting advanced strategic weapons.

Northrop won the ATB competition with Lockheed in October 1981 to develop and produce a fleet of 132 ATB aircraft (including 6 test aircraft, of which 5 would subsequently be reconfigured to combat capability). Because of major cost increases, in April 1990, Secretary of Defense Dick Cheney reduced the B-2 program to 75 aircraft; in January 1992 President Bush announced that the total B-2 buy was limited to a total of 20 aircraft including the flight tests units. Subsequently, a 21st aircraft was produced.

The aircraft encountered some technical difficulties and as late as July 1991, a report of the General Accounting Office stated that despite some progress, overall the

> B-2 contractors [Northrop, Boeing, LTV] are generally not meeting the manufacturing goals they established. They are continuing to experience significant problems in reducing labor hours, numbers of defects, engineering drawing changes and in completing work at major section assembly sites rather than transferring it to the final assembly site. Also, their production management programs are not fully effective.[6]

6 U.S. General Accounting Office, *B-2 Program Trends in Manufacturing* (Washington, D.C., July 1991), p. 3.

Design: The B-2 has a flying-wing design without a true fuselage and has no vertical tail structure, which reduces radar reflections. The wing's leading edges are swept back at approximately 40° to the tips, which are raked back toward the centerline. This unusual tip rake is believed to decrease the flying wing's tendency to yaw while rolling. The wing's sawtooth trailing edge is designed to cope with a flying wing's natural instability.

Two engine intakes are raked (the top edge being further aft than the bottom) and have sawtooth top edges and a narrow sawtooth fence mounted a few inches above the wing's surface just ahead of the intake openings. It is believed that these fences help to smooth the air flow into the intakes and may help to reduce the intake's radar signature. Four non-afterburning turbofan engines are buried in the wing.

Twin, internal, side-by-side bomb bays can accommodate a variety of conventional or nuclear weapons; the aircraft's maximum weapons load is 37,300 pounds consisting of up to 20 nuclear bombs. The conventional weapons load consists of 80 Mk 82 500-pound bombs.

The aircraft has two flat-plate antennas for the Hughes multimode, phased-array radar. Digital controls overcome the problems encountered in the previous U.S. flying wing bomber, the YB-49, which first flew in 1947.

The stealth traits of the B-2 include greatly reduced infrared, acoustic, electromagnetic, visual, and radar signatures. This reduction is accomplished by the aircraft's shape, materials, and signature suppression features.

There is cockpit space for a third crew member for special missions.

Designation: All 21 aircraft were designated B-2A.

Operational: The first flight was on 17 July 1989. The first B-2 was delivered to an operational training squadron in December 1993. The B-2A was first used in combat over Kosovo in 1999. Flying from Whiteman AFB, the eight combat-ready Block 30 B-2s—those with the most advanced stealth configuration and avionics upgrade—flew 45 combat sorties, attacking their targets at night and often in weather conditions that prevented other strike aircraft from attacking targets. On those 45 sorties the B-2s released about 700 guided weapons, and 2,000-pound guided bombs.

Subsequently, combat missions were also flown over Afghanistan and Iraq, with record-duration missions flown from the B-2A home base of Whiteman AFB, Missouri. Advance operating bases have been established at Andersen AFB on Guam and on Diego Garcia in the Indian Ocean.

A B-2A crashed on takeoff from Andersen AFB on 23 February 2008. The two pilots ejected safely. It was the first loss of a B-2 after more than 14,000 B-2A sorties totaling 75,000 flying hours.

Operational:	Dec 1993–current	
Design:	Northrop	
Aircraft data:	B-2A	
Crew:	2 (pilots)	
Engines:	4 General Electric F118-GE-100 turbofan; 17,300 lbst each	
Weights:	empty	160,000 lb
	max takeoff	336,500 lb
Dimensions:	length	69 ft
	wingspan	172 ft
	wing area	5,000 ft^2
	height	17 ft
Speed:	maximum	630 mph
Ceiling:	50,000 ft	
Range:	radius: 2,300 miles unrefueled with full combat payload	
	approx 6,000 miles with 37,300 lb weapons (high-altitude attack)	
	approx 5,000 miles with 37,300 lb weapons (high-low-high attack)	
	approx 7,600 miles with 24,000 lb weapons (high-altitude attack)	
	approx 1,150 miles with 24,000 lb weapons (high-low-high attack)	
Radar:	AN/APQ-181 multimode phased array	
Guns:	none	
Ordnance:	20 nuclear bombs B61 or 16 nuclear bombs B83	
or	80 500-lb conventional bombs	

B-29 Superfortress

The Boeing B-29 was the world's first nuclear-capable aircraft and the most advanced bomber flown by any nation in World War II. The B-29 carried out intensive attacks against Japanese industrial and urban targets in 1944–1945. The only nuclear weapons to be used in combat were delivered by B-29s on the Japanese cities of Hiroshima on 6 August 1945 (Little Boy), and on Nagasaki on 9 August 1945 (Fat Man).

The overall AAF project to organize, train, and equip a B-29 force to deliver the atomic bomb was given the code name Alberta. The aircraft were modified to carry an atomic bomb under the Silverplate program. In early

The Boeing B-29 Superfortress was the most-capable bomber aircraft of World War II. Although not as fast as the Ardo Ar 234 jet-propelled bomber, nor did it have the bomb capacity of the Avro Lancaster, the B-29 could fly higher, with a greater range, and had a more-potent defensive armament. B-29s served in the strategic bomber role until 1954, with KB-29 tankers and WB-29 weather reconnaissance aircraft serving longer, the latter until 1959. (U.S. Army Air Forces)

August 1945 there were 18 Silverplate-modified B-29s of the 509th Composite Group, on Tinian in the Mariana Islands, which had been specially trained for atomic bomb attacks; six of their crews were briefed to deliver nuclear weapons (the additional aircraft being employed in scientific and other support operations). By that time the unit had discarded their original Silverplate-modified aircraft.

After the war additional B-29s were modified for nuclear weapons delivery under the Silverplate program; in 1947 the code name was changed to Saddletree and would include other types of aircraft to be modified. (The number of B-29s fitted with a nuclear weapons capability are listed on page 3.) Official sources differ on exact numbers, not all were fully operational, and there were never fully trained crews available for all of the nuclear-capable aircraft.

Design: The B-29 was a four-engine heavy bomber, the first with pressurized compartments for the flight crew. It was also the first U.S. bomber built with a radar bombing system (initially the AN/APQ-13), which supplemented the Norden bombsight. With a maximum takeoff weight of 140,000 pounds, the B-29 was the largest bomber of World War II. Up to 20,000 pounds of bombs were carried in two internal bomb bays.

Silverplate aircraft could not carry conventional bombs. The so-called Pumpkin bomb in the shape of the Fat Man bomb was developed for training; it contained 5,500 pounds of high explosives. B-29s of the 509th carried out Pumpkin bombing missions in the Pacific: 12 missions over Japan on 20–29 July, prior to the atomic bomb attacks; 2 additional Pumpkin missions flown over Japanese cities on 8 August, between the two atomic bomb attacks; and 2 on 14 August, after the atomic bombings.

The basic B-29 armament consisted of 10 or 12 .50-cal machine guns in five remote-controlled power turrets. The forward dorsal turret had two or four guns; the after dorsal turret and two ventral turrets had two guns each, as did the tail turret; in early aircraft a 20-mm cannon was additionally fitted in the tail. The Silverplate aircraft had all guns and ammunition removed except for two .50-cal machine guns in the tail turret. Any one of four gunners could control any of the turrets.

Designation: The B-29D model was changed to B-50 (see separate entry).

Operational: The first XB-29 flew on 21 September 1942, and first delivery to an AAF unit occurred in July 1943.

The first B-29 combat mission was flown against Japanese-occupied Bangkok from staging bases in China on 5 June 1944. B-29s were briefly based in China in 1944 (flying one mission against Japan as well as against other Japanese targets in the Far East). Subsequently, heavy B-29 raids began against Japan from the Marianas in late 1944, a 13-hour round-trip flight. In addition to dropping 171,000 tons of conventional bombs during 14 months of war, B-29s sowed more than 12,000 aerial mines in Japanese and Korean waters to stop coastal shipping. When the war ended, the AAF had 40 groups of B-29s, of which 21 with almost 1,500 aircraft were at advance bases in the Pacific.

In January 1946 the AAF designated the 58th Very Heavy Bombardment Wing at March Field, California, as its atomic striking force; the wing consisted of three Bomb Groups—the 40th (March Field), 444th (Castle Field, California), and 509th (Roswell Field, New Mexico).[7] At that time the total U.S. strategic bomber force consisted of 148 B-29s.

On 30 June 1946, in Operation Crossroads, a B-29 dropped an atomic bomb on target ships arrayed at Bikini atoll. (Bikini and the two Japanese strikes were the only airdrops of nuclear weapons by B-29s.)

The last B-29 bomber was withdrawn from the SAC in late 1954; specialized variants, especially KB-29 tankers, survived in USAF service until 1959.

The Royal Air Force flew 89 B-29s with the name Washington (nonnuclear capable). The B-29 was copied and produced by the Soviet Union as the Tu-4, given the NATO code name Bull; some of those aircraft were transferred to China, which flew the aircraft in the research role into the early 1990s.[8]

Production: Total B-29 production was 3,628 aircraft by the Boeing, Bell, and Martin companies, with all but some 230 delivered by August 1945; the last aircraft was delivered in June 1946 (another 200 B-29 and 5,000 B-29C models were cancelled at the end of the war).

Operational:	July 1943–Nov 1954 in bomber role	
Design:	Boeing	
Aircraft data:	B-29	
Crew:	11 (pilot, copilot, bombardier, radar operator, navigator, engineer, radio operator, 4 gunner-observers)	
Engines:	4 Wright R-3350-57 or -57A piston; 2,200 hp each	
Weights:	empty	71,500 lb
	combat	101,082 lb
	max takeoff	140,000 lb
Dimensions:	length	99 ft
	wingspan	141 ft 3 in
	wing area	1,736 ft²
	height	27 ft 9 in
Speed:	maximum	399 mph at 30,000 ft
	cruise	253 mph
Ceiling:	39,650 ft	
Range:	1,975 mi with max bomb load	
Radar:	AN/APQ-13 bombing	
Guns:	2 .50-cal machine guns (tail position) in Silverplate aircraft 10 or 12 .50-cal machine guns in non-Silverplate aircraft	
	1 20-mm M2 cannon (tail position) in early non-Silverplate aircraft	
Ordnance:	1 nuclear bomb Little Boy, Fat Man, Mk 3, Mk 4, Mk 5, Mk 6 in Silverplate and Saddletree aircraft	
	20,000 lb conventional bombs, mines in non-Silverplate aircraft	

B-36 Peacemaker

The Convair B-36 was the largest combat aircraft ever built. Designed for conventional transoceanic bombing missions, it was the mainstay of the U.S. strategic bomber force during the 1950s. The B-36 was initiated to bomb European targets from bases in the United States because in 1941 the U.S. War Department believed that the United States would likely be drawn into the war and that all of Europe—including Britain—could be under Axis control. (Also developed under this transoceanic bomber program was the Northrop B-35 Flying Wing bomber, an early precursor to that firm's B-2 "stealth" bomber.)

In the spring of 1943, when it appeared that the Chinese government might collapse under Japanese attack, the B-36 was considered for long-range operations in the Pacific in the event of problems in the B-29 Superfortress

7 When established, the 509th was a Composite Group.
8 The Chinese aircraft included airborne early warning (AEW) variants.

The Convair B-36 Peacemaker was the largest combat aircraft ever built, dwarfing even the massive Tupolev Tu-20/95/142 Bear-series aircraft. The Air Force planned to procure thousands of B-36s; the total built, however, was 385. This is an RB-36D reconnaissance variant, which still could deliver nuclear weapons. (U.S. Air Force)

and B-32 Dominator heavy bomber programs. When the B-29 program achieved success, the B-36 was continued, but with lesser priority.

The size and cost of the B-36 made it the most-publicized bomber of the immediate postwar era. During World War II, AAF planners proposed a force of up to 3,740 B-36s. The Air Force leadership in 1949 planned a B-36 force of four bomber groups—120 aircraft plus 39 spares—and two groups of the RB-36 long-range reconnaissance variant. Although the B-36 was relatively slow, it could fly an Arctic route to reach the Soviet Union from bases in the United States without in-flight refueling.

The B-36 was the centerpiece for the 1948–1949 debate between the Navy and Air Force on the relative effectiveness of land-based and carrier-based bombers. There was also controversy over the aircraft within the USAF: General George C. Kenney, the first commander of SAC, declared, "The B-36 is a night bomber. I would not use it in daytime," and stated that the plane might be suitable for the Navy's anti-submarine mission. On 12 December 1946 he recommended the procurement of the B-36 only for test purposes and instead preferred the B-50 as a strategic bomber until jet-propelled bombers became available.

Kenney's successor as head of SAC, the outspoken and innovative General Curtis LeMay, supported the B-36, claiming, "We can get a B-36 over a target and not have the enemy know it is there until the bombs hit."[9] In an effort to overcome the perceived vulnerability of the B-36, the Air Force developed the diminutive F-85 Goblin or "parasite" fighter to be carried by the B-36 and released when enemy fighters threatened, and subsequently taken back aboard the bombers. The XF-85 was flight tested from a modified B-29, but never from a B-36. That scheme was quickly discarded (although SAC did operate its own land-based escort fighters until 1957). RF-84 Thunderflash reconnaissance aircraft also were

9 LeMay served as Commander and (from 1953) Commander-in-Chief Strategic Air Command from 19 October 1948, to 30 June 1957. He served as Chief of Staff of the Air Force from 30 June 1961, to 31 January 1965.

evaluated in a "parasite" role with 7 RB-36D bombers and 23 RF-84K aircraft being modified; after flight trials from December 1955 to February 1956 this scheme was also discarded.

Concern over B-36 survivability continued; an Air Force evaluation of strategic bombing noted: "The speed and operational radius of the B-36 left much to be desired, however. Because of its low speed, the B-36 was vulnerable to heavy attrition from fast, modern fighters. Relatively heavy losses, by comparison with World War II experience, could be tolerated within limits because of the destructive power of the A-bomb, but the unanswerable question was whether B-36 losses could be kept within tolerable limits."[10]

Efforts to improve B-36 survivability by increasing the aircraft's ceiling led to the "featherweight" program of reducing weights.

Design: The swept-wing aircraft was propelled by six large piston-pusher engines, supplemented in the B-36D and later models by four turbojets in underwing pods to provide "burst" speed as the big bomber neared its target.

The maximum bomb load was 72,000 pounds with a gross weight of 328,000 pounds, although the B-36D could carry two 42,000-pound Grand Slam bombs in the internal weapon bays. The B-36B was the first model configured to carry nuclear bombs. Eighteen B-36B aircraft were modified to also carry two VB-13 Tarzon guided bombs; those aircraft had the AN/APG-24 bomb/nav radar. Later, one aircraft was fitted to carry the GAM-63 Rascal guided missile; one B-36H was fitted as an aerial tanker; and one B-36H served as an aerial test bed for a nuclear reactor in support of the nuclear aircraft program, making 47 test flights (redesignated NB-36H). The B-36A aircraft were unarmed and were used mainly for training and then converted to the RB-36E configuration.

The aircraft had the heaviest defensive armament every carried by an aircraft: Up to 16 20-mm cannon in twin turrets—the four side-by-side dorsal and two side-by-side ventral turrets were retractable (to reduce drag); additional two-gun turrets were mounted in the nose and tail, all directed by a central control system. (There was no gunner position in the tail.) The featherweight program reduced the number of guns until the modified B-36H (III) had only the tail turret with two guns.

General Curtis E. LeMay, USAF. (U.S. Air Force)

The RB-36 variants were configured for strategic photo reconnaissance. The RB-36D had two of the four bomb bays fitted with 14 cameras; the third bomb bay had an additional fuel tank to increase endurance to 50 hours; and the fourth bay was fitted with electronic countermeasures gear. The RB-36D crew was increased to 22 men. (In June 1954 the RB-36s were given a primary bombing mission; they retained a limited, secondary reconnaissance role.)

As Soviet air defenses improved during the 1950s, an effort was made to increase the altitude of the RB-36 through the featherweight program in which most of the plane's guns and some other grear was removed to reduce weight; the crew of featherweight RB-36s was only 19 men. The featherweight aircraft could reach just over 45,000 feet, some 5,000 to 8,000 feet more than a standard aircraft. A number of programs were undertaken to upgrade the B-36 capabilities:

Saddletree—Modify aircraft to carry the Mk 3 and Mk 4 bombs. In 1951 there were 52 B-36Ds equipped to carry the Mk 4.

On Top—Provide the Mk 5 and Mk 6 nuclear bomb capability to B-36s. By the end of 1951 there were 121 Mk 6 modified B-36s. In July 1950 SAC established a requirement that B-36s carry more than one atomic bomb. Another phase fitted RB-36s to carry Mk 5, Mk 6, and Mk 13 weapons in bomb bay number 2. Still another On Top program provided

10 U.S. Air Force, *USAF Supporting Studies*, Vol. I of *A History of Strategic Arms Competition, 1945–1972* (Washington, D.C.: 1973), pp. 21–22.

atomic carrying capability in all four bays of 30 B-36s.

Caucasian—Improve delivery methods for B-36s carrying thermonuclear weapons. Caucasian began well before the United States had a deliverable thermonuclear weapon, or even before scientists understood the H-bomb concept. The proposed tentative size to guide the effort was a 50,000-pound bomb, 6 feet in diameter and 20 feet in length. Four B-36H aircraft were modified to carry the TX-14 bomb and, subsequently, the TX-16 bomb, both hydrogen weapon prototypes.

Barroom—Equipping B-36s to carry thermonuclear weapons begun in April 1953. After Operation Castle the TX-16 was discarded as a possible weapon and the program name was changed to Cauterize.

The maximum B-36 mission duration was almost 40 hours.

A single C-99 cargo version of the B-36 was built; it could carry up to 400 troops.

Many early B-36 models were updated; for example, 24 B-36B variants were upgraded to the B-36D configuration.

The YB-36G was a swept-wing, all-jet variant, which also carried the designation YB-60. Two aircraft were built (see separate entry).

Designation: The name Peacemaker was rarely used; the aircraft was generally known as simply the B-36.

Operational: The first XB-36 flew on 8 August 1946, almost five years after the prototype was ordered. SAC received the first (unarmed) B-36A on 26 June 1948, the first operational B-36B aircraft in November 1948, and the B-36D in August 1950. These aircraft were not considered fully operational until 1951 and were not free of mechanical problems until 1954.

The B-36F entered service in March 1951, the B-36H in 1952, and the final B-36J in 1953.

The unarmed B-36A aircraft were later converted to the RB-36E bomber/reconnaissance configuration.

The B-36 force peaked in October 1955 with ten heavy bombardment wings—six with B-36s and four with RB-36s. Each wing had a nominal assignment of ten planes in each of three squadrons—a total of 209 B-36 and 133 RB-36 aircraft. SAC became an all-jet bomber force when the last B-36J was retired on 12 February 1959.

No B-36 ever dropped a conventional bomb in combat. B-36s were used in six nuclear bomb drops in weapon tests.

Production: A total of 385 B-36 bombers were produced through 1954 plus one similar cargo aircraft (XC-99). The B-36C was to have had upgraded engines; the RB-36E aircraft were converted from B-36A models.

Aircraft	Built	First flight
XB-36	1	1946
YB-36	1	1947
B-36A	22	1947
B-36B	62	1948
B-36C	cancelled	
B-36D[a]	27	1949
RB-36D	24	1949
B-36F[a]	34	1950
RB-36F	24	
YB-36G	2	1952
B-36H[a]	83	1950
RB-36H	73	
B-36J	33	1953

a Includes one YB-36 prototype of each type that had a full military capability.

Operational:	26 June 1948–12 Feb 1959	
Design:	Convair/General Dynamics	
Aircraft data:	B-36H (III)	
Crew:	15 (pilot, copilot, engineer, navigator, bombardier, radar operator, radio operator, auxiliary crew member, 7 gunners)	
Engines:	6 Pratt & Whitney R-4360-53 piston; 3,800 hp each	
	4 General Electric J47-GE-19 turbojet; 5,010 lbst each	
Weights:	empty	164,522 lb
	combat	248,900 lb
	max takeoff	370,000 lb
Dimensions:	length	162 ft 1 in
	wingspan	230 ft
	wing area	4,772 sq[2]
	height	46 ft 9½ in
Speed:	maximum	428 mph at 38,400 ft
	cruise	250 mph
Ceiling:	5,100 ft	
Range:	7,200 miles	
Radar:	AN/APG-3 gunfire control	
	AN/APQ-24 bomb/nav	
Guns:	2 20-mm M24A1 cannon (600 rounds per barrel)	

The Boeing B-47 Stratojet introduced new concepts in U.S. bomber design, albeit based largely on German aviation technology. Beyond the strategic bomber role, RB-47 variants had an important place in U.S. reconnaissance operations, including flights over Soviet territory. This B-47E has the Strategic Air Command "band" and insignia on its nose. (U.S. Air Force)

Ordnance:	2 nuclear bombs Mk III, Mk 4, Mk 5, Mk 6, Mk 8, B17, B18, B24, B36, B39 (not every bomber was configured to carry every type)
or	86,000 lb conventional bombs

B-47 Stratojet

Aviation historian Bill Gunston observed that "very rarely through the history of aviation has a design team come up with a design so advanced technically as to appear genuinely futuristic."[11] The sleek, swept-wing, Boeing B-47 was such an aircraft. It was the first turbojet bomber to enter service with SAC as a "medium bomber" (replacing the B-29 and B-50), and it flew in large numbers in specialized reconnaissance configurations.

Designed by the Boeing Airplane Company, which had produced the renowned B-17 and B-29 bombers, the aircraft was one of several designs considered by the AAF at the end of World War II for the initial generation of U.S. jet-propelled bombers. After studying a number of configurations, Boeing engineers took advantage of German research into swept-back wings to produce the B-47 design.

The B-47 program was the largest U.S. post–World War bomber procurement. In both design and quantity, the B-47 was a milestone in the development of strategic bombers and gave the United States a potent nuclear strike capability.

Design: The basic B-47 design originated in the fall of 1945. It was a relatively large, swept-wing aircraft with six turbojet engines mounted in two twin and two single pods beneath the wings. The shoulder-mounted wings were swept back 35°.

The pilot and copilot sat in tandem in the narrow fuselage, with a navigator-bombardier in the nose, a total of 3 men compared to 10 for the B-17 and 11 for the B-29, both of which were smaller aircraft. Space and weight were instead allocated to fuel, with the B-47E carrying 14,610 gallons internally and 3,390 gallons in two wing drop tanks, totaling almost ten times the B-17 capacity and three times that of the B-29. The B-47 was fitted for in-flight refueling.

The internal bomb bay could accommodate two small or one large nuclear-thermonuclear weapon; the theoretical conventional bomb capacity was 1 12,000-pound, or 6 2,000-pound, or 18 1,000-pound conventional bombs.

Defensive armament consisted of a remote-controlled tail turret, initially twin .50-cal machine guns; twin 20-mm cannon were provided beginning with the B-47E (compared to about a dozen guns in the B-17 and B-29). The copilot's seat rotated to a rear-facing position to permit him to use the A-5 fire control system for the tail guns. (The B-47A, which did not enter squadron service, was unarmed.) A vertical camera was fitted in the tail of all aircraft

Thirty B-47s in the Ebb Tide program became DB-47Bs to carry the GAM-63 Rascal air-to-surface missile. That missile had a very stormy history before it was cancelled in 1958.

Several B-47 upgrade programs were undertaken, including projects High Noon, Turn Around, and Ebb Tide, which provided the capability of delivering a thermonuclear bomb. By the end of April 1956 1,100 B-47s were modified to carry thermonuclear bombs.

11 Bill Gunston, *Bombers of the West* (New York: Charles Scribner's Sons, 1973), p. 126.

An early program to have the B-47 operate as an unmanned bomb carrier was developed under the code name Brass Ring. As the MB-47, it would have to be air-refueled several times and manned until the last refueling. The crew would then bail out over friendly territory and the MB-47 would continue on automatic control to its preset target. There were problems with the guidance system, and as other ways of delivering H-bombs became available, Brass Ring faded and was cancelled on 1 April 1953.

Major variants included electronic warfare (EB-47), reconnaissance (RB-47), training (TB-47), and weather reconnaissance (WB-47) configurations.

Designation: The B-47 was begun as weapon system MX-584. The single B-47C had a different engine configuration and was redesignated XB-56.

Operational: The first flight of XB-47 was on 17 December 1947. The B-47B entered service with the 306th Bomb Wing on 23 October 1951. The definitive B-47E version made its first flight in 1953. The first B-47E entered operational service with 303rd Bomb Wing in April 1953.

The aircraft lacked the range to reach the Soviet Union from bases in the United States. Accordingly, several "permanent" SAC bases were established in forward areas, and large numbers of piston-engine KC-97 tanker aircraft were procured to provide in-flight refueling. The first B-47 wing (45 aircraft) to deploy overseas was sent to England on a 90-day rotational training mission in June 1953. B-47s set several endurance records: in August 1954 a B-47 made a nonstop flight of 17,000 miles in 35 hours, and in November 1954 B-47s made a nonstop flight from Davis-Monthan Air Force Base (Arizona) to England via North Africa, a flight of 21,000 miles in 47 hours, 35 minutes.

The last two Air Force B-47s went into storage on 11 February 1966. The last RB-47H was retired on 29 December 1967, and the last operational Stratojet—a WB-47—was retired in November 1969. In the 1980s the U.S. Navy flew (with contractor crews) three B-47s removed from storage—an NB-47H for engine test and two EB-47E aircraft to operate against U.S. warships in an electronic aggressor role.

There were no test drops of nuclear bombs by B-47s, and they were never used in conventional bombing missions in wartime.

Production: B-47 production totaled 2,042 aircraft, including several hundred EB-47 electronic warfare and RB-47 reconnaissance variants. The last B-47E was delivered on 18 February 1957. Production was undertaken by Douglas (274) and Lockheed (393) firms as well as by Boeing-Wichita. At one point the Air Force planned to procure more than 2,700 aircraft.

Two B-47s were modified to the B-47D configuration as test beds for the Curtiss Wright YT49-W-1 turboprop engine; the twin inboard engine pods were each replaced by a single YT49 engine. The single outboard jet engines were retained on both aircraft.

The single XB-56 was redesignated YB-47C (see separate entry); that aircraft was not built.

The "bomber" variants of the B-47 series were as follows:

Aircraft	Built	First flight
XB-47	2	1947
B-47A	10	1950
B-47B	399	1951
B-47E	1,341	1953

Operational:	1951–1966 in bomber role 1953–1969 in special-mission role	
Design:	Boeing	
Aircraft data:	B-47E-IV	
Crew:	3 (pilot, copilot, navigator-bombardier)	
Engines:	6 General Electric J47-GE-25 turbojet; 7,200 lbst each	
Weights:	empty	79,074 lb
	max takeoff	230,000 lb
Dimensions:	length	107 ft 1 in
	wingspan	116 ft
	wing area	1,428 ft^2
	height	28 ft
Speed:	maximum	607 mph at 16,300 ft
	cruise	500 mph
Ceiling:	39,300 ft	
Range:	radius	2,378 mi
Radar:	K-4A bomb/nav AN/APS-54 tail warning	
Guns:	2 20-mm M24A1 cannon (350 rounds per barrel)	
Ordnance:	2 nuclear bombs Mk 4, Mk 5, Mk 6, Mk 8, B15, B28, B36, B39, B41, B43, B53	
or	18,000 lb conventional bombs	

The Boeing B-50 Superfortress was an improved version of the B-29. Generally similar in appearance, B-50s had more powerful engines housed in larger nacelles. This B-50A has had its four fuselage gun turrets deleted, but retains it tail guns. Note the mid-fuselage observer-gunner domes. Beginning with the B-29 most U.S. bombers had remote-control gun turrets. (U.S. Air Force)

B-50 Superfortress

In 1944 the Army Air Forces asked the Boeing Aircraft Company to pursue an improved B-29 that could fly farther and faster. The aircraft was initially designated B-29D. At the end of World War II, despite the cancellation of almost 5,000 B-29s, the B-29D survived but with a reduction from the planned 200 aircraft to only 60.

The aircraft—redesignated B-50 in December 1945—was earmarked for the atomic role as a stopgap measure until a more capable aircraft was available. The modification program to enable them to carry nuclear bombs was more complicated, time consuming, and expensive than originally anticipated.

B-50A deliveries to SAC's 43rd Bombardment Wing, at Davis-Monthan AFB, Arizona, began in June 1948. The single B-50 prototype was a B-29A fitted with upgraded engines and redesignated XB-44. Boeing delivered the first B-50 in October 1947 and the last B-50D in December 1950. The single YB-50C prototype was changed to B-54A.

The B-50 production provided 346 bomber aircraft in addition to the single B-29D/B-50 prototype; 24 TB-50H training aircraft for navigator-bombardiers were also built, the last delivered in February 1953. Numerous bombers were converted to reconnaissance and tanker configurations.

Aircraft	Built	First flight
B-50A	79	1947
B-50B	45	1949
B-50D	222	1949

Design: In most respects the B-50 was an improved B-29 with more-powerful engines.

Operational: B-50A deliveries to an operational unit began in June 1948. The last B-50D bomber aircraft was taken out of service on 20 October 1955. Specialized B-50s continued in Air Force service for another decade.

On 2 March 1949, the *Lucky Lady II*, a B-50A of the 43rd Bomb Group, completed the first nonstop around-the-world flight, having covered 23,452 miles in 94 hours, 1 minute, departing from and returning to Carswell AFB, Texas. It was refueled in flight four times by KB-29M tankers.

Operational:	June 1948–Oct 1955 in bomber role	
Design:	Boeing	
Aircraft data:	B-50A	
Crew:	11 or 12 (pilot, copilot, bombardier, navigator, flight engineer, radio operator, 1 or 2 radar operators, 4 gunners)	
Engines:	4 Pratt & Whitney R-4360-35 piston; 3,500 hp each	
Weights:	empty	81,050 lb
	max takeoff	168,480 lb
Dimensions:	length	99 ft
	wingspan	141 ft 3 in
	wing area	1,736 ft^2
	height	32 ft 8 in
Speed:	maximum	385 mph at 25,000 ft
	cruise	235 mph
Ceiling:	37,000 ft	
Range:	4,650 miles with 10,000 lb bomb load	
Radar:	AN/APQ-24 bomb/nav	
Guns:	13 .50-cal MG (500 rounds per barrel)	
Ordnance:	1 nuclear bomb B3, B4, B5, B6	
or	20,000 lb conventional bombs	

The Boeing B-52 Stratofortress was the linear successor to the B-47 Stratojet. The B-52H variants now in service are expected to serve in the conventional and nuclear strike role for another decade. This is a now-discarded B-52G variant. Like the B-47, these aircraft were only fitted with guns in the tail position. (U.S. Air Force)

B-52 Stratofortress

The venerable B-52 remains the backbone of the U.S. strategic bomber force. It is a large, long-range strategic bomber that has been the most numerous bomber in the SAC and its successor, the Air Force's Air Combat Command, from 1992 to the present. The B-52 force peaked in 1962 with 639 aircraft in 22 heavy bomb wings (at which time the SAC bomber force also had 880 B-47s and 76 B-58 strategic bombers). The B-52H is expected to remain in first-line USAF service into the 2020s.

Development of the B-52 officially began with the award of a contract to Boeing on 28 June 1946, although the plane's roots can be traced to a June 1945 AAF directive calling for a heavy bombardment aircraft. Over the next few years the military characteristics for the plane were revised several times before the first production contract was awarded to Boeing on 14 February 1951.

Developed as a high-altitude strategic bomber, the improvement of Soviet surface-to-air missile systems in the late 1950s forced a shift to low-level tactics for strategic penetration of Soviet air defenses. These defenses also required that B-52s carry standoff, air-to-surface weapons and decoys in addition to gravity bombs.

Design: The B-52 has a distinctive configuration, with a blunt-nose fuselage and large, shoulder-mounted wings swept-back 37°. Eight turbojet engines (turbofan in the B-52H) are paired in pods mounted on pylons that project forward of the wing's leading edges. The early variants had tall vertical fins, with the B-52G/H having shorter fins for improved stability in high-speed, low-level flight. Two 700-gallon drop tanks were usually carried on dedicated pylons outboard of the outer pair of engines. Twenty-seven B-52B and all 35 B-52C aircraft were "convertible," able to carry a camera/electronics pack in their bomb bays in place of weapons.

Weapons are carried in two large bomb bays with a capacity of 27,000 pounds of gravity nuclear bombs or, when modified for conventional bombs, more than 40,000 pounds; weapons and decoys could also be carried on two wing pylons. Two AGM-28 Hound Dog air-to-surface missiles could be carried under the wings of the B-52D and later aircraft; the Hound Dog's J52 turbojet engines could be used for additional takeoff thrust to augment the B-52 engines. The GAM-87A Skybolt missile was intended for the B-52, with four to be carried on two wing pylons; that missile was cancelled.

GAM-72 Quail (unarmed) decoys were carried to help penetrate Soviet air defenses.

The first B-52s armed with the AGM-69A short-range attack missile (SRAM) became operational on 4 August 1972. Subsequently, 98 B-52G variants were armed with the AGM-129A air-launched cruise missile (ALCM). The B-52H has been fitted with the rotary bomb bay launchers to carry free-fall nuclear weapons, SRAMs, and ALCMs.

Defensive armament in early aircraft consisted of a tail gunner position with four .50-cal machine guns; in the B-52G the gunner was moved to the forward crew compartment and the guns remotely controlled. The B-52H was fitted with a 20-mm M61 multibarrel cannon, since removed from surviving aircraft.

The surviving B-52H aircraft have upgraded sensors and defensive systems.

Designation: B-52s are nicknamed "BUFF" for "Big Ugly Fat Fella."

Operational: The prototype YB-52 flew on 15 April 1952, before the first flight of the XB-52 on 2 October 1952. The first flight of a production B-52A occurred on 5 August 1954.

The first B-52 to enter SAC service was a B-52B reaching the 93rd Heavy Bomb Wing at Castle AFB, California, on 29 June 1955.

On 21 May 1956, a B-52B made the first airdrop of a U.S. thermonuclear weapon, releasing a B15 bomb over Bikini atoll. This was shot Cherokee of Operation Redwing; the detonation was measured at 3.4 megatons. A total of 30 nuclear bomb drops were made by B-52s in weapons tests through 1962; that amounted to 58 percent of the U.S. airdrops of nuclear weapons.

On 16–18 January 1957, three B-52B aircraft flew 24,325 miles nonstop around the world, the lead plane in 45 hours, 19 minutes. The bombers were refueled in flight five times by KC-97 tankers.

While forming the backbone of the SAC bomber force, from 1960 to 1968 B-52s carried out an airborne alert—Operation Chrome Dome—with at one point during the Cuban missile crisis up to 60 aircraft were airborne, armed with nuclear weapons, and periodically refueled in flight. The normal airborne alert rate varied, with an average of 10 B-52s. The airborne alerts ended after a series of accidents including a B-52 that collided with a KC-135 tanker near Palomares, Spain, on 17 January 1966 and a B-52G that crashed near Thule, Greenland, on 22 January 1968. Four thermonuclear bombs were carried by both of the B-52s involved in these accidents.

B-52D aircraft received the Big Belly modification in 1965 to 1967 to increase their conventional bomb load to 108 500-pound bombs (84 internal, 24 external) or 42 750-pound bombs for missions over South Vietnam. B-52F aircraft—not modified—also were used in conventional bombing operations in the Vietnam War (1965–1973). Later B-52s were used for conventional bombing in the Gulf War of 1991. On 16 January 1991, seven B-52G aircraft of the 2nd Bomb Wing took off from Barksdale AFB, Louisiana, and flew 14½ hours to launch 35 conventional ALCMs against Iraq from a position over Saudi Arabia (to avoid Iraqi air defenses); they then returned to Barksdale, the 34½-hour flight, requiring multiple refuelings from tanker aircraft. This was the longest combat mission in U.S. aviation history.

B-52D variants also were modified to carry CAPTOR naval mines and Harpoon anti-ship missiles.[12]

In 2008 the U.S. Air Combat Command (SAC's successor from 1992) operated 44 B-52H "combat coded" aircraft with a total of 93 in inventory. The first B-52H models were retired in August 2008.

Production: Boeing produced 744 B-52 aircraft from 1951 through June 1962. Although 13 B-52As were ordered, only 3 were built; accepted by the USAF in June–September 1954, they were returned to Boeing for use in test programs and never entered operational service.

The last plane produced was a B-52H, delivered to SAC on 26 October 1962.

Aircraft	Built	First flight
XB-52	1	1952
YB-52	1	1952
B-52A	3	1954
B/RB-52B[13]	50	1955
B/RB-52C[13]	35	1956
B-52D	170	1956
B-52E	100	1957
B-52F	89	1958
B-52G	193	1958
B-52H	102	1961

Operational:	June 1955–current	
Design:	Boeing	
Aircraft data:	B-52H (2008 data)	
Crew:	5 (pilot, copilot, radar navigator, navigator, ECM operator)	
Engines:	8 Pratt & Whitney TF33-P-3/103 turbofan; 17,000 lbst each	
Weights:	empty	185,000 lb
	max takeoff	488,000 lb
Dimensions:	length	159 ft 4 in
	wingspan	185 ft
	wing area	4,000 ft^2
	height	40 ft 8 in
Speed:	maximum	650 mph
	cruise	

12 CAPTOR = encapsulated torpedo; this was a mine that launched a Mk 46 lightweight torpedo against submarine targets.
13 Twenty-seven B-52B and all B-52C aircraft could be rapidly modified to the reconnaissance role.

Ceiling:	50,000 ft	
Range:	8,800 miles	
Radar:	AN/ASQ-38 bomb/nav	
Guns:	removed	
Ordnance:	*internal*	*external*
	4 B28, B43, B53, B57, B61, B83	2 Hound Dogs
or	8 SRAM	12 SRAM
or	8 ALCM	12 ALCM
or	50,000 lb conventional bombs, mines	

B-54

The B-54 was a further refinement of the B-29/B-50 design. Proposed by Boeing in early 1947, the aircraft—initially designated B-50C—was larger, with a single bomb bay and VDT (variable discharge turbine) engines, which were also being proposed for the B-36C aircraft.

President Truman's severe cuts in the fiscal year 1949 defense budget led to several program changes. The B-54 was cancelled on 18 April 1949, because of size limitations (e.g., suitable jet engine pods could not be provided), wider runways needed because of its outrigger landing gear, and the need to redesign the fuselage to provide for the new K-1 bombing system.

A single YB-50C was built but not completed. When the B-54 program was cancelled the Air Force had 14 B-54A bombers and 29 RB-54A reconnaissance aircraft on order. Of great significance to the program, General LeMay favored additional piston-engine B-36 bombers (with jet pods) and jet-propelled aircraft rather than the B-54.

Design: The B-54 design closely resembled the B-50 with external fuel tanks fitted to the wing. The most noticeable external difference between the B-50 and B-54 designs—other than size—was single-wheel outriggers under the B-54's outboard engines. Improved materials kept the empty weight about that of the B-50 series.

The R-4360-51 engines had water injection for takeoff with fuel injection to each of 27 cylinders; exhaust gases continually passed through a General Electric CHM-2 turbosupercharger. Two external wing tanks each would hold 1,500 gallons, more than twice that of the wing tanks on the B-50D.

Defensive armament was to consist of five remote-controlled turrets, two with four machine guns (forward dorsal and tail) and three with two guns, an arrangement similar to the B-29 and B-50 aircraft.

Designation: The aircraft's designation was changed from B-50C to B-54A in late 1948. Some documents refer to the aircraft as the YB-54.

Operational:	cancelled	
Design:	Boeing	
Aircraft data:	B-54	
Crew:	approx 10	
Engines:	4 Pratt & Whitney R-4360-51 VDT Wasp Major piston; 4,300 shp each wet; 4,000 shp each dry	
Weights:	empty	approx 81,000 lb
	max takeoff	230,000 lb
Dimensions:	length	111 ft
	wingspan	161 ft 2 in
	wing area	
	height	32 ft 8 in
Speed:	maximum	433 mph
	cruise	304 mph
Ceiling:	40,000 ft	
Range:	9,300 miles	
Radar:	bomb/nav radar	
Guns:	14 .50-cal machine guns	
Ordnance:	1 nuclear bomb	
or	conventional bombs	

B-55

Based on the B-47 Stratojet design, the Boeing XB-55 was the only U.S. Air Force turboprop-powered bomber. The aircraft was proposed to the Air Force in 1949 but did not progress past the blueprint stage because of the pending jet-propelled bomber designs. The aircraft, like the B-47, would have had a nuclear delivery capability.

Design: This was to have been a swept-wing aircraft, strongly resembling the Boeing B-47. The four Allison T40 turboprop engines—mounted on pylons beneath the wings—turned contrarotating propellers. This was the same engine planned for the Navy's A2D Skyshark and A2J Savage. (The Navy's R3Y Tradewind transport and P5Y patrol flying boats were the only U.S. military aircraft with the T40 to enter service.)

The B-55, unlike the B-47, was to have a relatively large crew and a heavy defensive armament of 12 20-mm cannon in three turrets.

Operational:	cancelled
Design:	Boeing

The Convair B-58 Hustler was a Mach 2 strategic bomber, the fastest bomber to serve in the Strategic Air Command. The delta-wing aircraft had no horizontal tail surfaces. It carried a nuclear weapon (and fuel) in an under-fuselage pod. That configuration limited the aircraft's payload flexibility and its career in SAC was relatively brief. (U.S. Air Force)

Aircraft data:	B-55	
Crew:	up to 10	
Engines:	4 Allison T40-A2 turboshaft; 5,643 shp each	
Weights:	empty	78,000 lb
	max takeoff	153,000 lb
Dimensions:	length	118 ft 10½ in
	wingspan	135 ft
	wing area	1,500 ft²
	height	33 ft 8 in
Speed:	maximum	490 mph
	cruise	452 mph
Ceiling:	42,000 ft	
Range:	5,300 miles	
Radar:	bomb/nav	
Guns:	12 20-mm cannon	
Ordnance:	1 nuclear bomb	
or	20,000 lb conventional bombs	

B-56 Stratojet

Boeing's 1950 proposal for the B-56 Stratojet was based on the B-47 design with four more-powerful engines. The aircraft's basic design was that of the B-47 but with four single-engine pods under the wing in place of the two double and two single pods of the B-47.

The B-56 was redesignated YB-47C.

The aircraft was not built.

Operational:	cancelled	
Design:	Boeing	
Aircraft data:	B-56	
Crew:	3	
Engines:	4 Allison J35-A-23 turbojet; 10,090 lbst each	
Weights:	empty	
	max takeoff	
Dimensions:	length	106 ft 10 in
	wingspan	116 ft
	wing area	1,428 ft²
	height	27 ft 10½ in
Speed:	maximum	600+ mph
	cruise	
Ceiling:	40,000+ ft	
Range:	3,000+ miles	
Radar:	bomb/nav	
Guns:	tail guns	
Ordnance:	2 nuclear bombs	
or	20,000 lb conventional bombs	

B-58 Hustler

The Convair B-58 was the first strategic bomber to enter service in the West with supersonic (Mach 2) performance. The aircraft was highly controversial: the official Air Force bomber history notes that the "B-58 established itself at home as one of the most expensive and controversial weapon systems . . . [but] it also attracted the world's attention as one of the most extraordinary airplanes. . . . The B-58 broke a great many speed records . . . and won almost every major aviation award."[14]

The B-58 evolved from an Air Force competition in 1949 to develop a supersonic strategic bomber and reconnaissance aircraft. Although the B-58 established several speed records, it was not well liked because of its comparatively short range and limited weapons load, and the small numbers procured required a disproportionate share of SAC support resources.

14 Knaack, *Post–World War II Bombers 1945–1973,* p. 394.

Although he had been an early proponent of supersonic bombers, General LeMay strongly opposed the aircraft because of its operational limitations. And, reportedly, after flying the B-58, he declared it was too small—"it didn't fit my 'arse.'"

The B-58 suffered technical problems during its test phases and operational service, among them unreliable bombing and navigation systems. The early aircraft had no low-level flight capability.

Design: The B-58 was characterized by its small size and advanced design, with a nuclear weapon carried in an external pod. It had an area-ruled ("Coke bottle") fuselage, a large delta wing, and no horizontal tail surfaces. (The competitive Martin B-68 design, cancelled in 1957, was also a delta-wing design.) The B-58's four podded engines were mounted directly on the wing, and not on pylons as in the B-47 and B-52 aircraft. The three-man crew sat in tandem.

In an effort to reduce aircraft size and improve performance, an internal weapons bay was deleted in favor of an external pod that could carry fuel, sensors, electronic countermeasures, and nuclear weapons. In the strategic strike role the entire 57-foot pod with 4,172 gallons of fuel and a nuclear weapon (W39Y1 and TX-53) could be released over the target area at speeds up to Mach 2 from 55,000 feet; the pod was designated MB-1C. The pod had a diameter of 60 inches and an empty weight of 2,500 pounds; with a W39Y1 bomb, the loaded weight was 8,550 pounds.

Early proposals also provided for an air-to-surface missile designated MA-1C.

A remote-controlled 20-mm rotary-barrel Gatling gun was fitted in the tail.

Under Project Hustle Up—undertaken from November 1962 to May 1964—the aircraft was fitted to carry an alternate payload of four nuclear bombs on wing pylons. Conventional weapons could not be carried by the B-58.

The single reconnaissance pod delivered for the RB-58 role was lost in an aircraft crash in 1958; that configuration was then cancelled, but briefly reinstated for the delivery of two more specialized pods before again being cancelled. Subsequently, a camera was fitted in the nose of ten pods (with 44 aircraft modified to carry the pod in the RB-58 role). The proposed B-58B and B-58C variants with enhanced performance and capabilities were cancelled; the B-58C was a Mach 2.4 aircraft proposed by Convair as an alternative to the B-70.

Designation: The first aircraft was designated XB-58 with the other test aircraft being designated YB/YRB-58 and, from the tenth aircraft, YB/YRB-58A. In practice all were referred to as YB-58.

Operational: The first XB-58 flight was on 11 November 1956; first delivery of a B-58A to a SAC squadron was on 1 August 1960, although the aircraft was not considered operational until August 1962.

The B-58 was deployed with two bomber wings, each with three squadrons; at peak strength, each wing had 40 aircraft.

In December 1965 Secretary of Defense McNamara directed that the B-58 force be phased out by 1970. The last B-58s were retired by January 1970, the aircraft having had a service life of less than nine years.

(LeMay's opposition to the B-58 led to consideration of operating the B-58 in a tactical role and in 1957 the head of the USAF Tactical Air Command expressed a desire for at least two B-58 wings. This role was not pursued.)

The B-58 set a large number of performance records; among the more impressive was a March 1962 flight from Los Angeles to New York in 2 hours, 59 seconds at an average speed of 1,214.65 mph; in October 1963 a B-58 flew the 8,028 miles from Tokyo to London in 8 hours, 35 minutes, 20 seconds, an average of 938 mph with five in-flight refuelings.

No B-58 dropped weapons in combat.

Production: Convair produced 116 B-58s with the last aircraft delivered in October 1962.

Of the 30 prototype and test aircraft, 15 were brought up to B-58A combat standards in 1961 and eight others were converted to trainers with tandem pilot seating (TB-58).

Aircraft	Built	First flight
XB-58	1	1956
YB/YRB-58	29	1956
B-58A	86	1959

Operational:	Aug 1962–Jan 1970	
Design:	Convair/General Dynamics	
Aircraft data:	B-58A	
Crew:	3 (pilot, navigator/bombardier, defensive systems operator)	
Engines:	4 General Electric J79-GE-5B turbojet; 15,600 lbst each	
Weights:	empty	54,794 lb
	max takeoff	163,000 lb
Dimensions:	length	96 ft 10 in

The Convair B-60 design—originally designated B-36G—provided eight turbojet engines in place of the standard Peacemaker's six piston-pusher engines. The B-60, however, was inferior in performance to the B-52 Stratofortress and, after the two YB-60 prototypes flew, the program was cancelled. (U.S. Air Force)

	wingspan	56 ft 10 in
	wing area	1,543 ft^2
	height	31 ft 5 in
Speed:	maximum	1,320 mph at 63,500 ft
	cruise	620 mph
Ceiling:	64,400 feet	
Range:	5,890 miles on high-altitude mission	
	5,970 miles on sea-level mission	
	7,185 miles maximum	
Radar:	AN/ASQ-42 bomb/nav	
Guns:	1 20-mm M61 multibarrel cannon (1,120 rounds)	
Ordnance:	1 nuclear bomb B39, B53 in pod	
or	4 nuclear bombs B43 externally on pylons	

B-59

The B-59 was proposed by Boeing in competition to the Convair B-58 Hustler as a supersonic (Mach 2) strategic bomber.

The design was not pursued.

Design: The aircraft was to have had short, swept-back wings in the shoulder position and a narrow, streamlined fuselage; the tail surfaces (with the horizontal surface midway up the tail fin) were also swept back. Four turbojet engines were to be buried in the wing roots.

The aircraft was to have a bicycle landing gear with small outriggers at the end of the wings that would retract up into small, streamlined pods (a variation of the arrangement used on the firm's B-47 Stratojet).

The aircraft would have an internal bomb bay; a single 30-mm tail gun would provide defense against enemy fighters attempting to catch the aircraft.

Operational:	cancelled	
Design:	Boeing	
Aircraft data:	B-59	
Crew:		
Engines:	4 General Electric J73-X24A turbojet; approx 9,300 lbst each; 14,000 lbst with afterburner	
Weights:	empty	63,200 lb
	max takeoff	148,300 lb
Dimensions:	length	123 ft 4 in
	wingspan	81 ft 4 in
	wing area	1,650 ft^2
	height	25 ft 4 in
Speed:	maximum	1,200 mph
Ceiling:	51,000 ft	
Range:	2,740 miles	
Radar:	bomb/nav radar	
Guns:	1 30-mm cannon	
Ordnance:	1 nuclear bomb	
or	14,000 lb conventional bombs	

B-60

The B-60 was a swept-wing, all-jet development of the B-36 Peacemaker for the strategic bombing role. The design was proposed by Convair in 1950 as the B-36G, a swept-wing aircraft, initially propelled by six turboprop engines and then by eight J57 turbojets, the latter paired in underwing pods (as in the Boeing B-52 Stratofortress).

Two prototypes were built, although only the first one flew because of delays in delivering the engines for the second aircraft. The first prototype flew some 40 hours.

The program was cancelled in January 1953 in favor of the faster B-52.

Design: The aircraft generally resembled the B-36 but with smoother lines, a more pointed nose, and the wings swept back 35°, leading to a 24-foot reduction in wing span.

A two-gun tail turret was to be fitted.

The cost to produce and test the two YB-60s was approximately $15 million.

A transport version of the aircraft was also considered.

Designation: The two prototype aircraft were initially designated YB-36G and were redesignated YB-60 in 1951.

Operational: The first flight was on 18 April 1952.

Operational:	cancelled	
Design:	Convair/General Dynamics	
Aircraft data:	B-60	
Crew:	5 to 10	
Engines:	8 Pratt & Whitney J57-P-3 turbojet; 8,700 lbst each	
Weights:	empty	153,016 lb
	max takeoff	280,000 lb
Dimensions:	length	171 ft
	wingspan	206 ft
	wing area	5,239 ft^2
	height	
Speed:	maximum	508 mph at 39,250 ft
Ceiling:	44,650 ft	
Range:	approx 8,000 miles	
Radar:	bomb/nav radar	
Guns:	2 20-mm cannon	
Ordnance:	2 nuclear bombs	
or	approx 80,000 lb conventional bombs	

B-68

The B-68 was the Martin entry in the Air Force competition of 1952 for a supersonic nuclear bomber that led to the Convair B-58 Hustler. Martin was awarded a development contract in September 1956, with this design being chosen in preference to a competing Douglas design. However, the Martin project was cancelled early in 1957 at the preliminary design stage.

Design: The B-68 had a delta-wing design with two engine nacelles mounted directly on the underside of the wing and the third engine buried in the fuselage with "cheek" position air intakes. The aircraft was to be capable of Mach 2.5 at altitude.

Designation: The designation B-68 was subsequently reassigned to the Titan ICBM (later SM-68).

Operational:	cancelled	
Design:	Martin	
Aircraft data:	B-68	
Crew:	2 (pilot, weapon systems operator)	
Engines:	3 General Electric J79-GE-1 turbojet; 10,500 lbst each; 16,000 lbst with afterburning	
Weights:	empty	
	max takeoff	
Dimensions:	length	approx 58 ft
	wingspan	approx 45 ft
	wing area	
	height	approx 18 ft
Speed:	maximum	1,650 mph at altitude
Ceiling:	50,000+ ft	
Range:		
Radar:	bomb/nav	
Guns:	none	
Ordnance:	1 nuclear bomb	

B-70 Valkyrie (later RS-70)

The North American B-70 Valkyrie was developed as a supersonic, high-altitude strategic bomber to succeed the B-52. However, the B-70 was essentially stillborn because of (1) the high procurement and maintenance costs for an aircraft that large and complex, (2) the plane's flight conditions were very poor at other than design-point performance of Mach 3 at 70,000 feet, and (3) the increasing high-altitude bomber defenses of the Soviet Union.

Development of the B-70 began in late 1954 when General LeMay requested the development of a successor to the B-52 bomber with as high a speed as possible. Following several industry proposals, North American Aviation was awarded a contract for the aircraft's development in December 1957. The design evolved rapidly.

In November 1959, during a meeting to discuss defense programs, President Eisenhower told Air Force chief of staff General Thomas D. White that "the B-70 left him cold in terms of making military sense." The president also

15 Quoted in Knaack, *Post–World War II Bombers 1945–1973*, p. 568.

The North American B-70 Valkyrie had a very long and convoluted development—and never entered operational service. The massive, delta-wing bomber was overtaken by events as Soviet air defenses made high-altitude bombing missions impracticable. Failing to "sell" the B-70 as a strategic bomber, the Air Force proposed it as a strategic reconnaissance (RS-70) and transport aircraft. (U.S. Air Force)

noted that if the B-70 were allowed to reach the production stage, it would not be available for eight or ten years, by which time missiles would be the primary strategic retaliatory weapon.[15] On 29 December 1959, the Air Force reluctantly made the decision to procure only a single prototype XB-70 for technology development. The final decision on the B-70 was postponed until the 1960s and a new administration.

The B-70 program underwent a resuscitation a few months later as the 1960 presidential campaign rekindled interest in strategic weapons in light of the alleged bomber and ICBM "gaps" vis-à-vis the USSR. In August 1960 the Eisenhower administration increased the program to 13 prototype and test aircraft, making the B-70 a weapon system moving toward production.

On 28 March 1961, President Kennedy recommended to Congress that the program be continued to explore the problems of Mach 3 flight with an aircraft "potentially" useful as a bomber.

The first XB-70 had still not flown when, in 1961, the Air Force redesignated the aircraft as a "reconnaissance-strike" vehicle (RSB-70 and, subsequently, RS-70). Under the RS-70 concept, 60 aircraft would be procured to (1) carry out reconnaissance of enemy targets during a missile attack, (2) examine targets, and (3) if required, attack them immediately with air-to-surface missiles. Desperate to keep the B-70 effort alive, the Air Force also looked into the possibility of employing the B-70 as a transport aircraft. General White told a congressional committee, "As a troop carrier with atomic Army weapons, where cost of operation is not the prime consideration that it is with the airlines, I am sure it can successfully carry about 60 people or, say, six of the Honest John Army weapons with their serving crews."[16] At Mach 3 speeds the flight time to carry these atomic missile launchers from the U.S. west coast to Taiwan, an example given by General White, would be four hours.

Even within the Air Force there were serious questions being asked about proceeding with the B-70. Missile supporters argued instead for more ICBMs and the tactical commanders for the F-108, a Mach 3 fighter-interceptor. Earlier, Secretary of the Air Force Dudley Sharp admitted to Congress: "I was not so strong for the B-70 as the Chief of Staff, and we had some very heart-to-heart discussions about this, because I do not like to be at differences with the Chief of Staff of the Air Force. There are points on

16 Gen. Thomas D. White, USAF, testimony before the Committee on Appropriations, House of Representatives, 22 January 1960.

17 Secretary of the Air Force Dudley C. Sharp, testimony before the Appropriations Committee, House of Representatives, 22 January 1960.

both sides and I think they are very valid points. I wanted to point out it was a controversial issue and there were two sides to it."[17]

The penultimate blow came in April 1961, when Secretary of Defense McNamara directed that the program be reduced back to three prototypes without military systems installed. McNamara explained his opposition to the reconnaissance-strike concept: "Quite apart from the technical feasibility of developing, producing, and deploying such a system within the time frame proposed by the Air Force (which we do not think possible), there are better ways, when one considered both cost and effectiveness, to obtain both of these capabilities." After explaining these points, McNamara said, "The RS-70, by carrying air-to-surface missiles, would provide only a very small increase in over-all effectiveness. In my judgment, this increase is not worth the large additional outlay of funds estimated at more than $10 billion above the $1.35 billion already approved [and allocated to the B-70]."[18]

In late 1962 the USAF proposed a force of 60 RS-70A aircraft to be operational by 1969. In the end, only the two prototypes were procured and flown.

Design: The B-70 was the heaviest bomber aircraft ever to fly. It was a precursor of supersonic transport designs with a large delta wing, small canards forward, a needle nose, and the turbojet engines nested under the wing. The wingtips folded down during high-speed flight to increase control.

A tubular weapons bay for gravity bombs was to have been fitted between the massive engine intakes. The aircraft was to carry one Class A (25,000-pound) or two Class B (20,000-pound) nuclear weapons. No defensive guns were to be fitted.

Designation: This was the last aircraft designated in the bomber series that had originated with the Army Air Corps in 1924. The next aircraft in this series was the supersonic reconnaissance aircraft SR-71; the next U.S. bomber aircraft was the B-1 Lancer, designated under the scheme instituted in 1962.

Operational: The first prototype XB-70A flew for the first time on 21 September 1964; the second prototype flew in July 1965; the latter aircraft was destroyed in a midair collision with an F-104 fighter during a photo shoot on 8 June 1966 while on its 46th flight. The first XB-70 was transferred to the National Aeronautics and Space Administration (NASA) and flew until 4 February 1969, making a total of 83 flights before being retired to a museum.

The third prototype, to have been the XB-70B with an integrated bombing-navigation system, was cancelled in 1964.

Operational:	cancelled	
Design:	North American Aviation	
Aircraft data:	B-70A	
Crew:	4 (pilot, copilot, navigator-bombardier, electronics specialist)	
Engines:	6 General Electric YJ93-GE-3 turbojet; 28,000 lbst each	
Weights:	empty	231,215 lb
	max takeoff	521,792 lb
Dimensions:	length	185 ft 10 in
	wingspan	105 ft
	wing area	6,297 ft^2
	height	30 ft 8 in
Speed:	maximum	1,979 mph at 75,550 ft
	cruise	1,252 mph at 35,000 ft
Ceiling:	75,500 ft	
Range:	3,400 miles	
Radar:	AN/ASQ-28 bomb/nav	
Guns:	none	
Ordnance:	2 nuclear bombs	

FB-111A (later F-111G)

The FB-111 was the strategic bomber variant of the F-111 multipurpose aircraft (see separate entry for F-111). Development of the FB-111 variant as a penetration strategic bomber was announced on 10 December 1965, by Secretary of Defense McNamara, two days after he revealed that he was accelerating retirement of 345 B-52C/D/F Stratofortress and 80 B-58 Hustler bombers from SAC. (The B-52G/H aircraft were to be retained.)

The FB-111s were assigned to SAC until June 1990 when the Air Force began modifying the 51 surviving aircraft for the tactical/theater strike role (changed to F-111G).

Design: The principal differences between the F-111A and FB-111A were the engines with the FB-111 having a higher gross weight, additional electronics, and 3½-foot

18 Secretary of Defense Robert S. McNamara, "U.S. Defense Program 1964–1968," testimony before the Armed Services Committee, House of Representatives, 31 January 1963.

The General Dynamics FB-111A program was an effort by Secretary of Defense McNamara to keep the F-111/TFX production lines open after the Navy withdrew from the program. The F-111/FB-111 was the world's first operational aircraft with variable-sweep (swing) wings. (U.S. Air Force)

extensions on both wingtips. There were six wing pylons; the outermost pylon on each wing was fixed so that it created the least drag at a 26° sweep angle; these pylons were used for fuel tanks and were jettisoned with the tanks when the latter were empty. Weapons could not be carried on those pylons. The four other pylons could carry nuclear bombs, AGM-69 SRAM missiles, 24 conventional 750-pound bombs, cluster bombs, or 600-gallon drop tanks. Two nuclear bombs or SRAMs could be carried in the internal weapons bay.

Proposed updates of the FB-111A were designated FB-111B and FB-111C; both were cancelled in 1979. The follow-on FB-111G was to be a "stretched" variant of the FB-111A. The FB-111H was to be a larger aircraft with an enlarged bomb bay to accommodate five nuclear bombs plus four on wing pylons and up to six conformally mounted on fuselage stations (providing for a maximum of 12 SRAM missiles).

No Gatling gun was fitted in FB-111 aircraft.

Designation: The FB-111A was changed to F-111G when the aircraft were assigned to a tactical strike role.

Operational: The first flight of FB-111A (F-111A conversion) was on 30 July 1967; the first flight of a production of FB-111A was on 13 July 1968. Although the first FB-111A was delivered on 8 October 1969, to the 340th Bomb Wing, the first aircraft were not fully combat ready until October 1971. These aircraft were never used in com-bat; the F-111 saw extensive action in the Vietnam War.

Throughout its service as a strategic bomber, the FB-111s were based at Pease AFB, New Hampshire, and Plattsburgh AFB, New York. The maximum number of operational FB-111s available to SAC was 70.

In the early 1990s the 51 surviving FB-111A aircraft were transferred to the Air Force Tactical Air Command, redeployed to Cannon AFB, New Mexico, and redesignated F-111G. The last active F-111G was retired in July 1993.

Production: In December 1965 Secretary McNamara announced the procurement of 263 FB-111s—210 to equip 14 squadrons and 53 for training, development, and attrition. The procurement goal was reduced to 76 aircraft on 19 March 1969, with the last delivered in June 1971, including the prototype modified from an F-111A. (One FB-111A was lost before delivery to the Air Force, thus the maximum force was 75 aircraft assigned to four operational squadrons.)

Additional FB-111 procurement was deferred because of higher-than-expected costs.

The last FB-111A was delivered on 3 June 1971.

Operational:	Oct 1969–July 1993 (see operational notes)	
Design:	General Dynamics	
Aircraft data:	FB-111A	
Crew:	2 (pilot, navigator-bombardier)	
Engines:	2 Pratt & Whitney TF30-P-7 turbofan; 20,350 lbst each	
Weights:	empty	47,481 lb
	combat	70,380 lb
	max takeoff	114,300 lb
Dimensions:	length	75 ft 6½ in
	wingspan	70 ft at 16° sweep
		34 ft at 72.5° sweep
	wing area	550 ft²
	height	17 ft

Speed:	maximum	1,450 mph at 50,000 ft
	cruise	510 mph at 23,200 ft
Ceiling:	50,000 ft	
Range:	6,150 miles on hi-lo-lo-hi mission	
	7,500 miles on hi-hi mission	
Radar:	AN/AJN-16 bomb/nav	
	AN/APQ-114 attack	
	AN/APQ-134 terrain following	
Guns:	none	
Ordnance:	2 nuclear bombs B43, B57, B61 internal	
	4 nuclear bombs B43, B57, B61 external or 2 SRAM missiles internal	
	4 SRAM missiles external	
or	18,000 lb conventional bombs	

PATROL AIRCRAFT

The P2V-3C Neptune and P6M Seamaster "patrol" aircraft were, respectively, converted and built specifically for strategic attack roles. Other U.S. Navy patrol aircraft that could carry nuclear weapons are described in Chapter 4.

Aircraft		Service	IOC	Type
P2V-3C	Neptune	Navy	1949	carrier/land
P6M	Seamaster	Navy	cancelled	seaplane

P2V-3C Neptune

The P2V-3C was the U.S. Navy's first aircraft capable of delivering a nuclear weapon. One P2V-2 and 11 P2V-3 aircraft were modified to the P2V-3C configuration to carry the Mk 8 atomic bomb and take off from the Navy's three large *Midway*-class aircraft carriers. Takeoffs were made with the aid of rocket-assistant canisters (JATO "bottles"). However, the aircraft could not land on board (they were loaded on board at pierside by cranes).

The Lockheed P2V Neptune was the U.S. Navy's principal land-based maritime patrol/anti-submarine aircraft for two decades after World War II (see separate listing for basic P2V/P-2 aircraft).

The first carrier takeoff by a Neptune was made by a P2V-2 from the USS *Coral Sea* (CVB 43) on 28 April 1948. Several additional takeoffs were made from the three *Midway*-class carriers by P2V-3C and standard P2V aircraft (fitted with auxiliary rocket boosters) to simulate atomic bomb delivery missions. Although the landing of a P2V-3C on a carrier (with arresting hook) was proposed, and practiced regularly ashore, no carrier landing was attempted. A hook-equipped P2V-3C made 16 arrested landings ashore at the Naval Air Station Patuxent River, Maryland.

Design: The 12 P2V-3C aircraft were modified to carry an Mk 8/ Little Boy–type atomic bomb in their inter-

The Lockheed P2V-3C Neptune was heavily modified from the standard maritime patrol/anti-submarine aircraft flown by the U.S. Navy and several other nations. The dozen modified aircraft provided the Navy with a limited and somewhat impractical nuclear strike capability. But the program gave the Navy a "foot in the door" to develop a realistic nuclear strike capability. (Lockheed Aircraft Corp.)

A Lockheed P2V-3C Neptune—with rocket boosters—takes off from the carrier *Franklin D. Roosevelt* (CVB 42) during a nuclear exercise. The large aircraft could operate only from the Navy's three *Midway* (CVB 41)-class large carriers. Carrier landing trials were proposed, but never undertaken. (U.S. Navy)

nal bomb bay. An AN/APA-5 low-altitude bombing system and AN/APS-31 radar were installed; and the dorsal gun turret and certain antennas were deleted to reduce drag. Twin 20-mm guns were retained in the tail position.

No jet pods were fitted to these aircraft.

Operational: The 12 P2V-3C conversions were completed in 1949; they were initially assigned to Navy Composite Squadron (VC) 5. The squadron had been established the previous year to evaluate naval nuclear strike aircraft and develop the doctrine and tactics for carrier-based operations of nuclear strike aircraft. (The squadron also was assigned AJ-1 Savage aircraft in 1949—at the end of the year VC-5 had 12 P2V-3C and 6 AJ-1 aircraft.) Subsequently, the 12 modified Neptunes were divided between squadrons VC-5 and VC-6.

Several distance records for carrier-launched aircraft were set in P2V-3C operations, and on 21 April 1950, a P2V-3C took off from the *Coral Sea* with a record gross weight of 74,668 pounds.

The first overseas operation of P2V-3C aircraft occurred in February 1961 when three Neptunes and six Savages of VC-5 deployed to Port Lyautey, French Morocco. The AJ-1s operated from *Midway*-class carriers in the Mediterranean; the Neptunes operated from shore bases, ready to be loaded aboard carriers (by crane) in the event of war—or more likely fly strikes against the Soviet Union from NATO land bases.

The P2V-3C aircraft were operational as nuclear strike aircraft from 1949 until early 1952, after which a few were flown in the training role.

Operational:	1949–1952	
Design:	Lockheed	
Aircraft data:	P2V-3C	
Crew:	approx 7 (pilot, copilot, navigator, bombardier, radio operator, engineer, gunner)	
Engines:	2 Wright R-3350-26W piston; 3,200 hp each	
Weights:	empty	approx 33,500 lb
	max takeoff	approx 74,700 lb
Dimensions:	length	75 ft 4 in
	wingspan	100 ft
	wing area	1,000 ft^2
	height	28 ft 6 in
Speed:	maximum	354 mph
Ceiling:	17,000 ft	
Range:	4,500 miles with 10,000 lb bomb load	
Radar:	AN/APA-5 bombing AN/APS-31 search	
Guns:	2 20-mm cannon (400 rounds per barrel)	
Ordnance:	1 nuclear bomb Mk 8	

P6M Seamaster

Designed by the Martin Company, the P6M was intended primarily for long-range minelaying and reconnaissance missions, with a secondary nuclear strike capability. In the

A Martin P6M-2 Seamaster flying boat taxis on Chesapeake Bay. The large, multi-mission aircraft was the world's fastest seaplane. It was to be employed primarily in long-range minelaying and reconnaissance missions, with a secondary nuclear strike role. The P6M program was cancelled to help fund the Polaris submarine/missile program. (U.S. Navy)

The first prototype Martin XP6M-1 Seamaster in flight, showing the aircraft's swept-back wings, four shoulder-mounted turbojet engines, and wingtip stabilizing floats. The aircraft had a rotary bomb bay, permitting it to be loaded with conventional bombs or a nuclear bomb while resting on the water. Mission scenarios included refueling the aircraft in the water by submarine tanker. (Martin)

minelaying role, the Seamaster was to mine straits the approaches to Soviet submarine bases, denying them access to the sea and preventing submarines already at sea from returning for replenishment. The P6M Seamaster was a top-performing aircraft for its time and the last flying boat developed for the U.S. Navy.

The P6M program began in July 1951 when the Navy requested 12 aircraft manufacturers to submit design proposals for a high-speed, minelaying flying boat. Martin, which had built the highly successful PBM Mariner and P5M Marlin flying boats, won the design competition. The Navy initially ordered two XP6M-1 test aircraft and six YP6M-1 flight demonstration aircraft.

Subsequently 24 production P6M-2 aircraft were ordered on 29 August 1956. Engineering problems increased the costs of the aircraft and caused further delays in the program. Although the technical problems were solved, the costs and need to fund the Polaris submarine missile program led to the 24-plane contract being reduced to 18 aircraft on 21 June 1957, and then to only eight P6M-2 aircraft on 24 November 1958. At that time, the Navy announced a Seamaster patrol squadron would be formed with those eight units to explore the capabilities of the aircraft.

Four P6M-2 production aircraft were completed before cancellation of the program on 24 August 1959. It was one of several Navy projects halted to help pay for the acceleration of the Polaris program. By the time of the cancellation four of the YP6M-1s had been partially disassembled and a fifth had been destroyed in static tests; all surviving aircraft were scrapped.

Cancellation of the P6M marked the end of flying boat development by the U.S. Navy. (The development of military flying boats continued in Japan and the Soviet Union.)

Design: A graceful-looking aircraft, the Seamaster's flying-boat hull had an extremely high length-to-beam ratio with the wings swept back 40°. It had a high T-tail, and fixed wing-tip stabilizing floats were fitted to the aircraft. The four turbojet engines were mounted in pairs above the shoulder-high wing.

A payload of up to 30,000 pounds of mines, bombs, or cameras and flash bombs was to be carried on a rotary bomb-bay door that formed part of the hull bottom. During trials the aircraft successfully released stores at a speed of 500 mph.

Two 20-mm cannon in a remote-controlled tail turret, beneath a gunfire control radome, comprised the aircraft's defensive armament.

The Seamaster was a large aircraft with the P6M-2 production models expected to weigh some 85,000 pounds empty and almost 190,000 pounds loaded for rough-water takeoffs. The aircraft was designed for operation in six- to eight-foot waves.

The Seamaster was to remain in the water for as long as six to eight months, being maintained, refueled, and rearmed by seaplane tenders and by submarines. Weapons were loaded through the top of the fuselage, enabling the aircraft to be rearmed while in the water. Similarly, engines could be replaced while the plane was afloat. Clandestine refueling of the aircraft in forward areas from submarines was proposed, with the fleet submarine *Guavina* being converted to a tanker configuration (AOSS 362) to evaluate at-sea refueling of flying boats.

Operational: The first XP6M-1 flew on 14 July 1955; that aircraft crashed into the Potomac River on 7 December 1955, because of a structural failure. The second XP6M-1 crashed near New Castle, Delaware, on 9 November 1956, because of control problems. Modifications were made to the six YP6M-1 aircraft, and there were no further accidents in the flight-test program; the first YP6M-1 flew on 1 July 1957.

The 12 completed P6M aircraft had logged a total of 536 hours of flight time when the project was halted.

Operational:	cancelled	
Design:	Martin	
Aircraft data:	P6M-1	
Crew:	5 (pilot, copilot, navigator-mine-layer, radio operator, gunner)	
Engines:	4 Allison J71-A-6 turbojet; 13,000 lbst each	
Weights:	empty	84,685 lb
	max takeoff	160,000 lb in rough water
		190,000 lb in sheltered water
Dimensions:	length	134 ft 4 in
	wingspan	102 ft 7 in
	wing area	1,900 ft^2
	height	33 ft 9 in
Speed:	max	645 mph at 5,000 ft and sea level
	cruise	540 mph
Range:	2,100 miles on mining mission	
	2,740 miles on reconnaissance mission	
Ceiling:	43,500 ft	
Radar:	AN/ASB-1 search	
Guns:	2 20-mm cannon (500 rounds per barrel)	
Ordnance:	1 nuclear bomb Mk 28	
or	30,000 pounds conventional bombs, mines	

An airman checks an ALCM—air-launched cruise missile—at the RAF base at Fairford, England; a B-52H Stratofortress from the 20th Bomb Wing towers over the missile in this 1999 photo. The plane's tail gun position has been replaced by electronic warning/countermeasures equipment. (U.S. Air Force)

CHAPTER FOUR

Tactical Aircraft

BOMBER AIRCRAFT

Aircraft		Service[a]	IOC	Type
B-45	Tornado	AF	1948	land
B-46	—	AF	cancelled	land
B-48	—	AF	cancelled	land
B-57	Canberra	AF	1954	land
B-59	—	AF	cancelled	land
B-60	—	AF	cancelled	land
B-66	Destroyer	AF	1956	land

a AF = Air Force

B-45 Tornado

The North American B-45 was the first U.S. multiengine jet bomber. The project was initiated by the Army Air Forces in 1944 (also see B-48). Although intended from an early date to carry an atomic bomb as well as conventional bombs, because of the limited nuclear weapons data available even within the AAF, the plane initially could not carry atomic weapons without major modifications. Accordingly, the B-45 was reassigned to the tactical (conventional) strike role in September 1947.

However, in the early 1950s a total of 56 B-45s were modified to carry the smaller atomic bombs then becoming available. Initially 41 aircraft were extensively modified under Project Back Breaker followed by 15 aircraft that received more extensive modifications under Project Second Call. (One of the Back Breaker aircraft was destroyed by fire; hence, only 40 of those aircraft became operational.) The 40 Back Breaker–modified B-45A aircraft were sent to England in May–June 1952 while still suffering support and mechanical problems. They remained at British bases until 1958 when they were replaced by the B-66.

The newer Project Second Call improvements were also retrofitted to the 40 Back Breaker aircraft. The Second Call program was not completed until March 1954.

Design: The B-45 was a straight-wing aircraft with the wing mounted in the shoulder position, aft of the fighter-

The North American B-45 Tornado was the only one of several multi-engine tactical bomber programs initiated by the Army Air Forces in 1944 that became operational. The straight-wing, four-turbojet aircraft had a surprisingly large bomb–carrying capacity. The serial number 8001 of this B-45C indicates that it was the first USAF aircraft ordered in the Fiscal Year 1948 program. (U.S. Air Force)

like bubble cockpit that housed the pilot and copilot; the bombardier sat in a glazed nose housing. (The fourth crewman, the gunner, was in the tail position.)

The dissimilar engines were paired in nacelles under each wing.

The standard B-45 bomb bay was capable of accommodating the World War II–era Grand Slam (22,000-pound) and Tall Boy (12,000-pound) bombs. Maximum bomb load was 22,000 pounds with up to 8,000 pounds of conventional bombs normally carried (4,000-pound or smaller bombs.)

The defensive armament was a tail gun turret with twin .50-caliber machine guns.

Operational: The first flight of XB-45 was on 17 March 1947. The first B-45A was assigned to an operational squadron in November 1948. The first B-45A Back Breaker–modified aircraft entered service in June 1952.

The B-45 remained in the Air Force's tactical strike inventory until 1958 when it was replaced by the B-66 Destroyer. No B-45 dropped a bomb in combat.

The RB-45C photo-reconnaissance variant was the only B-45 to serve with the Strategic Air Command (SAC), from August 1950 to December 1953. British-manned RB-45C aircraft overflew the Soviet Union several times in 1952.

In 1951–1952 B-45s made two nuclear bomb drops in weapon tests at the Nevada nuclear test site.

Production: B-45 production totaled 142 aircraft. Additional aircraft were cancelled. Only 56 B-45A aircraft were nuclear capable.

Aircraft	Built	First flight
XB-45	3	1947
B-45A	96	1948
B-45C	10	1949
RB-45C	33	1950

Operational:	Nov 1948–mid-1958 nuclear bomber variant	
Design:	North American Aviation	
Aircraft data:	B-45A Back Breaker	
Crew:	4 (pilot, co-pilot, navigator-bombardier, gunner)	
Engines:	2 General Electric J47-GE-7	
or	-13 turbojet; 5,500 lbst each and 2 General Electric J47-GE-9	
or	-15 turbojet; 5,000 lbst each	
Weights:	empty	47,022 lb
	max takeoff	92,745 lb
Dimensions:	length	75 ft 4 in
	wingspan	89 ft
	wing area	1,175 ft^2
	height	25 ft 2 in
Speed:	maximum	566 mph at sea level
	cruise	462 mph
Ceiling:	46,400 ft	
Range:	1,160 miles	
	radius	460 miles with 10,000 lb bombs
Radar:	AN/APQ-24 bomb/nav	
Guns:	2 .50-cal M-3 machine guns (600 rounds per barrel)	
Ordnance:	1 nuclear bomb Mk 5, Mk 7, Mk 8	
or	8,000 lb conventional bombs	

B-46

The B-46 was the Convair proposal for the Air Force's early jet design competition. The AAF initiated the procurement on 17 January 1945, and three experimental XB-46 aircraft were subsequently ordered.

In the fall of 1945 the AAF became interested in a Convair proposal for a jet-propelled attack aircraft (designated XA-44) and considered cancellation of the B-46 program in favor of the attack aircraft. Convair proposed completing one XB-46 and producing two XA-44s in place of the original agreement for three of the B-46 aircraft. The AAF concurred with the proposal, but only one XB-46 was built.

The B-46 program was formally cancelled in August 1947. Although cancelled before atomic bomb dimensions were fully available, it was assumed that the B-46 would be nuclear capable.

Design: The B-46 was a straight-wing, streamlined aircraft with twin-engine nacelles faired into the wings. It featured a conventional landing gear. The pilot had a fighter-type canopy and a tail (unmanned) gun position was provided.

The bomb bay could accommodate bombs up to 22,000 pounds (see B-45 design comments).

Operational: The sole XB-46 prototype first flew on 2 April 1947, and the Air Force accepted delivery of the aircraft on 12 November 1947. After tests, the aircraft made its last known flight in July 1950.

Operational:	cancelled
Design:	Convair

The Convair proposal in the AAF's early jet design competition was the four-turbojet B-46. The development of the aircraft was convoluted as Convair was also proposing a twin-turbojet attack aircraft—the A-44. In the event, only this single XB-46 prototype was produced. The B-36 Peacemaker was the last Convair "bomber" produced in large numbers. (U.S. Air Force)

Aircraft data:	B-46	
Crew:	3 (pilot, copilot, navigator-bombardier)	
Engines:	4 General Electric J35-C3 turbojet; 4,000 lbst each	
Weights:	empty	48,000 lb
	max takeoff	94,400 lb
Dimensions:	length	105 ft 10 in
	wingspan	113 ft
	wing area	$1,285^2$
	height	27 ft 11 in
Speed:	maximum	490 mph at approx 40,000 ft
	cruise	440 mph
Ceiling:	approx 40,000 ft	
Range:	radius	690 miles
Radar:	bomb/nav	
	AN/APG-27 gunfire control	
Guns:	2 .50-cal machine guns (600 rounds per barrel)	
Ordnance:	1 nuclear bomb (tentative)	
or	22,000 lb conventional bombs	

B-48

The Martin B-48 was a first-generation jet bomber initiated by the AAF in 1944 (also see B-45 and B-46 entries). The B-48 was considered from an early date to be capable of carrying an atomic bomb as well as conventional bombs. After many vicissitudes, and with the B-48 design lagging behind the competitive North American B-45 and Convair B-46 designs, a contract was awarded to the Glenn L. Martin Company on 13 December 1946, for two aircraft.

The aircraft encountered major delays, caused in part by the engines. With the first flight of the Boeing B-47 and the B-48 in 1947, the Air Force decided not to procure the B-48. Indeed, the directly competitive B-45 was produced in only limited numbers.

Design: The B-48 was a bulky but streamlined aircraft with midmounted straight wings and six turbojet engines encased in a squared-off pod beneath each wing. The pilot and copilot sat in tandem under a fighter-type canopy mounted far forward; the navigator-bombardier was seated in the nose. The bomb bay doors retracted up into the fuselage.

This was the first U.S. bomber aircraft with a bicycle-type landing gear with a tandem main undercarriage and small "outrigger" wheels retracting into the engine pods. The wings were too thin for a conventional undercarriage.

Like the B-45 and B-46, the aircraft was to carry up to 22,000 pounds of bombs, including the 22,000-pound Grand Slam weapon.

It was planned to provide twin machine guns in a remote-controlled tail turret (not fitted in prototypes).

Operational: The first prototype XB-48 flew on 22 June 1947, followed by the second XB-48 on 16 October 1948.

The first aircraft flew until the fall of 1949 and the second until the fall of 1951.

Production: Only the two prototypes were produced before program cancellation on 31 March 1949. Martin had proposed several variations in an effort to save the aircraft, including the fitting of four Navy-developed XT-40A turboprop engines in a revised engine arrangement.

The XT-40, which was also proposed for several Air Force aircraft, was not a successful program.

Martin's B-48 jet-bomber proposal was unusual in having six turbojet engines fitted in two "groups" under the wings. The low thrust of early turbojets led to four and six engines being used for these early bombers. The aircraft also had a bicycle-type tandem main landing gear with small "outrigger" wheels that retracted into the engine pods. (U.S. Air Force)

Operational:	cancelled	
Design:	Martin	
Aircraft data:	B-48	
Crew:	3 (pilot, copilot, navigator-bombardier)	
Engines:	6 General Electric J35-GE-7 (TG-180-B1) turbojet in first prototype; J35-GE-9 (TG-180-C1) in second prototype; 3,820 lbst each	
Weights:	empty	58,500 lb
	max takeoff	92,600 lb
Dimensions:	length	85 ft 10 in
	wingspan	108 ft 4 in
	wing area	1,330 ft^2
	height	26 ft 6 in
Speed:	maximum	522 mph at 35,000 ft
	cruise	415 mph
Ceiling:	43,000 ft	
Range:	radius 500 miles with max bomb load	
Radar:	bomb/nav radar	
	AN/APG-27 gunfire control	
Guns:	2 .50-cal machine guns	
Ordnance:	1 nuclear bomb	
or	22,000 lb conventional bombs	

B-57 Canberra

The Canberra was one of the most versatile aircraft of the post–World War II era, having a nuclear strike capability in the British B-series aircraft and the U.S. B-57 models. The U.S. aircraft also flew in the tactical strike, reconnaissance (including high-altitude "spyplane" configuration), weather, and training roles.

The U.S. Air Force selected the British Canberra for operation in the night/tactical intruder role, a need accelerated by the start of the Korean War in June 1950. Two RAF aircraft were flown to the Glenn L. Martin Company, in Baltimore, Maryland, to serve as development models.[1] The Canberra was selected for production by the Air Force on 26 February 1951, and 250 were ordered the following March. However, tests immediately revealed flaws and one British aircraft crashed, causing major delays in the program; the first U.S.-built Canberra did not fly until seven days before the end of the Korean conflict.

Still, in U.S. service the Canberras proved to be an excellent strike and reconnaissance aircraft, seeing extensive service in the Vietnam War as well as in other areas of the world in the strategic reconnaissance role.

Design: The Canberra design featured low aspect–ratio wings with a distinctive rectangular center section and double-taper outer panels. Twin engine turbojet nacelles were embedded in the wings at midspan.

The U.S. aircraft were fitted with a revolving bomb-bay door that rotated 180° to eliminate drag caused by opening conventional bomb bay doors. The armament in U.S. planes initially consisted of eight .50-cal machine guns (300 rounds per barrel) fitted in the wings; in the 85th B-57B and later aircraft these were replaced by four 20-mm cannon. These later aircraft also had fittings for underwing fuel tanks plus eight weapon pylons.

1 The first aircraft, a Canberra B.2, flew across the Atlantic on 20 February 1951; it was the first jet aircraft to complete an unrefueled flight across the Atlantic.

The English Electric Canberra was one of the most versatile aircraft of the Cold War era, being produced in the United States by the Glenn L. Martin Company as the B-57 (with the British name being adopted). This B-57B painted in "jet black" had tandem seating for its two-man crew. This was the principal variant produced for the USAF. (U.S. Air Force)

The U.S. planes had a crew of two, compared to three in some British variants. The B-57C differed from the B-57B only in having dual controls in the rear cockpit.

The RB-57 reconnaissance variant was begun early in the program, and those aircraft were fitted with a variety of cameras that could be rapidly interchanged for day and night operation. Some RB-57A aircraft were modified for electronic countermeasures (becoming EB-57A). The improved RB-57D and RB-57F spyplanes with enlarged wings could operate at altitudes greater than 70,000 feet. The Canberra also flew in the WB-57 weather reconnaissance role.

Operational: The first flight of B-57A was on 20 July 1953, with all eight B-57A aircraft relegated to tests and special projects. The RB-57A became operational in mid-1954, and the first Canberra "bomber" to enter Air Force service was the B-57B, joining tactical bomb wings in late 1954. The B-57B made its combat debut as a night/all-weather strike aircraft in late 1957 (with upgraded aircraft designated B-57G).

Bomber variants served in Vietnam until 1972 and then were turned over to the Air National Guard; those were phased out by 1974. Specialized B-57s remained in Air Force service another few years. In the mid-1990s an ex-USAF WB-57F was reactivated and operated by the National Science Foundation and National Center for Atmospheric Research for high-altitude research.

British Canberras flew in first-line service until 2006.[2]

Production: A total of 403 B-57s of all types were produced by Martin; the bomber variants are listed below. The last B-57C was delivered in May 1956. Also produced were 67 RB-57A and 20 RB-57D reconnaissance aircraft, and 68 B-57E target tow aircraft. Early RB-57s were upgraded to the RB-57F configuration.

A total of 878 Canberras of all types were produced in Britain; a further 48 B-20 variants were built in Australia. Some British-built bomber aircraft were produced specifically for other air forces, and several other countries subsequently flew former RAF aircraft.

Aircraft	Built	First flight
B-57A	8	1953
B-57B	202	1954
B-57C	38	1954

Operational:	June 1954–1974 in bomber role	
Design:	English Electric/Martin	
Aircraft data:	B-57B/Night Intruder	
Crew:	2 (pilot, bombardier-navigator)	
Engines:	2 Wright J65-W-5 turbojet; 7,220 lbst each	
Weights:	empty	25,811 lb
	max takeoff	54,476 lb
Dimensions:	length	65 ft 6 in

2 The final RAF Canberras in operational service were PR.9 reconnaissance aircraft. That type first flew in 1960 (the first Canberra flight was in 1949). The last PR.9 was retired from first-line service on 23 June 2006.

The Douglas B-66 Destroyer was adapted—with numerous changes—from the Navy's carrier-based A3D/A-3 Skywarrior. Most Destroyers flew in the electronic reconnaissance (EB-66), photo-reconnaissance (RB-66), and weather reconnaissance (WB-66) roles. This is one of the few B-66B bomber variants. Note the remote-control 20-mm tail turret beneath the MD-1A fire control radar. (U.S. Air Force)

	wingspan	64 ft
	wing area	960 ft^2
	height	14 ft 10 in
Speed:	maximum	585 mph at 17,400 ft
		500 mph at sea level
	cruise	487 mph
Ceiling:	39,900 ft	
Range:	radius	1,020 miles
Radar:	AN/APS-54 warning	
	S-4 bomb/nav	
Guns:	4 20-mm M39 cannon (200 rounds per barrel)	
Ordnance:	1 nuclear bomb Mk 7	
or	6,000 lb conventional bombs, rockets in internal weapons bay and on eight wing pylons	

B-66 Destroyer

The Douglas B-66 was a light tactical bomber and day/night reconnaissance aircraft developed to succeed the B-26 Invader, which had entered AAF service in 1944.[3] The B-66 was intended as an interim aircraft until the B-68 came into service (the latter plane being cancelled in 1957).

The Air Force selected the not-yet-flown Navy A3D Skywarrior for these bomber/reconnaissance roles in 1952. The aircraft was to be an "off-the-shelf" procurement and the first time the Air Force procured a Navy-developed aircraft. However, the Air Force directed a major redesign of the wing and fuselage; provided for new engines and ejection seats for the three-man crew; deleted various naval features such as folding wings, strengthened undercarriage, and arresting hook; and provided for a different radar and gun system. Except for external appearance, little of the naval aircraft remained in the B-66.

Furthermore, of 294 aircraft produced by Douglas, only one-quarter were bomber aircraft, with the remaining B-66s built for specialized reconnaissance missions. Later 13 of the bombers were converted to electronic countermeasures (ECM) aircraft, and two others were modified for high-altitude parachute drops of Apollo and Gemini space capsules.

Design: The B-66 was a twin-turbojet aircraft with a swept wing mounted in the shoulder position; the engines were mounted in pods on wing pylons. Originally fitted with a remote-controlled 20-mm tail turret, most bombers had an ECM package in place of the guns.

Weapons were carried in an internal bomb bay.

The principal variants of the B-66 were the RB-66A/B photo-reconnaissance aircraft, EB-66B/E ECM aircraft (conversions), RB-66C electronic reconnaissance/ECM aircraft, and WB-66D weather reconnaissance aircraft. Two WB-66D variants were reconfigured to test laminar flow control systems (changed to X-21A).

Operational: The first flight of RB-66A was on 28 June 1954; the first flight of B-66B was on 4 January 1955. IOC of the B-66B occurred in March 1956. The bomber aircraft were deployed overseas primarily to bases in England. (Only after conversion to EB-66B under the Brown Cradle program did those aircraft reach Vietnam.)

Production: A total of 294 B-66s were produced—72 bombers, 186 RB-66 variants, and 36 WB-66 variants. An additional 69 B-66B aircraft were cancelled. (The five prototypes were designated RB-66A; they were used exclu-

3 The Invader entered AAF service as the A-26, the wartime B-26 being the Martin Marauder. The Invader was redesignated B-26 in June 1948 because the Air Force eschewed the use of the attack designation.

sively for flight testing.) The last B-66 was delivered in October 1957.

Operational:	Mar 1956–1963 bomber variant	
Design:	Douglas	
Aircraft data:	B-66B	
Crew:	3 (pilot, bombardier-navigator, gunner)	
Engines:	2 Allison J71-A-13 turbojet; 10,200 lb each	
Weights:	empty	42,549 lb
	max takeoff	76,600 lb
Dimensions:	length	75 ft 2½ in
	wingspan	72 ft 6 in
	wing area	780 ft^2
	height	23 ft 7 in
Speed:	maximum	630 mph at 6,000 ft
	cruise	525 mph
Ceiling:	36,500 ft	
Range:	radius	900 miles
Radar:	AN/APS-54 bomb/nav	
	MD-1A fire control	
Guns:	2 20-mm M-24A-1 cannon (500 rounds per barrel)	
Ordnance:	1 nuclear bomb Mk 5, B28	
or	15,000 lb conventional bombs	

FIGHTER-ATTACK AIRCRAFT

Aircraft[a]			Service[b]	IOC	Type
F-110	F-4	Phantom	AF	1964[c]	land
F-15E		Strike Eagle	AF	1988	land
F-16		Fighting Falcon	AF	1979	land
F/A-18		Hornet	N-MC	1981	carrier
F-35A		Lightning	AF	2013[d]	land
F-84F		Thunderstreak	AF	1954	land
F-84G		Thunderjet	AF	1951	land
F-86F		Sabre	AF	1952	land
F-86H		Sabre	AF	1954	land
F-89		Scorpion	AF	1954	land
F-100		Super Sabre	AF	1955	land
F-101		Voodoo	AF	1957	land
F-102		Delta Dagger	AF	1956	land
F-104		Starfighter	AF	1958	land
F-105		Thunderchief	AF	1958	land
F-106		Delta Dart	AF	1959	land
F-107		—	AF	cancelled	land
F-108		Rapier	AF	cancelled	land
F-111		Aardvark	AF	1968	land
F-117A		Nighthawk	AF	1983	land
F9F-8B	AF-9J	Cougar	N-MC	1954	carrier
F2H	F-2	Banshee	N-MC	1951	carrier
F3H-2N	F-3C	Demon	N	1956	carrier
FJ-4	AF-1E	Fury	N-MC	1954	carrier
F7U-3		Cutlass	N	1953	carrier

a Aircraft designations in second column indicate 1962 redesignations.
b AF = Air Force; N = Navy; MC = Marine Corps.
c The Navy–Marine Corps variant—which was not nuclear capable—became operational in 1962.
d Estimated.

F-1 Fury (see FJ Fury)

F-2 Banshee (see F2H Banshee)

F-3 Demon (see F3H Demon)

F-4 Phantom (formerly F-110)

The Phantom was one of the most successful fighter-attack aircraft of the post–World War II era and was produced in larger numbers than any other U.S. postwar military aircraft except for the P-80/F-80 Shooting Star and F-86 Sabre/FJ Fury series. During the 1960s and 1970s the Phantom was the most numerous fighter aircraft in the U.S. Air Force, Navy, and Marine Corps, and was flown by ten foreign nations. Only Air Force variants—both fighter-attack and reconnaissance—could deliver nuclear weapons.

The McDonnell Douglas F-4 Phantom was the outstanding multipurpose fighter of the Cold War era. It proved successful in several roles in U.S. Air Force, Navy, and Marine Corps service in the Vietnam War as well as in Middle East conflicts flown by the Israeli Air Force. USAF Phantoms were nuclear capable, but not the Phantoms flown by the Navy and Marine Corps. (U.S. Air Force)

The Navy selected the F4H Phantom over the F8U-3 Crusader as its standard carrier-based fighter in 1958. Subsequently, in 1962, Secretary of Defense Robert McNamara directed that the Air Force procure the aircraft in place of additional F-105 Thunderchief production. See production notes for Air Force model designations.

The Phantom is no longer in U.S. service.

Design: The Mach 2.24 aircraft (F-4E variant) was noted for its maneuverability, large payload, and excellent carrier operating characteristics. The Phantom was a twin-engine, swept-wing fighter with box-like "cheek" air intakes; the wings were swept 45°. The aircraft had a large nose radar housing and (in most models) a distinctive infrared sensor pod fitted directly beneath the radar. There was tandem seating for the two crewmen under a single canopy.

Several air-to-air missiles could be carried semirecessed under the fuselage, and various combinations of weapons, sensors, and fuel tanks could be carried on six wing pylons plus two fuselage attachment points (available when semirecessed missile positions are not used). The Air Force RF-4C variants could carry a single nuclear weapon on a centerline station; no other ordnance capability was provided.

Only the F-4E variant had an internal cannon; external gun pods could be carried on other aircraft. This was the first U.S. Navy fighter aircraft to be built without guns. The missile armament for Air Force Phantoms included the Sidewinder, Sparrow, Shrike, and Falcon.

Air Force models had limited dual controls in the rear seat (a feature not found in naval aircraft). Navy–Marine Corps aircraft had folding wings for carrier stowage (a feature retained on Air Force aircraft).

The Air Force operated the Phantom in the Wild Weasel (anti-radar) and photo-reconnaissance roles.[4] (The U.S. Marine Corps flew the RF-4B variant in small numbers; the Navy did not have reconnaissance variants. The RF-4B had no weapons capability.)

Designation: Initially designed AH-1 in development (the letter "H" at the time indicating McDonnell), the aircraft entered Navy–Marine Corps service as the F4H Phantom II. The Air Force began acquisition of the Phantom as the F-110. All Phantoms were redesignated in the F-4 series on 1 October 1962.

Operational: The first flight of XF4H-1 was on 27 May 1958. Navy squadron introduction was in 1961, and Air Force F-4C IOC in 1963. The Air Force F-4D became operational in June 1966, and the F-4E in October 1967.

The Phantom was flown extensively in combat by the U.S. services in the Vietnam War and by the Israeli Air Force in several Middle East conflicts. All five U.S. fighter aces of the Vietnam War (two Navy, three Air Force fliers—in two-man crews) scored their kills of MiG-type aircraft in Phantoms; of the 57 "kills" by U.S. aircraft in that conflict, 36 plus several "probables" were made by Phantoms. U.S. Air Force F-4G Wild Weasel aircraft were used in the 1991 war in the Persian Gulf, the last operational use of that plane by U.S. forces.

The U.S. Navy's last active Phantom was retired from squadron service in 1986, and the last Naval Reserve unit gave up Phantoms in 1987; the last U.S. Marine active Phantom was retired in 1987, and the last reserve Phantom in 1992. The last Phantoms—F-4G Wild Weasel aircraft—were retired in 1996.

Production: McDonnell (subsequently McDonnell Douglas) produced 5,050 aircraft through 25 October 1979—1,244 for the U.S. Navy (A/B/G/J models), 2,870 for the USAF (C/D/E models), and 936 for foreign users; additional Air Force aircraft were subsequently transferred to other nations.

In addition, Mitsubishi built 125 F-4EJ aircraft in Japan plus 11 aircraft assembled from parts produced by McDonnell Douglas (included above in the number produced for foreign users).

The following aircraft were produced for the U.S. Air Force:

Aircraft	Built	First flight
F-110A F-4C[a]	583	1963
YF-4C	2	1963
YRF-4C	2	
RF-110A RF-4C[a]	503	1964
F-4D[a]	793	1965
YF-4E	1	1965
F-4E[a]	988	1967
F-4G	12	1963

a Nuclear-capable aircraft

4 Wild Weasel aircraft carried sensors and missiles to detect, jam, and attack hostile ground radars.

A pair of McDonnell Douglas F-15E Strike Eagles are shown during high-speed maneuvers. The Mach 2.5 multipurpose fighter is the most capable fighter aircraft in U.S. military service today. Although procured in relatively small numbers by the U.S. Air Force, it is also flown by several allied air forces. (U.S. Air Force)

Operational:	July 1961–1992 U.S. Navy and Marine Corps	
	Oct 1964–1996 U.S. Air Force	
Design:	McDonnell	
Aircraft data:	F-4E	
Crew:	2 (pilot, weapons system officer)	
Engines:	2 General Electric J79-GE-17 turbojet; 17,900 lbst each in afterburner	
Weights:	empty	30,328 lb
	max takeoff	61,795 lb
Dimensions:	length	63 ft
	wingspan	38 ft 5 in
	wing area	530 ft^2
	height	16 ft 6 in
Speed:	max 1,220 mph at 40,000 ft	
	cruise	582 mph
Range:	1,050 miles	
	combat radius 413 miles	
Ceiling:	57,200 ft	
Radar:	AN/APQ-120 fire control	
Guns:	1 20-mm M61A1 Vulcan multibarrel cannon (640 rounds)	
Ordnance:	2 nuclear bombs B28, B43, B61	
or	4 nuclear bombs B57	
or	16,000 lb (7,258 kg) conventional bombs, missiles	

F-9 Cougar (see F9F-8B Cougar)

F-15E Strike Eagle

The F-15 is a long-range, high-performance (Mach 2.5) fighter-interceptor used by the U.S. Air Force and several other air forces. The F-15E variant is an air superiority aircraft as well as an all-weather/night, deep-penetration strike aircraft. The F-15 is the only Air Force fighter able to carry and deliver air-to-surface weapons at supersonic speeds.

In the mid-1960s the U.S. Air Force sought industry proposals for an advanced tactical fighter for the post-1975 period. On 23 December 1969 McDonnell Douglas was declared winner of the competition and an initial order was placed for 20 research and development aircraft—18 F-15A and 2 TF-15A Eagles. The F-15E Strike Eagle evolved from the earlier models with only that aircraft configured to carry nuclear weapons. The earlier F-15A through D models cannot carry nuclear weapons, although they can carry up to 23,600 pounds of bombs and missiles.

The F-15E was begun as a company-funded, upgraded air-to-surface aircraft based on the two-seat F-15B aircraft. The design was selected by the Air Force in February 1984 to augment and eventually replace F-111 Aardvark and F-4 Phantom in the strike role. The F-15E can carry 24,500 pounds of ordnance, three times the bomb load of the World War II B-17 and only 500 pounds less that the F-111F.

Design: The aircraft has a shoulder-mounted wing with swept leading edges; the twin engines are on each side of the fuselage with squared-off engine intakes. The aircraft has distinguishing twin tail fins and rudders.

The F-15's excellent maneuverability and acceleration are achieved through its high engine thrust-to-weight ratio and wing loading. It was the first U.S. operational aircraft whose engine thrust exceeded the plan's loaded weight, permitting it to accelerate even while in a vertical climb. The F-15 design provides a high thrust-to-weight ratio (1.17:1 for the F-15A compared to 0.88:1 for the F-4E Phantom) and a low wing loading (65.8 lbs versus 76.5 lbs).

All bombs and missiles are carried externally on 18 pylons.

Operational: The first flight of F-15A was on 27 July 1972; F-15A IOC was in November 1974. The first flight of F-15E was on 11 December 1986; F-15E IOC was in December 1988.

Earlier F-15 models are flown by the Israeli, Japanese, Saudi, Singapore, and South Korean air forces; more than 300 F-15s have been delivered or are on order for those nations. Japan also produces the F-15.

Production: The Air Force acquired 209 F-15E aircraft through fiscal year 1991; 8 additional attrition replacements were subsequently procured.

Operational:	Dec 1988–current	
Design:	McDonnell Douglas	
Aircraft data:	F-15E	
Crew:	2 (pilot, weapon systems officer)	
Engines:	2 Pratt & Whitney F100-PW-229 turbofan; 29,000 lbst each	
Weights:	empty	32,000 lb
	max takeoff	81,000 lb
Dimensions:	length	63 ft 9¾ in
	wingspan	42 ft 9¾ in
	wing area	608 ft^2
	height	18 ft 6 in
Speed:	maximum	1,875 mph
Ceiling:	60,000 ft	
Range:	1,935 mi	
Radar:	AN/APG-63(V)1 bomb/nav (replacing AN/APG-70)	
	AN/AWG-27 weapons control	
Guns:	1 M61A1 Vulcan multibarrel cannon (500 rounds)	
Ordnance:	2 nuclear bombs B61	
or	24,500 lb conventional bombs	

F-16 Fighting Falcon

The F-16 is a lightweight, single-engine, supersonic (Mach 2) fighter capable of air-to-air and air-to-ground roles. It is flown by the U.S. Air Force and several NATO air forces. The F-16 complements the F-15 in the air superiority role and replaced the F-4 Phantom in the air-to-ground role. The F-16C/D aircraft have numerous improvements over the original F-16A/B including a reconfigured cockpit, improved radar, and increased computer capacity.

The General Dynamics YF-16 was named winner of the Air Force's lightweight fighter. On 13 April 1972, the Air Force ordered two YF-16 prototypes and two YF-17 prototypes. The first flight of the YF-16 occurred on 2 February 1974, and the YF-17 made its first flight on 9 June 1974. General Dynamics YF-16 won the competition and began full-scale development on 13 January 1975. The YF-17 evolved into the F/A-18 Hornet flown by the Navy and Marine Corps.

Design: The F-16 has slightly swept-back wings with long root extensions to improve supersonic maneuvering. The engine intake is below the cockpit, giving the aircraft a distinctive profile.

The aircraft carries two Sidewinder air-to-air missiles on its wingtips with additional stores carried on four wing pylons and a centerline attachment station.

The two-seat F-16B has tandem seating with a student position in the front seat and pilot in the rear seat; the F-16D is also a twin-seat aircraft.

Operational: The first flight production of the F-16A was on 8 December 1976. F-16A IOC was in January 1979. The U.S. Navy currently operates ten F-16A and four F-16B aircraft in the adversary training role with Fighter Weapons School (Top Gun).[5]

Twenty-five other nations fly various models of the F-16. Nuclear-capable F-16A/B variants were flown by Belgium, Greece, The Netherlands, and Turkey. Those aircraft were certified to carry the B57 and B61 nuclear bombs.

Production or assembly also takes place in Belgium, The Netherlands, South Korea, and Turkey.

Production: More than 4,000 F-16s have been produced. The U.S. Air Force took delivery of 2,231 F-16s, the last in March 2005.

Operational:	Jan 1979–current
Design:	General Dynamics
Aircraft data:	F-16C
Crew:	1 in F-16A/C
	2 (pilot, student or weapon systems officer) in F-16B/D

5 The U.S. Navy previously flew 26 modified F-16C Fighting Falcon fighters in the adversary training role; these were 22 single-seat F-16N and 4 two-seat TF-16N aircraft delivered in 1987–1988. These aircraft were grounded beginning in 1991 because of structural cracking in the center fuselage and were disposed of.

Engines:	1 General Electric F-110-GE-100/129 turbofan; or 1 Pratt & Whitney F100-PW-200/220/229 turbofan; 29,588 lbst	
Weights:	empty	16,794 lb
	max takeoff	37,500 lb
Dimensions:	length	49 ft 5 in
	wingspan	32 ft 8 in
	wing area	300 ft^2
	height	16 ft
Speed:	maximum	1,520 mph
Ceiling:	50,000 ft	
Range:	radius	575 miles
Radar:	AN/APG-68 multifunction	
Guns:	1 20-mm M61A1 Vulcan multibarrel cannon (515 rounds)	
Ordnance:	2 nuclear bombs B57-1/2, B61-3/4/10	
or	21,850 lb conventional bombs, missiles	

General Dynamics F-16 Fighting Falcon was selected by the Air Force for the lightweight fighter role (the competitor being the Northrop F-17 design). The F-16 has served several nations in the fighter and attack roles. This USAF F-16, on a training flight over South Korea, is carrying bombs, drop tanks, and wingtip-mounted Sidewinder air-to-air missiles. (U.S. Air Force)

F/A-18 Hornet

The Hornet is a carrier-based strike-fighter flown in significantly larger numbers than any other fixed-wing aircraft by the U.S. Navy and Marine Corps. Only the C/D models have a nuclear capability and only Marine aircraft deployed in a nuclear strike role prior to 1992, when all Navy and Marine Corps aircraft gave up that role. Several other nations fly land-based F/A-18s.

Development of the F/A-18 was the result of pressure from Congress for the Navy to procure a lightweight fighter to complement the larger, more expensive F-14 Tomcat in carrier air wings. Congress directed that the Navy select the winner of the Air Force's lightweight fighter competition between the General Dynamics YF-16 and Northrop YF-17 prototypes. The Air Force selected the YF-16 for production, and the Navy selected the YF-17 but made major modifications, leading to the F/A-18 being developed jointly by McDonnell Douglas and Northrop.

In the U.S. Navy the F/A-18 replaced the F-14 and A-7E Corsair; in the Marine Corps the F/A-18 replaced the F-4S Phantom, A-6E Intruder, A-4M Skyhawk, RF-4B Phantom, and OA-4 Skyhawk.

The McDonnell Douglas–Northrop F/A-18 Hornet is flown by the U.S. Navy and Marine Corps as their principal fighter/attack aircraft, with the EA-18G Growler variant entering service as an electronic attack/jamming aircraft. The Hornet is produced in several variants, with both single- and two-seat models. The aircraft was derived from the Northrop F-17 design. Shown here is a Navy F/A-18C variant. (U.S. Navy)

Design: A twin-engine, single- or two-seat aircraft, the F/A-18 is characterized by its high maneuverability, the ability to operate in either the fighter or attack role with the push of a button, and comparatively low maintenance requirements. The initial versions were the F/A-18A single-seat and F/A-18B two-seat aircraft. The F/A-18C is an improved single-seat aircraft, and the F/A-18D is a two-seat aircraft with a weapons officer in the rear seat. The enlarged, more-capable F/A-18E (single seat) and F/A-18F (two-seat) Super Hornet aircraft are in production for the U.S. Navy; the Marine Corps is not procuring this variant.

The EA-18G Growler is a two-seat derivative configured for electronic attack/countermeasures; it replaces the EA-6B Prowler.

The twin-engine aircraft is distinguished by swept leading wing edges, and twin tail fins. It has wingtip rails for Sidewinder missiles and three fuselage and four wing stations for weapons and sensor/guidance pods. A variety of bombs, missiles, and rockets can be carried, including up to four 2,000-pound bombs.

Some Marine F/A-18D variants are configured for the advanced tactical air reconnaissance system (ATARS), which provides real- and near-real-time photo reconnaissance, with a limited data-link capability to transmit images to ground and shipboard facilities.

Wings fold on all models for carrier stowage.

Designation: The initial Navy order for 11 development aircraft used the designation YF-18, and production aircraft were to be F-18. However, in 1978 the designation was changed to F/A-18 for "strike-fighter." The F/A designation violates the official U.S. military aircraft designation policy.

Operational: The first flight of F/A-18A was on 18 November 1978. U.S. Navy–Marine Corps F/A-18A IOC was in February 1981 (transition squadron); Marine F/A-18A IOC was in March 1983.

Production: The U.S. Navy–Marine Corps accepted 1,048 F/A-18A/B/C/D variants (including prototypes):

Aircraft	Built	First flight
F/A-18A	380	1978
F/A-18B	41	1978
F/A-18C	466	1987
F/A-18D	161	1988

In addition, through 1 January 2008, the U.S. naval services accepted 148 F/A-18E and 193 F/A-18F Hornets. Almost all F/A-18A/B models have been retired from U.S. service.

Another 409 F/A-18A/B/C/D aircraft were produced for seven other nations with additional Hornets produced/assembled in Australia, Finland, and Switzerland. In March 2008 the Australian government decided to procure 24 F/A-18E/F variants, the first foreign sale of the Super Hornet. A Northrop-proposed F-18L land-based variant has not been procured.

Operational:	Feb 1981–current	
Design:	McDonnell Douglas and Northrop (now Boeing)	
Aircraft data:	F/A-18C	
Crew:	1 in F/A-18A/C 2 (pilot, radar intercept officer) in F/A-18B/D	
Engines:	2 General Electric F404-GE-400 turbofan; 16,000 lbst each; F404-GE-402 engine with 17,700 lbst each in FY 1992 and later aircraft	
Weights:	empty	23,050 lb
	max takeoff	56,000 lb
Dimensions:	length	56 ft
	wingspan	37 ft 6 in 40 ft 5 in with Sidewinder missiles
	wing area	400 ft^2
	height	15 ft 3½ in
Speed:	maximum	1,185 mph at 37,000 ft
Ceiling:	50,000 ft	
Range:	radius	230 miles in fighter role with 1 hour on station (2 Sidewinders + 4 AMRAAMs) 350 miles in fighter escort role (2 Sidewinders + 2 AMRAAMs) 320 miles in strike role (2 Sidewinders + 4 1,000-lb bombs)
Radar:	AN/APG-73 multimode	
Guns:	1 20-mm M61 Vulcan multibarrel cannon (570 rounds)	
Ordnance:	2 nuclear bombs B57, B61 + 2 Sidewinders	

or 2 Sidewinders + 4 Sparrows/AMRAAMs in fighter role
or 2 Sidewinders + 17,000 lb bombs, missiles, rockets in attack role

F-35A Lightning

The Lockheed Martin F-35 is a multirole aircraft developed for use by the U.S. Air Force, Navy, and Marine Corps. It was the winner of the U.S. Joint Strike Fighter (JSF) competition, with the Lockheed Martin X-35 being judged superior to the Boeing X-32 flight demonstration aircraft (see Designation section).

Because the F-35A variant will be a replacement for the F-16 and F-117 strike aircraft, it is believed that some or all of that variant will be fitted to carry the B61 nuclear bomb.

The JSF was planned to replace up to eight U.S. aircraft types:

The Lockheed Martin F-35A Lightning II prototype in flight. The so-called joint strike fighter (JSF) is planned for use by the U.S. Air Force (F-35A), Marine Corps (F-35B), and Navy (F-35C) as well as several foreign services. The F-35 was developed from the Lockheed Martin X-35 technology demonstrator, selected in competition with the Boeing X-32 aircraft. (U.S. Air Force)

- *F-35A conventional variant* for U.S. Air Force replacing the F-15 Eagle, F-16 Fighting Falcon, and F-117A stealth strike aircraft.
- *F-35B STOVL variant* for U.S. Marine Corps and Royal Navy replacing the AV-8B Harrier and RN Harrier. The F-35B may also replace the Royal Air Force's GR.Mk 7/9, which are the only British fixed-wing aircraft currently operated from the country's three VSTOL carriers.
- *F-35C carrier variant* for U.S. Navy replacing the F-14 Tomcat and F/A-18C/D Hornet.

Design: The F-35 series provides for a supersonic (approximately Mach 1.5) aircraft capable of fighter, attack, and related missions. The basic F-35/JSF concept is a family of three distinct variants that have a 70 to 90 percent aircraft commonality.

The Lockheed Martin X-35 was chosen over the competing Boeing X-32 primarily because of Lockheed's lift fan STOVL design, which is considered superior to Boeing's thrust-vectored approach. The lift fan, which is lowered by the aircraft engine through a clutched driveshaft, was technically feasible, but the Department of Defense concluded that Lockheed had the technology in hand. The lift fan has significant excess power that could be critical given the weight gain that all fighter aircraft tend to experience.

All three versions of the aircraft have the same fuselage and internal weapons bay, common outer mold lines, identical swept-back wings, and comparable tail configurations. All have the same core engine, which is based on the Pratt & Whitney F119 engine derived from the F-22 Raptor. The weapons bay will be an unusual feature for contemporary carrier-based aircraft, with previous naval fighters and light/medium attack aircraft carrying their weapons on external pylons and attachment points. In contrast, previous Air Force tactical strike aircraft—largely developed to carry nuclear weapons—tend to have internal weapon bays. The internal bay usually permits higher speeds and enhances stealth. In the Marine Corps and British F-35C variant the weapons bay may be smaller because of the STOVL requirement.

The aircraft is intended to have a radar cross section (RCS) slightly larger than the radar signature of the F/A-22 Raptor at a fraction of the cost. The reduction is achieved through shaping, construction materials, and skin coatings. An equivalent signature is the radar reflection from a perfect electrically conducting metal sphere the size of a golf ball.[6]

Designation: In 2002 then–Secretary of the Air Force James Roche announced that the JSF would be designated F-35 because the technology demonstration aircraft for the project was the X-35. That research aircraft was not a prototype JSF; rather, its experimental designation is

6 See David A. Fulghum, "JSF Reflection Is Golf Ball-Sized," *Aviation Week & Space Technology* (15 February 1999), pp. 27, 30.

akin to that of the XV-15 tilt-rotor aircraft, which was the technology demonstrator for the V-22 Osprey series. According to the Department of Defense aircraft designation system, the next U.S. fighter-type aircraft should have been designated F-24. (The F-23 was the McDonnell Douglas competitive design to the Lockheed F-22 Raptor high-performance fighter.)

The use of sequential numbers for each aircraft within a given category is spelled out in the official DoD instructions on aircraft designations.[7] The instruction specifically states that the system "is mandatory for use by all DoD Components."

Historical: Past efforts to develop multiservice tactical aircraft have not been marked by success. The most notable example was the 1960s effort to produce the tactical fighter experimental (TFX) for the Air Force, Navy, and Marine Corps, and, subsequently, for the strategic bomber role. The resulting F-111/FB-111 had a lackluster career as a tactical strike (not fighter) and strategic bombing aircraft; the Navy rejected their F-111B variant as being unsuitable for carrier operation, and that rejection ended possible Marine Corps procurement. The proposed British procurement of the F-111 was stillborn, although the Royal Australian Air Force acquired the F-111 and still flies that aircraft.[8]

Similarly, several land-based tactical aircraft proposed for naval (carrier) use have also been less than successful, with most of the examples of such efforts having been British. In sharp contrast, carrier-based aircraft that have been adopted for land operation have had marked success, among them the F4H/F-4 Phantom, AD/A-1 Skyraider, A3D/A-3 Skywarrior (as the B-66 Destroyer), A4D/A-4 Skyhawk, and A-7 Corsair.

Operational: The first flight of X-35A was on 24 October 2007; F-35A was on 15 December 2006. IOC for F-35A will be approximately 2013.

Production: The Air Force originally planned to procure 1,763 F-35A aircraft. That number is being reduced. The total Department of Defense F-35/JSF buy in 2008 was listed as 2,458 aircraft. The Royal Navy is purchasing the F-35B VSTOL variant, and several other nations have options.

Operational:	2013 (estimated)	
Design:	Lockheed Martin	
Aircraft data:	F-35A	
Crew:	1	
Engines:	1 Pratt & Whitney F135 turbofan; approx 35,000 lbst	
Weights:	empty	approx 22,500 lb
	max takeoff	approx 50,000 lb
Dimensions:	length	50 ft 6 in
	wingspan	35 ft
	wing area	460 ft^2
	height	15 ft
Speed:	maximum	Mach 1.5+
	cruise	
Ceiling:		
Range:		
Radar:		
Guns:	1 25-mm GAU-12 multibarrel cannon	
Ordnance:	1 nuclear bomb B61 + 2 Sidewinder missiles	
or	conventional bombs + 2 Sidewinder missiles	

F-84F Thunderstreak

The F-84F was a fighter-bomber capable of carrying nuclear bombs. Based on the F-84 Thunderjet fighter, development of a swept-wing F-84 led to the F-84F, the first swept-wing model of the series. The F-84F variant could carry heavier stores and was a faster aircraft.

The F-84F was developed from the F-84G, which had followed the F-84E in the development sequence of the Thunderjet-Thunderstreak series (see separate entry for the F-84G). The F-84F was proposed by Republic Aviation in November 1949, and the following month the firm was authorized to complete one F-84E as a swept-wing F-84F. The F-84F was a far more capable aircraft than the F-84G, the only other nuclear-capable variant of this fighter-bomber.

The RF-84F Thunderflash was a swept-wing, photo-reconnaissance aircraft that did not have a bombing capability. (Some YRF-84F aircraft were modified to be

7 Department of Defense, Office of the Under Secretary of Defense (Acquisition and Technology), "Model Designation of Military Aerospace Vehicles," DOD 4120.15-L (13 March 1996), p. 6. This was the last published edition of the document; it is available on the Internet.

8 Major variants of the General Dynamics F-111 Aardvark were the EF-111 electronic countermeasures aircraft, FB-111 strategic bomber, and RF-111 reconnaissance aircraft. The EF-111 was the last variant flown by the U.S. Air Force and was replaced by the Navy-Marine EA-6B Prowler.

Republic F-84F Thunderstreak in JATO takeoff. (U.S. Air Force)

carried aloft by RB-36D strategic bomber/reconnaissance aircraft.)

Design: The F-84F was generally similar to the previous F-84 configurations but was larger and had a greater maximum takeoff weight; the F-84F wings and tail were swept back 40°.

The aircraft suffered major engine deficiencies with the original Allison J35-A-25; it was replaced by a succession of engines until all operational F-84F aircraft were fitted with the J65-W-3 based on the British Sapphire engine. There were also weight and other technical problems with the aircraft, some of which were never fully solved. F-84F production was briefly halted in the fall of 1954 and in early 1955 all aircraft were grounded because of engine problems.

Twin 450-gallon drop tanks were normally carried under the wing roots of the F-84F, replacing the wingtip tanks used in earlier models. It was fitted for in-flight refueling.

Weapons were carried on wing pylons. Four machine guns were fitted in the nose and two in the wings.

Two F-84F aircraft were converted to XF-84H to test a supersonic propeller driven by a gas turbine engine, and one F-84F was converted to a YF-84J to evaluate a more powerful engine; neither design entered production.

Designation: Development began with the designation YF-96A; the designation YF-84F was assigned on 8 September 1950 with the name Thunderstreak being assigned to F-84 swept-wing variants.

Operational: The first flight of YF-84F was on 3 June 1950; the F-84F modified prototype with British Sapphire engine first flight was on 14 February 1951; and the first flight of a production F-84F was on 22 November 1952.

The first F-84F aircraft entered operational service with the Air Force in January 1954. The plane was assigned both to SAC and Tactical Air Command (TAC) with the first aircraft assigned to combat units going to SAC's 506th Strategic Fighter Wing at Dow AFB, Maine.

SAC operated F-84s in the fighter-escort role from 1950 until 1957, reaching a peak strength of 568 F-84/RF-84 aircraft in 1955. From late 1951 until June 1957 various models of the F-84 were the only fighter-type aircraft operated by SAC.[9]

The F-84F remained in first-line Air Force service until January 1958. As early as August 1954 SAC began transferring F-84Fs to the Air National Guard; the last SAC and TAC aircraft went to the Air National Guard (ANG) by January 1958, except for TAC training aircraft. During the Berlin crisis of 1961–1962, four ANG wings of F-84Fs were brought on active duty with some units being deployed to Europe. After the crisis 222 F-84Fs were retained by TAC to temporarily equip two new tactical fighter wings. In July 1964 TAC returned its last F84-Fs to the ANG. The last of those aircraft were discarded in 1972.

Production: Republic Aviation in Long Island, New York, and the General Motors Corp. plant in Kansas City, Kansas, produced the F-84F. Production totaled 3 YF-84F and 1,496 F-84F models for the U.S. Air Force and 852 F-84F models for foreign air forces. (This did not include RF-86F production; the RF-84F had nose-mounted cameras and the engine intakes moved to the wing roots.)

The last U.S. F-84F was delivered in February 1957, and foreign production ended in August 1957. (See F-84G entry for total F-84 production.)

Operational:	Jan 1954–July 1964 active Air Force
Design:	Republic Aviation
Aircraft data:	F-84F
Crew:	1
Engines:	1 Westinghouse J-65-W-3 turbojet; 7,200 lbst

9 SAC ceased operating fighter-type aircraft on 1 July 1957. However, in 1958 two fighter interceptor squadrons with 54 F-86 Sabres that were based in Spain were assigned to SAC control; on 1 July 1960 those squadrons were transferred from SAC to the operational control of U.S. Air Force Europe.

Weights:	empty	13,420 lb
	max takeoff	27,000 lb
Dimensions:	length	43 ft 4¾ in
	wingspan	33 ft 7 in
	wing area	325 ft^2
	height	14 ft 4½ in
Speed:	maximum	690 mph at 35,000 ft
	cruise	535 mph at 35,000 ft
Ceiling:	44,300 ft	
Range:	radius	375 miles
Radar:	AN/APG-30 gunfire control	
Guns:	6 .50-cal M-3 machine guns (300 rounds per barrel)	
Ordnance:	1 nuclear bomb Mk 7	
or	6,000 lb conventional bombs	

F-84G Thunderjet

The F-84G was a fighter-bomber developed from the basic F-84 series with a limited nuclear capability, making this variant the first U.S. Air Force fighter with a nuclear strike capability. The F-84 was an air-superiority fighter, conceived by Republic Aviation as a turbojet successor to its P-47 Thunderbolt piston fighter of World War II fame.

The Air Force gave Republic the go-ahead to develop the P-84 on 11 November 1944; the prototype XP-84 flew in early 1946 and, despite considerable technical problems, the aircraft entered production. The F-84A through F-84E and F-84G were primarily fighter aircraft with the last having a limited attack capability.

F-84s were employed mainly in the fighter-escort role with the F-84F/G variants additionally employed in the fighter-bomber role. The aircraft lacked agility for effectiveness in fighter-versus-fighter combat

Design: The F-84 had a conventional aircraft design with a large nose air intake and straight wings mounted at the midfuselage position. The aircraft had a straight-wing configuration.

The F-84G was the first Air Force fighter with a built-in flight refueling capability.

Designation: The designation was changed from P-84 to F-84 on 11 June 1948.

Operational: The first flight of XP-84 was on 28 February 1946. First squadron delivery of F-84B was in mid-1947, and first F-84G squadron delivery was in August

Republic F-84G Thunderjet. (U.S. Air Force)

1951 to SAC's 31st Fighter-Escort Wing at Turner AFB, Georgia.

In 1952–1953 SAC integrated nuclear weapons into strategic fighter wings flying the F-84G. By mid-1952 four fighter-escort wings had a total of 225 F-84G aircraft capable of delivering the Mk 7 bomb; by June 1953 those wings had 302 F-84Gs assigned with a fifth F-84G strategic fighter wing activated at the end of the year. [10]

More F-84s of various types were flown by SAC than any other fighter aircraft. The F-84G served in SAC until August 1955 and was fully retired from Air Force service in mid-1960.

F-84G aircraft were flown extensively in the Korean War, as were other F-84 variants.

An F-84E flew the first nonstop, trans-Atlantic flight by a jet fighter aircraft on 22 September 1950; the flight, from Manston, England, to Limestone (later Loring) AFB, Maine, took 10 hours, 2 minutes, and required three in-flight refuelings.

The F-84G was employed in Project ZELMAL (Zero Length Launch and Mat Landing) employing rocket boosters for zero-length takeoff and arrested landing on a pneumatic landing mat. These experiments took place in 1954.

Production: Republic Aviation and General Motors built 7,524 F-84 and RF-84 aircraft including 3,515 produced specifically for foreign air forces. F-84G production—the largest of any F-84 variant—totaled 789 for the USAF and 2,236 for foreign air forces; the last F-84G was completed in July 1953 (for foreign use).

Operational:	Aug 1951–1960
Design:	Republic Aviation
Aircraft data:	F-84G

10 SAC fighter wings were changed from Fighter Escort Wings to Strategic Fighter Wings in January 1953.

Crew:	1	
Engines:	1 Allison J-35-A-29 turbojet; 5,600 lbst	
Weights:	empty	11,095 lb
	max takeoff	23,525 lb
Dimensions:	length	38 ft 1 in
	wingspan	36 ft 6 in
	wing area	260 ft^2
	height	12 ft 7 in
Speed:	maximum	620 mph
	cruise	480 mph
Ceiling:	40,500 ft	
Range:	2,000 miles	
Radar:		
Guns:	6 .50-cal M-3 machine guns (300 rounds per barrel)	
Ordnance:	1 nuclear bomb Mk 7	
or	6,000 lb conventional bombs	

F-86F Sabre

The F-86 was developed as an air superiority fighter. It was produced in larger numbers than any other Western aircraft of the Cold War era. The F-86F/H variants were fighter-bombers with some aircraft configured to carry nuclear bombs. (See separate F-86H entry.)

The design was initiated in the fall of 1944 as a straight-wing turbojet aircraft derived from the Navy's FJ-1 Fury, then under development by North American Aviation. As the results of German wartime research into swept-wing aircraft became available to the U.S. aviation industry, a 35° wing sweep was adopted for the F-86. During a test flight in the spring of 1948, the aircraft exceeded Mach 1 in a shallow dive, the first U.S. fighter to do so.

North American F-86F Sabre. (U.S. Air Force)

In the fighter role, the aircraft entered combat in the Korean War in December 1950 and fought what was probably history's first jet-versus-jet duel with a MiG-15 on 17 December. Although the MiG-15 was slightly superior to the F-86A in performance, Air Force pilot training was superior. The ratio of MiG kills to F-86 losses during the Korean War strongly favored the Sabre.

Design: The F-86 was a low-wing aircraft with all wing and fin surfaces swept. The pilot's cockpit was located well forward of the wing, above the large nose air intake. The aircraft was fitted with four wing pylons for bombs, rockets, and drop tanks. Several aircraft were converted to the RF-86F photo-reconnaissance configuration.

The F-86F-35 subseries featured the low altitude bombing system (LABS) that was capable of lofting a nuclear bomb.

Designation: Originally designated P-86, the designation was changed to F-86 on 11 June 1948.

Operational: The first flight of the XP-86A was on 1 October 1947; the first flight of P-86A was on 20 May 1948. Squadron deliveries began in February 1949.

The first flight of the F-86F was on 19 March 1952; squadron deliveries began in March 1952, with aircraft arriving in the Korean war zone within three months. The F-86F scored many of the MiG kills of 1952–1953, having been the premier U.S. Air Force fighter of the Korean War. The last F-86Fs were phased out of active Air Force squadrons by the end of 1955 (replaced by the F-86H). The F-86F served in the Air National Guard for another few years.

F-86s were operated primarily by the Air Defense Command and Tactical Air Command; Strategic Air Command operated approximately 80 F-86A aircraft in the fighter escort role in 1949–1950 and 54 later F-86s in 1958–1960 (see entry for F-84F Thunderstreak).

Production: Total F-86 deliveries to the U.S. Air Force by North American Aviation numbered 5,893 aircraft; North American also produced 460 aircraft for allied air forces. Another 1,815 F-86s were built in Canada, 112 in Australia, 221 in Italy, and 300 in Japan—a total of 8,681 aircraft.

The U.S. Air Force accepted 1,959 F-86F variants, of which 265 aircraft were the nuclear-capable F-86F-35 series; another 2 were TF-86F, the only two-seat Sabres. The last F-86F was delivered in October 1955.

Operational:	1952–1955 as nuclear strike aircraft	
Design:	North American Aviation	
Aircraft data:	F-86F	
Crew:	1	
Engines:	1 General Electric J47-GE-27 turbojet; 5,910 lbst	
Weights:	empty	10,950 lb
	max takeoff	20,650 lb
Dimensions:	length	36 ft 7 in
	wingspan	39 ft
	wing area	313 ft^2
	height	14 ft 9 in
Speed:	maximum	690 mph at 35,000 ft
	cruise	
Ceiling:	45,000 ft	
Range:	radius	290 miles
Radar:		
Guns:	6 .50-cal M-3 machine guns	
Ordnance:	1 nuclear bomb Mk 12	
or	2,000 lb conventional bombs	

F-86H Sabre

The F-86H variant of the North American F-86 Sabre was a fighter-bomber configured for tactical nuclear strike. This variant was slightly larger than the F-86F, the other nuclear-capable variant of the F-86. (See F-86F entry for program details.)

The F-86H was the last production variant of the Sabre.

Design: With a more powerful engine than the J47 in the F-86F, this aircraft had a larger fuselage, larger tail plane, greater fuel capacity, heavier gun armament (with the 116th production aircraft having 20-mm cannon vice .50-caliber machine guns), and other improvements compared to the F-86F. The large number of changes led the Air Force to order two prototypes before starting series production.

The aircraft encountered numerous technical problems plus delays in J73 engine deliveries, compounded by the lack of spare engine parts.

Operational: The first YF-86H flew on 9 May 1953, and the first production F-86H flew on 4 September 1953.

The F-86H entered service with TAC in 1954. The plane was used exclusively by air defense and tactical air forces, not by SAC.

The Air Force quickly disposed of the trouble-plagued F-86H in favor of the F-100C Super Sabre. The last active F-86Hs were transferred to the Air National Guard in June 1958. During the Berlin crisis of 1961–1962 one ANG wing of these aircraft was reactivated. The F-86Hs were fully retired in the fall of 1970.

North American F-86H Sabre. (U.S. Air Force)

Production: The Air Force accepted two YF-86H and 471 F-86H Sabres, the last in October 1955.

Operational:	1954–1958 as nuclear strike aircraft	
Design:	North American Aviation	
Aircraft type:	F-86H	
Crew:	1	
Engines:	1 General Electric J73-GE-3D turbojet; 8,920 lbst	
Weights:	empty	13,836 lb
	max takeoff	21,800 lb
Dimensions:	length	38 ft 10 in
	wingspan	39 ft 1 in
	wing area	313 ft^2
	height	15 ft
Speed:	maximum	680 mph at sea level
		600 mph at 35,000 ft
	cruise	
Ceiling:	47,200 ft	
Range:	radius	420 miles
Radar:	AN/APG-30 gunfire control	
Guns:	4 20-mm M-39 cannon	
Ordnance:	1 nuclear bomb Mk 12	
or	2,720 lb conventional bombs	

F-89J Scorpion

The Northrop F-89J was the Air Force's first nuclear-armed interceptor. The F-89J was a modified F-89D whose new features included provisions for two unguided air-to-air MB-1 Genie rockets (later redesignated AIR-2A).

The F-89 suffered extensive problems and was replaced in the Air Defense Command by the F-102.

Design: The F-89 had straight, shoulder-mounted wing configurations with two turbojet engines mounted under the fuselage. Four cannon and air-intercept radar were fitted in the elongated nose.

Two Genie rockets could be carried beneath the aircraft's wings in addition to conventional Falcon missiles and unguided air-to-air rockets.

Conventional folding-fin rockets were fired from permanently mounted wingtip pods.

Operational: The first flight of XF-89 was on 16 August 1948, the first flight of YF-89D was on 23 October 1951, and first flight of F-89J was on 15 March 1956.

The first F-89J modified from an F-89D entered operational service in January 1957. By mid-1958 the Air Defense Command had a peak inventory of 286 F-89J interceptors serving in 11 active squadrons. Nine ANG squadrons flew F-89Js with Genie missiles until 1969.

Production: Of the 682 F-89Ds produced, 350 were modified and redesignated F-89Js. Total F-89 production was 1,052, all for the U.S. Air Force.

Operational:	Jan 1957–1969
Design:	Northrop
Aircraft data:	F-89J
Crew:	2 (pilot, radar operator)
Engines:	2 Allison J35-A-35 turbojet; 7,400 lbst each

Weights:	empty	25,194 lb
	max takeoff	45,575 lb
Dimensions:	length	53 ft 7 in
	wingspan	59 ft 8½ in
	wing area	650 ft²
	height	17 ft 6 in
Speed:	maximum	635 mph at 10,600 ft
	cruise	558 mph at 46,500 ft
Ceiling:	43,500 ft	
Range:	1,600 miles	
	radius	500 miles
Radar:	AN/APG-40 search	
	MG-12 fire control	
Guns:	none	
Ordnance:	2 nuclear AIR-2A Genie air-to-air rockets	
	4 Falcon air-to-air missiles	
	104 2.75-in folding-fin air-to-air rockets	

F-100 Super Sabre

The F-100 evolved from the F-86 Sabre and was the first Air Force fighter of the Century series. It is credited with being the world's first operational fighter capable of supersonic speed in level flight. The F-100A/C were tactical fighters, and the F-100D/F were tactical fighter-bombers. All four variants were nuclear capable.

A Northrop F-89J Scorpion with two MD-1 Genie nuclear armed rockets under the wings; the aircraft's wingtip pods carry 2.75-inch folding-fin rockets. Genie was one of two U.S. air-to-air nuclear weapons. At the height of the Cold War hundreds of U.S. and Canadian air-defense fighters were armed with nuclear Genies and Falcon missiles. (U.S. Air Force)

North American F-100A Super Sabre. (U.S. Air Force)

Originally a redesign of the F-86 air-superiority fighter, the F-100 was initiated in 1949 as a tactical fighter. However, the F-100C and later variants had a significant conventional as well as nuclear weapons delivery capability.

The development of the F-100 was slowed by emphasis on the F-84 and F-86 fighter production during the Korean War. The F-100's excellent performance led to it becoming a stalwart of the U.S. tactical air forces and several foreign air forces during the 1950s and 1960s.

The F-100 suffered severe maintenance problems and was involved in numerous accidents; almost one-third of all F-100s—more than 700 planes—were lost in accidents.

Design: The F-100 was a low-wing aircraft; the leading edge of the wing was sharply swept at 45° and was fitted with slats. All tail surfaces were swept with a taller fin and rudder fitted in the F-100D/F. A nose air intake supported the single turbojet engine buried in the after fuselage.

There were six wing pylons for bombs, missiles, rockets, and fuel tanks; the F-100D/F had an additional centerline pylon. All models could carry a single nuclear bomb. The single-seat F-100A/C/D had four 20-mm cannon; the two-seat F-100F variant had two guns.

The F-100D/F could carry a "buddy store" to permit an F-100 to refuel another F-100 in flight. Seven F-100F aircraft were modified for the Wild Weasel anti-radar role in the Vietnam conflict.

The F-100 demonstrated zero-length launching on 7 June 1957 at Edwards AFB, California. It was boosted aloft by a 150,000-pound thrust Astrodyne rocket. The final F-100D production aircraft incorporated fittings for the zero-length launches from atomic bomb–resistant shelters.

The F-100B was to have been a fighter-bomber as well as a day/night air-superiority fighter; it was changed to YF-107A prior to first flight on 10 September 1956; that program was not pursued (see separate entry). The single two-seat TF-100C (unarmed) served as prototype for the two-seat F-100F. The latter aircraft had dual controls and full combat systems, developed as an advanced trainer because of the alarming rate of F-100 flying accidents. (Still, 74 F-100F variants—one-fourth of those produced—were lost in major accidents.)

Six F-100A aircraft were secretly converted to the photo-reconnaissance role. They were fitted with cameras, wing extensions (increasing their altitude to 53,000 feet), and additional internal and external fuel. They were not redesignated as recce aircraft (i.e., RF-101).

Operational: The first flight of YF-100A was on 25 May 1953; squadron introduction was in September 1954, although the wing did not achieve an IOC until September 1955. The delay was caused by accidents, structural failures, and performance problems.

The F-100C entered service in July 1955, F-100D in November 1955, and F-100F in January 1958. The F-100-D/F aircraft were flown extensively in the strike role in the Vietnam War.

Almost all F-100s were gone from the active Air Force inventory by mid-1972. At that time the Air National Guard had 550 F-100s; the ANG operated those aircraft for several years. Three other nations flew the F-100, and an effort was initiated to produce the aircraft in France but did not materialize.

The first YF-100A achieved a world record speed of 755 mph in 1953; two years later an F-100C set a record at 822 mph.

Production: North American delivered 2,249 F-100s between 1953 and 1959 for U.S. Air Force use.

Aircraft	Built	First flight
YF-100A	2	1953
F-100A	203	1953
F-100C	475	1954
TF-100C	1	1956
F-100D	1,274	1956
F-100F	294	1957

An additional 248 F-100Fs were built by North American Aviation for foreign air forces.

Operational:	Sep 1955–mid-1972 in nuclear strike role
Design:	North American Aviation

Aircraft data:	F-100D	
Crew:	1 in F-100A/C/D	
	2 (pilot, second pilot) in F-100F	
Engines:	1 Pratt & Whitney J57-P-21A turbojet; 16,000 lbst in afterburner	
Weights:	empty	20,638 lb
	max takeoff	38,048 lb
Dimensions:	length	49 ft 4 in
	wingspan	38 ft 10 in
	wing area	400 ft^2
	height	16 ft 2½ in
Speed:	maximum	920 mph at 35,000 ft
	cruise	585 mph
Ceiling:	48,800 ft	
Range:	radius	650 miles
Radar:	AN/ASG-17 fire control	
	AN/AJB-1B LABS	
Guns:	4 20-mm M-39E cannon (200 rounds per barrel)	
Ordnance:	F-100A/C	1 nuclear bomb Mk 7
	F-100D	1 nuclear bomb B28, B43, B57
	F-100F	2 nuclear bomb Mk 7, B28, B43, B57, B61
or	5,000 lb conventional bombs, missiles, rockets, except F-100D could carry 7,040 lb	

F-101 Voodoo

The F-101 was designed as a long-range bomber escort and penetration fighter for SAC. With the availability of high-speed, high-altitude jet bombers, escort fighters became unnecessary. Accordingly, the F-101A/C variants served primarily in the tactical nuclear strike role and the F-101B/F in the air defense role with nuclear missiles and rockets. The F-101A/C were nuclear-only aircraft, incapable of delivering conventional bombs.

Prototypes for the F-101 were the XF-88 and XF-88A, which first flew in 1948; however, funding shortages led to cancellation of the project in August 1950. The requirement was resurrected in January 1951 to provide an escort for the B-36 strategic bomber; several designs were considered, and an improved version of the F-88 was selected and redesignated F-101. But SAC cancelled its requirement for the aircraft in 1954 (before the first flight of the F-101).

The RF-101 photo-reconnaissance variant was flown extensively in the Cuban missile crisis of 1962 and in Vietnam from 1961 to 1965.

Design: The swept-wing F-101 had two turbojet engines mounted in the after fuselage, with air intakes in the midmounted wing roots. The wing was swept back 35°. The high tail assembly revealed a family similarity to the McDonnell F3H Demon and F4H Phantom.

The F-101B/F carried air-to-air missiles and rockets mounted on the rotary door of an internal weapons bay. In the interceptor role they normally carried two 450-gallon drop tanks mounted under the wing roots.

The F-101A/C could carry a nuclear bomb on a centerline pylon—Mk 7, B28, or B43; they were armed with four 20-mm cannon for "close-in" combat.

The F-101A/C were single-seat aircraft; the F-101B had a crew of two. The F-101F aircraft was similar to the F-101B with dual controls.

The F-101B carried nuclear Genie rockets in an internal weapons bay and conventional AIM-4 Falcon missiles externally.

The McDonnell F-101 Voodoo was designed to escort SAC bombers in penetrating advanced Soviet air defenses. However, that role was overtaken by USAF bomber performance, with the speed and range of turbojet bombers making escorts impractical. The F-101s served in several other roles, gaining prominence in the RF-101 photo-reconnaissance role during the Cuban missile crisis. This is a two-seat F-101B. (U.S. Air Force)

Designation: The F-101 was developed for the SAC escort role with the weapon system designation WS-105A and the RF-101 as the WS-105L. The air defense F-101B was WS-217A.

Operational: Design of an advanced penetration fighter began in 1945, leading to the F-88 and subsequent F-101. The prototype XF-88 first flew on 20 October 1948. The F-101A first flew in 1954 and entered service in May 1957 with the 27th Fighter Bomber Wing of the Strategic Air Command; the wing was transferred to the Tactical Air Command on 1 July 1957.

The F-101C entered service in September 1957 and the F-101B in January 1959 with the Air Defense Command. The B model was also flown by the Canadian Forces with the first of 56 CF-101B/F air-defense fighters entering Canadian service in November 1961. In 1970–1971 the 46 surviving CF-101 aircraft were traded to the U.S. Air Force for 56 refurbished and modernized Voodoos (Operation Peach Wings), the last of which was delivered to Canada in 1985.

The F-101 also served in the Taiwanese air force.

The F-101A/C aircraft ended first-line service with the U.S. Air Force in 1970 and the F-101B/F in 1971; they remained with Air National Guard until the mid-1980s.

An F-101C established a world speed record of 1,207 mph on 12 December 1957.

Production: McDonnell produced 805 F-101s following the two XF-88 prototypes.

Aircraft	Built	First flight
F-101A	77	1954
F-101B	387[a]	1957
F-101C	47	1957
F-101F	93	

a 58 of these aircraft were converted to F-101F with post-production kits; 66 of these aircraft were transferred to Canada.

In addition to fighter aircraft, 201 RF-101 photo-reconnaissance variants were built, with the first of two YRF-101A prototypes flying on 10 May 1956; these aircraft had a lengthened nose and were fitted with cameras for day and night photography. Another 29 F-101A and 32 F-101C fighters subsequently were converted to the reconnaissance configuration for use by the ANG.

The last F-101 aircraft were produced in March 1961.

Operational:	May 1957–1971	
Design:	McDonnell	
Aircraft data:	F-101B	
Crew:	2 (pilot, radar operator)	
Engines:	2 Pratt & Whitney J57-P-55 turbojet; 16,900 lbst each in afterburner	
Weights:	empty	28,492 lb
	combat	41,490 lb
	max takeoff	52,400 lb
Dimensions:	length	67 ft 6 in
	wingspan	39 ft 7 in
	wing area	368 ft^2
	height	18 ft
Speed:	maximum	1,095 mph at 35,000 ft
	cruise	545 mph
Ceiling:	51,000 ft	
Range:	1,520 miles	
	radius	690 miles
Radar:	MG-13 fire control	
Guns:	none	
Ordnance:	2 nuclear AIR-2A Genie air-to-air rockets	
	2 Falcon air-to-air missiles	

F-102 Delta Dagger

The F-102A Delta Dagger was a supersonic, all-weather fighter-interceptor. The Hughes AIM-26A Falcon nuclear-armed missile was designed specifically for the F-102A, which was the only U.S. fighter aircraft to carry an air-to-air nuclear missile. (The nuclear MB-1 Genie was an unguided rocket.) The F-102 was the world's first supersonic delta-wing aircraft to enter service with any air force.

This was the first U.S. Air Force fighter designed from the outset with an all-missile/rocket armament. From 1960 the AIM-26A nuclear Falcon was added to the F-102A's arsenal. The aircraft also carried the conventional AIM-4 Falcon missile and unguided air-to-air rockets.

Design: The F-102 was the first delta-wing aircraft based on the experimental XF-92A, which Convair built in 1948, to enter Air Force service. The aircraft had the cinched or "Coke-bottle" waist, the application of the "area rule" concept. It had no horizontal tail surfaces.

Up to six conventional Falcon missiles or two nuclear missiles or one nuclear and three conventional missiles could be carried in an internal weapons bay. Folding-fin 2.75-inch rockets were also carried in a fuselage pack. Aircraft modified to carry a mix of conventional/nuclear Falcons could carry only 12 air-to-air rockets.

Convair/General Dynamics F-102 Delta Dagger. (U.S. Air Force)

During weapon trials, on 8 July 1955, a YF-102A fired 6 Falcon missiles and 24 rockets in less than ten seconds.

Designation: The aircraft was developed under the designation WS 201A.

Operational: The first YF-102 flight was on 24 October 1953. Problems with the early aircraft (ten YF-102s were built) led to major design changes, and the first YF-102A flew on 20 December 1954.

The F-102A entered squadron service with the Air Defense Command in April 1956. At peak deployment 26 ADC squadrons were equipped with F-102s.

The last active F-102s were retired in 1973 with the ANG continuing to fly the F-102, albeit without their nuclear capability.

At the height of the Cuban missile crisis, on 27 October 1962, two F-102A fighters from Eielson AFB, Alaska, took off with nuclear Falcons to assist a U-2 spyplane that was being approached by Soviet MiGs as the spyplane strayed toward Soviet Siberia. There was no engagement.

F-102s were deployed to South Vietnam almost continuously from March 1962 until the end of 1969 to provide an air-defense capability should North Vietnamese or Chinese aircraft enter South Vietnamese air space. From May 1963 until August 1964 they were withdrawn to Clark field in the Philippines because of overcrowding at Tan Son Nhut airfield in South Vietnam. In addition F-102 detachments periodically operated from other South Vietnamese bases and from airfields in Thailand. There were no engagements with hostile aircraft.

Production: Convair built 1,000 F-102s—10 YF-102, 879 F-102A, and 111 TF-102A trainer variants.

Operational:	Apr 1956–1973 in nuclear intercept role	
Design:	Convair/General Dynamics	
Crew:	1	
	2 in TF-102A (pilot, student)	
Engines:	1 Pratt & Whitney J57-P-23A turbojet; 16,000 lbst in afterburner	
Weights:	empty	19,350 lb
	max takeoff	31,276 lb with 2 215-gallon drop tanks
Dimensions:	length	68 ft 4 in
	wingspan	38 ft 1½ in
	wing area	695 ft^2
	height	21 ft 2½ in
Speed:	maximum	825 mph
	cruise	600 mph
Ceiling:	service	54,000 ft
	combat	51,800 ft
Range:	1,350 miles	
	radius	400 miles
Radar:	MG-3 fire control	
Guns:	none	
Ordnance:	6 AIM-4 Falcon air-to-air missiles	
or	2 AIM-26 Falcon air-to-air missiles	
or	1 AIM-26 Falcon air-to-air missiles	
plus	24 2.75-in folding-fin air-to-air rockets	

F-104 Starfighter

The F-104 was a single-seat, daylight fighter-interceptor and fighter-attack aircraft. It represented a radical departure in aircraft design as Lockheed developed a relatively lightweight, simplified Mach 2.2. fighter.

The U.S. Air Force F-104C tactical strike version had a nuclear capability; later F-104 variants with a nuclear strike capability were flown by seven NATO air forces as

A prototype Lockheed XF-104 Starfighter shows the clean lines and stub wings of this high-performance, multirole fighter, developed as an alternative design to the more-complex Century-series fighters. Still, the aircraft had a brief career in U.S. markings while serving in several NATO air forces—with a nuclear strike capability in seven of their air forces. (U.S. Air Force)

well as the U.S. Air Force. The majority of the aircraft—almost 90 percent of those produced—went to allied air forces.

Design: The F-104 was derived, in part, from the Lockheed F-90 design, which flew in 1949. The F-104 featured a long, sleek fuselage with a circular cross section dominated by a pointed nose. There were "cheek" mounted engine inlets, well aft of the cockpit. The relatively short and extremely thin wings had slightly swept (18°6') leading edges and trailing edges. The horizontal tail surfaces were set high on the tail fin. The thin wings required the tricycle landing gear to retract into the fuselage. The F-104C carried its nuclear weapon in an internal weapons bay. Drop tanks and weapons were carried under the wings and fuselage, and wingtip rails for Sidewinder air-to-air missiles were an option. An internal rotary cannon was fitted.

The F-104B/D were two-seat trainers with limited armament. The F-104G was an all-weather fighter-bomber with a reinforced structure produced for foreign service by Lockheed and overseas firms (see Production section, below).

Operational: The first XF-104 flight was on 28 February 1954; first flight of a production F-104A was on 26 January 1958. The F-104A entered service as a fighter interceptor with Air Defense Command in September 1958.

From November 1965 to July 1967, U.S. F-104Cs were sent to Southeast Asia where they logged 2,269 combat sorties against targets in North and South Vietnam with conventional weapons.

Pakistani F-104s were in combat against Indian aircraft in the 1965 and 1971 conflicts.

Taiwanese F-104G aircraft engaged Chinese MiG-19s over the island of Quemoy in January 1967.

While the F-104 remained in the U.S. Air Force inventory for only a brief period, it constituted the backbone of NATO's tactical nuclear forces being flown by 15 other nations from the mid-1960s. The first time non-U.S. NATO F-104Gs apparently began to qualify to carry U.S. B28/B43/B57 nuclear bombs was in 1965. The seven NATO air forces with U.S. nuclear certification for the F-104 were the Belgian, Canadian, West German, Dutch, Greek, Italian, and Turkish.

The F-104 was the first aircraft to simultaneously hold world records for both absolute speed and altitude (1,404.19 mph and 91,249 feet), both accomplished in May 1958. On 14 December 1959 an F-104C reached a then-record altitude of 103,389 feet.

Production: The production of all models/variants of F-104s totaled 2,578 aircraft, of which 296 were for U.S. Air Force service (i.e., 11 percent). The U.S. Air Force procurement was

Aircraft	Built	First flight
XF-104	2	1954
YF-104[a]	15	1955
F-104A	165	1956
F-104B	26	1957
F-104C	77	1958
F-104D	21	1959

a Some official records list these aircraft as simply F-104A.

The Royal Canadian Air Force received 200 Canadair-built CF-104s and 38 Lockheed CF-104Ds to equip eight squadrons assigned to NATO, with two fighter wings in France and two in West Germany. But because of French policy that no foreign nuclear weapons could be placed on French soil, the Canadian squadrons subsequently transferred to West Germany. The CF-104 carried the B28/B43/B57, and the CF-104D the B57.

Other foreign orders consisted of 1,127 F-104G variants with 139 produced by Lockheed, 140 by Canadair, and the balance by several European firms. The Italians produced an upgraded variant known as the F-104S as an interceptor and as a fighter-bomber. In the latter role it could carry up to 7,500 pounds of bombs on nine external hardpoints. Fiat produced 206 aircraft for Italy and 40 for Turkey.

West Germany received 915 F-104s; The Netherlands received 120 F-104Gs; Greece initially received 35 F104-Gs, and later 19 additional F-104Gs were transferred from other air forces. Italy received 125 F/RF-104Gs, and Turkey initially received 32 F-104G aircraft, later supplemented by a large number of discarded F-104s from other NATO air forces.

Japan acquired 209 F-104G variants designated F-104J; 3 were built by Lockheed, 29 assembled by Mitsubishi with Lockheed-produced components, and 177 built by Mitsubishi. These aircraft had no strike capability but were used only as interceptors.

Operational:	Sep 1958–1967	
Design:	Lockheed	
Aircraft data:	F-104C	
Crew:	1	
	2 in F-104B/D (pilot, student)	
Engines:	1 General Electric J79-GE-7 turbojet; 15,800 lbst in afterburner	
Weight:	empty	12,760 lb
	max takeoff	27,853 lb
Dimensions:	length	54 ft 10 in
	wingspan	21 ft 11 in
	wing area	196.1[2]
	height	13 ft 6 in

Speed:	maximum	860 mph at sea level
		1,324 mph at 35,000 ft
	cruise	585 mph
Ceiling:	43,200 ft	
Range:	radius	490 miles air superiority
		600 miles long-range escort
Radar:	AN/ASG-14T-2 fire control	
Guns:	1 20-mm M61 Vulcan multibarrel cannon (725 rounds)	
Ordnance:	1 nuclear bomb Mk 28, Mk 43, B57	
or	2,000 lb conventional bombs, rockets	
or	4 Sidewinder air-to-air missiles	

F-105 Thunderchief

The F-105 Thunderchief was an all-weather, supersonic, long-range strike aircraft. Designed specifically as a tactical strike aircraft, the F-105 had an internal weapons bay for carrying nuclear weapons; the early models of the F-105 could not accommodate conventional weapons. The aircraft had no air-to-air capability despite its fighter designation.

Republic Aviation developed the aircraft as a company venture, proposing it to the Air Force as a Mach 1.5 successor to the F-84F Thunderstreak, which was not yet operational. The F-105 was planned to have an air-to-air as well as nuclear strike role. The Air Force approved development of the aircraft in 1952 with series production following.

Design: This was the world's largest and heaviest single-engine "fighter" aircraft. It was a large, midwing aircraft with the wings swept 45°. The aircraft featured internal and external payloads, a large combat radius, and

The Republic F-105 Thunderchief was a large, powerful nuclear strike aircraft that was used extensively in the Vietnam War in a conventional bombing role. The aircraft was nicknamed "Thud" for Thunderchief, although during the conflict it was said to be the sound that the aircraft—397 were shot down—made when they struck the ground. This is a bomb-laden F-105D variant. (U.S. Air Force)

the capability of speeds in excess of Mach 2 in level flight. (The F-105 bomb bay was larger than that of the four-engine B-17 Flying Fortress of World War II.)

When employed in the conventional bombing role in the Vietnam War, the bomb bay was fitted with fuel tanks and up to 12,000 pounds of external ordnance; additional fuel tanks were carried on five fuselage and wing pylons.

The F-105B could carry either three B28 or B43 nuclear bombs (one internally and two on wing pylons); the F-105D/F could carry a single bomb internally. The F-105D and later variants could carry air-to-surface missiles, beginning with the AGM-12C Bullpup-B nuclear weapon. F-105s could also be fitted with "buddy" refueling stores to permit one F-105 to refuel another in flight.

The F-105F/G variants had a second crewmen; these aircraft (as well as the F-105D) carried the Shrike or Standard anti-radar missiles in the Wild Weasel role. These configurations retained the internal bomb bay and could accommodate a nuclear bomb.

Specialized variants were the photo-reconnaissance (RF-105B) and Wild Weasel anti-radar aircraft (F-105F). Beginning in 1966, about 55 existing F-105D/F aircraft were modified to an improved F-105G Wild Weasel configuration.

An internal 20-mm multibarrel gun was fitted in all variants.

Designation: The F-105 was assigned the Air Force designation WS-306A during development.

The aircraft was generally referred to as the "Thud," short for Thunderchief. But during the Vietnam War, this was said to be the sound an F-105 made when it crashed. (F-105 losses were high during that conflict.)

Operational: The first YF-105A flight was on 22 October 1955. Squadron introduction of the F-105B was in 1958, F-105D in 1961, and F-105F in 1963. Despite the May 1958 service introduction, until April 1960 no aircraft were combat ready because of technical and maintenance problems. The F-105 continued to suffer mechanical problems through most of its career.

F-105s were used extensively in the Vietnam War as conventional bombing aircraft from 1964 to 1972; Wild Weasel–configured F-105s remained in combat until 1973.

Production: Republic Aviation produced 833 F-105s through January 1965:

Aircraft	Built	First flight
YF-105A	2	1955
YF-105B	10	1956
F-105B[a]	65	1956
TF-105B	3	1959
F-105D[a]	610	1959
F-105F[a]	143	1963

a Nuclear-capable aircraft

The large-scale procurement of the planned RF-105B reconnaissance variant was cancelled in July 1956 (with the prototypes designated TF-105B and then JF-105 as test aircraft); a two-seat F-105C combat proficiency trainer was cancelled in 1957, as was a two-seat F-105E in 1959.

Operational:	Aug 1958–Feb 1984	
Design:	Republic Aviation	
Aircraft data:	F-105D	
Crew:	1	
	2 in TF-105B (pilot, student)	
	2 in F-105F/G (pilot, weapon systems officer)	
Engines:	1 Pratt & Whitney J75-P-19W; 26,500 lbst with afterburner	
Weights:	empty	26,354 lb
	takeoff	50,511 lb
Dimensions:	length	64 ft 5 in
	wingspan	34 ft 11 in
	wing area	385 ft^2
	height	19 ft 8 in
Speed:	maximum	1,415 mph
	cruise	585 mph
Ceiling:	48,700 ft	
Range:	radius	890 miles
Radar:	AN/ASG-19 fire control	
Guns:	1 20-mm M61 Vulcan multibarrel cannon (1,028 rounds)	
Ordnance:	1 nuclear bomb, B43, B28	
or	3,000 lb conventional bombs, missiles, rockets	

F-106 Delta Dart

The General Dynamics/Convair F-106 was a single-seat, all-weather, supersonic interceptor flown by the Air Force and Air National Guard. The aircraft evolved from the same firm's F-102 Delta Dagger. It was in service in the air defense role for almost 30 years, from 1959 to 1988.

The F-106 was the last U.S. aircraft to carry the Genie nuclear air-to-air weapon.

Design: The F-106 had a low delta wing with a 60° leading-edge sweep. The needle-nose, area-ruled fuselage

Convair/General Dynamics F-106A Delta Dart. (Montana Air National Guard)

resembled that of the F-102 but was stretched to accommodate the more powerful J75 engine. The F-106B carried a second crewmen with a reduction in internal fuel (i.e., 64 gallons less in the fuselage fuel tank).

Two 358-gallon external tanks could be fitted on pylons below the wing leading edges; they could be retained when empty because they did not restrict speed or lower load factor limits.

The fire control system consisted of the Hughes MA-1 electronic fire-control system, which was linked digitally to the semiautomatic ground environment (SAGE) air defense network. The radar provided automatic search and track, interception, and firing of the missiles. Infrared search and track (IRST) was added later.

Missiles were carried in an internal weapons bay. In the 1970s a 20-mm rapid-fire cannon and ammunition were fitted in place of the Genie capability.

Three additional variants were proposed: the F-106C (two prototypes built), which would have had the JT4B-22 engine, a new fuselage, larger radar antenna, and other changes; the F-106D with additional improvements; and the F-106X, the basic F-106 aircraft with a larger radar antenna and modified fire-control system with a "look-down" capability.

Operational: The first F-106A flight was on 26 December 1956, YF-106B was on 9 April 1958, and F-106B was in October 1958. IOC for F-106A was in May 1959 with the Air Defense Command; F-106B was in July 1960. The last F-106 was retired from the Air National Guard in 1988.

In 1959 an F-106 set a world speed record of 1,525.9 mph.

Production: A total of 340 F-106s were produced: 273 F-106A, 2 YF-106B, 63 F-106B, and 2 YF-106C variants. The last aircraft were delivered in December 1960.

Operational:	Oct 1959–Aug 1988	
Design:	Convair/General Dynamics	
Aircraft data:	F-106A (with Genie deletion)	
Crew:	1 2 in F-106B (pilot, radar operator)	
Engines:	1 Pratt & Whitney J75-P-17 turbojet; 24,500 lbst with afterburner	
Weights:	empty	24,861 lb
	max takeoff	41,831 lb
Dimensions:	length	70 ft 8¾ in
	wingspan	38 ft 3½ in
	wing area	697.8 ft^2
	height	20 ft 3 in
Speed:	maximum	1,265 mph
	cruise	594 mph
Ceiling:	52,000 ft	
Range:	2,000 miles	
	radius	725 miles with drop tanks 417 miles with internal fuel only
Radar:	MA-1 fire control	
Guns:	1 20-mm M61A1 Vulcan multi-barrel cannon (625 rounds)	
Ordnance:	4 AIM-4 Falcon air-to-air missiles	

F-107

This was an advanced derivative of the Super Sabre series, originally designated F-100B. However, the North American design evolved into a dual-mission aircraft: all-weather interceptor and fighter-bomber. Maximum speed was to be Mach 2.25.

Design: The F-100B design effort began in 1953. As the design evolved, all commonality with the F-100 disappeared except for the basic wing and horizontal tail platforms. A bifurcated engine intake was mounted above and immediately behind the cockpit. The foreshortened fuselage had a needle nose. The flight control system used mechanically linked spoilers and slot-type deflectors in place of conventional ailerons. Also, both the horizontal and vertical tail surfaces were a single-piece slab type.

Armament was to consist of four 20-mm cannon in the fuselage with conventional and nuclear weapons carried semirecessed on the centerline. Six wing hardpoints for pylons were to be used for ordnance but not for external fuel tanks.

On 16 May 1954 work on the interceptor project was terminated in favor of emphasizing the fighter-bomber variant.

Operational: The first of four prototype F-107A aircraft first flew on 10 September 1956. The F-107 program was terminated in 1957 with the decision to proceed with the F-105 fitted with the J75 engine.

Production: Nine aircraft were ordered from North American but this was subsequently reduced to four prototypes. Only the first three aircraft were completed. (All three prototypes flew in 1956.)

Operational:	cancelled	
Design:	North American Aviation	
Aircraft data:	F-107A	
Crew:	1	
Engines:	1 Pratt & Whitney YJ75-P-9 turbojet; 24,500 lb thrust in afterburner	
Weights:	empty	22,696 lb
	loaded	39,755 lb
Dimensions:	length	61 ft 10 in
	wingspan	36 ft 7 in
	wing area	376 ft^2
	height	19 ft 8 in
Speed:	maximum	890 mph at sea level
	1,295 mph at 36,000 ft	
Ceiling:		
Range:	790 miles	
Radar:	fire control	
Guns:	4 20-mm M-39E cannon	
Ordnance:	1 nuclear bomb	
or	conventional bombs	

F-108 Rapier

The F-108 Rapier was initiated as a long-range interceptor and to serve as an escort for the B-70 Valkyrie strategic bomber. A letter contract for development of the aircraft was awarded in June 1957. The specifications called for a fighter with a combat speed of at least Mach 3 and maneuverable at 70,000 feet.

The project was cancelled on 23 September 1959, prior to the construction of any aircraft.

Design: The initial F-108 design greatly resembled the B-70 bomber. During the design process the forward canard (wings) were deleted, as were the upper portions of the wing-mounted vertical stabilizers. The wing was changed from a standard delta to a double-delta configuration with a sweep of about 65°.

The internal weapons bay was to hold two or more nuclear or conventional air-to-air missiles.

The following dimensions are based on the mock-up configuration.

Designation: The aircraft was developed under the long-range interceptor, experimental (LRIX) program.

Production: The Air Force in 1957 planned to acquire 480 F-108s.

Operational:	cancelled	
Design:	North American Aviation	
Aircraft data:	F-108	
Crew:	2 (pilot, weapon systems officer)	
Engines:	2 General Electric J93-GE-35 turbojet; 30,000 lbst each in afterburner	
Weights:	empty	
	max takeoff	approx 100,000 lb
Dimensions:	length	89 ft 2½ in (plus nose boom)
	wingspan	57 ft 5 in
	wing area	
	height	22 ft 1½ in
Speed:	maximum	Mach 3 (for 10 minutes)
	cruise	1,150 mph
Ceiling:	70,000 feet	
Range:		
Radar:	air intercept	
Guns:	4 20-mm cannon	
Ordnance:	several nuclear and conventional air-to-air missiles	

F-111 Aardvark

The F-111 was a U.S. effort to develop a multimission, multiservice aircraft. Although a conceptual failure, the F-111 served in U.S. Air Force operational squadrons for 28 years. The F-111 was the world's first variable-sweep ("swing") wing combat aircraft, with a maximum speed of Mach 2.5. All F-111 variants were nuclear capable. The aircraft also formed the basis for the FB-111 strategic bomber (see pages 106–8).

The design had been initiated by the Air Force to provide a successor to the F-105 Thunderchief as a tactical strike aircraft. Secretary of Defense McNamara ordered development of the TFX (tactical fighter, experimental) in 1962, directing that the existing Air Force proposal for a tactical strike aircraft to be developed for multiple mission use by the Air Force, Navy, and Marine Corps.

General Dynamics F-111A Aardvark. (U.S. Air Force)

The process to develop the F-111 became the most controversial U.S. aircraft development in history. Aviation historian Bill Gunston wrote:

> McNamara had himself conceived the concept of a major multi-service weapon system, and brusquely banged together the heads of generals and admirals to get it. With his personal identification with the TFX he had exceptional motivation for making the correct contract award, and the belief . . . that he chose capriciously is false. Yet his actions laid himself and his closest Pentagon colleagues open to the vicious attack they received. For a Defense Secretary personally to order the development of a major weapon system that the two user services had said was technically impossible was unprecedented. For the contract award of this system to require four rounds of evaluation, taking almost the whole of one year, was unprecedented. For the final choice to be not the proposal enjoying the almost unanimous support of the user services was unprecedented. For this surprising choice to be seemingly taken as a result of one or two casual conversations, and supported by *no written document* except a five-page memo was unprecedented.[11]

General Dynamics was named winner of the TFX competition (over the lower-cost Boeing proposal) on 24 November 1996, and two weeks later the Senate began an investigation into the TFX program. Nevertheless, the GD effort continued and the TFX was developed, becoming the F-111. The requirement for the close air support mission was dropped at an early date, and the Navy withdrew from the program in August 1968 because of the aircraft being severely overweight and considered unsuitable for carrier operation. In addition, the aircraft fell short of its supersonic dash requirement, was underpowered, and suffered other deficiencies.

It was procured by the U.S. Air Force as a tactical strike aircraft and, at Secretary McNamara's direction, as the FB-111 strategic bomber. The Air Force variant had no air-to-air capability.

The Australian government procured 24 F-111C aircraft (as Canberra replacements), but the planned British procurement of 46 F-111K and 4 TF-111K variants was cancelled in January 1968.

Design: The TFX/F-111 had shoulder-mounted, variable-geometry wings that sweep from a fully extended position (16°) for landing, takeoff, and cruise, to a "tucked-in" position (72½°) for high-speed flight. There was almost no gap between the wings and tail-planes at full sweep. The aircraft had side-by-side seating for the two-man crew, based on a Navy requirement (which also provided a wide nose to house a large radar antenna for the Navy's long-range intercept requirement).

The aircraft was designed to carry two nuclear weapons in an internal weapons bay; subsequently, wing pylons were added to carry nuclear or conventional weapons at subsonic speeds. Up to eight wing pylons could be fitted, four of which were retained when the wings are swept back. Heavy weapons were usually carried on the four inboard pylons, which could each accommodate six 1,000-pound bombs—that is, 24 bombs. Installing the rapid-fire gun displaced one of two internal positions for nuclear weapons (i.e., right-hand position).

Laser bomb guidance was fitted to many Air Force aircraft. (Australian F-111C aircraft can carry the Harpoon anti-ship missile.)

The aircraft had a night/all-weather capability.

Several reconnaissance variants were proposed, but only a single aircraft was built to that configuration (RF-111A). That aircraft was to have had a removable sensor pod. A more sophisticated RF-111D was planned but cancelled because of budget constraints. However, 42 F-111A aircraft were converted to the EF-111A Raven configuration

11 Bill Gunston, *Attack Aircraft of the West* (New York: Charles Scribner's Sons, 1974), pp. 183–84.

and fitted with Wild Weasel electronic countermeasures systems; they were delivered from 1981–1985. The EF-111A had no weapons capability.

Plans to arm surviving F-111Es with the nuclear AGM-131B short-range attack missile, tactical (SRAM-T) ended when that missile program was terminated in 1991.

Designation: The aircraft was changed from TFX to F-111 in December 1961. The F-111 development aircraft were F-111A and F-111B, with no X or Y prefixes being employed.

In 1989 the 51 surviving FB-111 strategic aircraft were modified for the dual (tactical/theater) strike role and were redesignated F-111G.

The name Aardvark was an unofficial name for the F-111 throughout its service career. On the day that the last F-111F was retired, the Air Force made the popular name official.

Operational: The first flight of F-111A was on 21 December 1964. First squadron delivery was on 28 April 1968 to the Tactical Air Command. The F-111E entered service in September 1969, the F-111F in January 1972, and the F-111D in September 1972.

The Royal Australian Air Force received 24 F-111C aircraft in 1973. Subsequently four U.S. F-111A aircraft were refitted as RF-111A reconnaissance aircraft for Australian service.

Forty-eight F-111A aircraft were used in bombing strikes against North Vietnam from a base in Thailand in 1968 and 1972–1973; these were single-plane missions flown at night. The 1968 deployment was aborted because of high losses; the planes were more successful in 1972–1973.

Sixteen F-111 aircraft and two EF-111 aircraft flew a strike against Libya from bases in England on 15 April 1986 (Navy carrier aircraft carried out simultaneous strikes against Libya). Eighteen F-111E and 66 F-111F aircraft carried out strikes against Iraq in the 1991 Gulf War, with 18 EF-111A aircraft also having a major role in that conflict.

During the Cold War F-111s were on nuclear quick-reaction alert (QRA) for sustained periods at two British air bases, RAF Upper Heyford and RAF Lakenheath.

The last F-111 strike aircraft was retired in 1996; the last EF-111 was retired in 1999.

Production: Planned procurement goals for the F-111 series changed continuously from 1961 to 1976, reaching a peak estimate in March 1964 of 2,411 aircraft—1,706 for the U.S. Air Force and 705 for the U.S. Navy plus 22 development aircraft.

Total F-111 tactical aircraft production for U.S. service was 462 aircraft, the last delivered in November 1976. These included 17 F-111A development aircraft for the Air Force, five F-111B development aircraft, and two production aircraft for the Navy. Two British F-111K aircraft that were almost completed became YF-111A in U.S. records; they never flew.

An additional 100 aircraft of this basic design were produced, 24 F-111C variants for Australia and 76 FB-111 aircraft as strategic bombers. Thus, the grand total for all F/FB-111 variants was 562 aircraft.

Aircraft	Built	First flight
F-111A[a]	156	1964
RF-111A	1	1967
YF-111A		—
(ex-F-111K)[b]	2	
F-111B	7	1965
F-111D[a]	96	1970
F-111E[a]	94	1969
F-111F[a]	106	1971

a Nuclear-capable aircraft.
b Never flown.

Operational:	April 1968–1996 in USAF strike role	
Design:	General Dynamics	
Aircraft data:	F-111A	
Crew:	2 (pilot, weapon systems officer)	
Engines:	2 Pratt & Whitney TF30-P-3 turbofan; 18,500 lbst each in afterburner	
Weights:	empty	46,172 lb
	combat	59,620 lb
	max takeoff	91,300 lb
Dimensions:	length	75 ft 6½ in
	wingspan	63 ft extended (16°); 32 ft swept back (72.5°)
	wing area	525 ft²
	height	17 ft
Speed:	maximum	1,451 mph at 56,800 ft
	cruise	745 mph
Ceiling:	56,200 ft	
Range:	radius	1,260 miles
Radar:	AN/APQ-110 terrain-following radar	
	AN/APQ-113 attack radar	

Guns:	1 20-mm M61A1 Vulcan multi-barrel cannon (2,050 rounds)
Ordnance:	4 nuclear bombs B43, B57, B61 external,
or	2 nuclear bombs B43, B57, B61 internal,
or	24,000 lb conventional weapons external

Lockheed F-117A Nighthawk. (U.S. Air Force)

F-117A Nighthawk

This aircraft was developed under a highly secret program—code-named Have Blue—that began in 1973 when the Department of Defense asked five manufacturers to offer a design for a stealth strike aircraft. In 1976 Lockheed proposed a faceted approach that led to the F-117A. The project was extremely successful in its ability to keep information about the plane secret. The F-117A had been flying for more than seven years before the official announcement of its existence was made on 10 November 1988, when a single photograph was released by the Pentagon. No further visual depictions were released until 21 April 1990, when there was a public unveiling at Nellis AFB, Nevada.

The production contract for the F-117A was awarded to Lockheed in 1978.

Design: The F-117A design employed flat, angled fuselage and wing panels to defect radar energy in a few sharply defined directions. The aircraft also used advanced surface materials and exhaust reduction features to attain stealth. The design presented several aerodynamic problems that were compensated for through the extensive use of computer-supported, quadruple redundant, fly-by-wire controls. The aircraft had two, nonafterburning engines that provided subsonic speeds.

Lockheed F-117A Nighthawk. (Lockheed)

It could carry one B61 nuclear bomb or 5,000 pounds of conventional (laser-guided) bombs in an internal weapons bay. Although called a "fighter" the F-117 was an attack aircraft; it had no air-to-air capability.

The pilot had a triangular front windshield and trapezoidal side windows.

Designation: No fighter-series designations in the bloc F-112 through F-116 are known to have been formally assigned.

Operational: The first F-117A flight was on 18 June 1981. The 37th Tactical Fighter Wing flew the F-117A.

The F-117 entered combat on 20 December 1989, during Operation Just Cause—the U.S. removal of Panamanian dictator Manuel Noriega. Two Nighthawks each dropped 2,000-pound laser-guided bombs that struck next to the Rio Hato barracks in Panama, stunning troops there and allowing US troops to overrun the base with minimal casualties to both sides.

Less than a year later, on 17 January 1991, ten F-117s helped to open Operation Desert Storm, the assault against Iraq following Saddam Hussein's invasion of Kuwait. Twelve F-117s in a second wave destroyed other high-priority targets—a total of 26 centers critical to Iraq's ability to mount a coordinated defense. No F-117s were fired on with most aircraft departing the target areas before defenses had time to react. Protected from radar and visual detection, the F-117s flew 40 percent of all Coalition air strikes against Iraqi targets in the first three nights of the conflict, although stealth fighters comprised less than 3 percent of the total air force. Because of the intense air defenses around the Iraqi capital of Baghdad, F-117s were the only U.S. aircraft to attack the central portions of the city. (U.S. Navy ship-launched Tomahawk missiles were the only other coalition weapons to strike targets in Baghdad.) In total, F-117s flew approximately 1,300 sorties and scored 1,600 direct hits before flight operations ended on February 28, 1991. Forty-two F-117s were committed to Operation Desert Storm.

The F-117 was again in combat in 1999 in support of NATO Operation Allied Force against the former Yugoslavia. Again, F-117s led the way, dropping bombs in the first air strike on 24 March 1999. F-117s continued to fly during the conflict, with the first Allied aircraft lost to enemy action occurring on the night of 27–28 March when an F-117 was shot down, apparently by an S-125 Neva (NATO SA-3 Goa) surface-to-air missile some 25 miles west of Belgrade. The pilot was rescued in the only combat loss of an F-117. (The only other Allied combat loss was a U.S. Air Force F-16 Falcon.) The Air Force deployed 24 F-117A aircraft to Aviano Air Base in Italy and Spangdahlem Air Base in Germany in support of Operation Allied Force.

Production: Under the Have Blue program Lockheed built two prototypes that were about 40 percent smaller than the F-117 and had tail surfaces that were canted inward. The first of these flew in December 1977. Under the code name Senior Trend, the program entered full-scale development in November 1977.

A total of 59 aircraft were procured with the first delivery in 1982; the 59th and final F-117A was delivered in July 1990.

The first 10 F-117s were retired in December 2006; the last of 55 surviving aircraft were retired in April 2008.

Operational:	Oct 1983–Apr 2008	
Design:	Lockheed	
Aircraft data:	F-117A	
Crew:	1	
Engines:	2 General Electric F-404 turbofan; 10,800 lbst each	
Weights:	empty	approx 30,000 lb
	max takeoff	52,500 lb
Dimensions:	length	65 ft 11 in
	wingspan	43 ft 4 in
	wing/body area 1,070 ft^2	
	height	12 ft 5 in
Speed:	maximum	645 mph at sea level
	cruise	625 mph at 35,000 ft
Ceiling:	45,000 ft	
Range:	radius	575 miles
Radar:		
Guns:	1 nuclear bomb B61	
or	5,000 lb conventional weapons	

F9F-8B Cougar (later AF-9J)

The F9F-6 Cougar derivative of the F9F Panther introduced the swept-wing configuration to operational Navy fighters. Used as a day fighter as well as light attack and photo-reconnaissance aircraft, the Cougar series reached the F9F-8 variant. Cougars were standard aircraft for the majority of the Navy's carrier-based fighter squadrons from the mid-1950s to 1960, and with reserve units for a few years longer.

Some F9F-8 aircraft were modified for tactical nuclear strike, being fitted with LABS and various arming and control devices. These F9F-8Bs were phased out of the fleet in 1959 and were replaced by A4D Skyhawks. Both the Navy and Marine Corps flew the F9F-8B. The F9F-8T two-seat trainer was also nuclear capable.

Design: The Panther/Cougar design provided for an attractive aircraft, with the Cougar series having a 35° wing sweep. In addition to a swept-wing design, the aircraft featured leading-edge slats, wing fences, and spoilers, which replaced ailerons.

The F9F-8 was lengthened eight inches from the straight-wing F9F-7 to provide for additional fuel. The F9F-8T two-seat training variant was 26 inches longer than the F9F-8, and the F9F-8P reconnaissance aircraft was approximately the same length as the F9F-8T. (The single-place F9F-8P was fitted with seven cameras in an elongated nose.)

The external stores capacity for the F9F-8B was one nuclear bomb and one 150-gallon fuel tank on wing pylons.

Cougar wings folded for carrier stowage. A nose boom was fitted for in-flight refueling.

A planned F9F-9 variant with the J65-W-6 engine was not produced.

Designation: The nuclear-capable aircraft were designated F9F-8B (changed to AF-9J in 1962). Some of these aircraft later were used as trainers and designated TAF-9J.

Operational: The first flight of the Cougar prototype was on 20 September 1951; first flight of F9F-8 was on

Grumman F9F-8B (AF-9J) Cougar. (U.S. Navy)

18 January 1954; first flight F9F-8T was on 29 February 1956. F9F-6 service introduction was in November 1951.

A small number of F9F-8T trainers were flown from land bases by the Argentine Navy.

Production: The total production of all Cougar variants was 1,985 plus three prototypes (converted F9F-5 Panthers), the last delivered in February 1960. Of those 601 were F9F-8s delivered from January 1954 to March 1957. The number completed to the F9F-8B configuration is not known. Four hundred F9F-8T trainers were produced.

Operational:	F9F-8	Oct 1954–1959
	F9F-8T	1957–1971 with nuclear capability
Design:	Grumman	
Aircraft data:	F9F-8	
Crew:	1	
Engines:	1 Pratt & Whitney J48-P-8A/8C turbojet; 7,250 lbst	
Weights:	empty	11,866 lb
	combat	17,345 lb
	max takeoff	24,763 lb
Dimensions:	length	41 ft 9 in
	wingspan	34 ft 6 in
	wing area	337 ft^2
	height	12 ft 3 in
Speed:	maximum	640 mph at 5,000 ft
	cruise	510 mph
Ceiling:	42,500 ft	
Range:	1,045 miles	
Radar:	AN/APG-30A fire control	
Guns:	4 20-mm M-3 cannon (190 rounds per barrel)	
Ordnance:	1 nuclear bomb Mk 7, Mk 12	
or	2,000 lb conventional weapons	

F2H Banshee (later F-2)

The F2H was the first U.S. Navy jet-propelled fighter to be produced in large numbers, developed slightly ahead of the competitive F9F Panther.

The F2H-2, which first saw action from the *Essex* (CV 9) in the Korean War in 1951, was slightly faster than the Panther, but its drawbacks were its size, having a wingspan a fraction under 45 feet (7 feet more than the Panther) and, at 40 feet long, almost 3 feet longer than the Panther. This led to modifications of elevators in the *Essex* class to accommodate the Banshee and other large aircraft.

McDonnell F2H (F-2) Banshee. (U.S. Navy)

Design: The Banshee was a low-wing, carrier-based fighter. Beginning with the F2H-2 the straight wing was fitted with jettisonable wingtip tanks that were usually retained except in an air-to-air combat situation; they each held 170 gallons.

The F2H-1 had no provisions for bombs or drop tanks.

McDonnell produced 25 F2H-2B aircraft configured for the nuclear strike role. Their wings were strengthened to carry a nuclear bomb beneath the left wing.

The F2H-3 was the basic F2H-2 provided with a night/all-weather capability plus provisions for a nuclear bomb. The later aircraft were slightly heavier and longer and were fitted with eight pylons for carrying weapons.

Thus, a total of 275 Banshees were configured for nuclear weapons delivery—25 F2H-2B and 250 F2H-3 aircraft.

Designation: The Banshee prototypes were changed from the original Navy designation XF2D-1 to XF2H-1 on 28 August 1947. The surviving F2H-3 aircraft—employed in utility roles—were changed to F-2C in 1962.

All surviving F2Hs were changed to F-2 in the redesignation of 1 October 1962.

Operational: The first XF2H-1 flight was on 11 January 1947. IOC for F2H-1 was in March 1949. The F2H-2 first flew on 18 August 1949, and entered squadron service in August 1951; the first flight of an F2H-2B took place in July 1952. The F2H-3 first flew on 29 March 1952, and entered service in mid-1952.

The F2H-2B aircraft were operational from 1951 to late 1958 and were assigned to several carriers while serving in detachments from Composite Squadrons 3 and 4. The F2H-3 was configured for nuclear weapons delivery from 1952 to late 1955 and served in fighter squadrons.

Operating from aircraft carriers, the F2H Banshee was in combat over Korea in the fighter role from 23 August 1951. The last F2H aircraft were retired from fleet squadrons in September 1959.

Beginning in late 1955, 39 F2H-3 aircraft were transferred to the Canadian Navy for carrier operations. The ex-*Powerful* was commissioned as HMCS *Bonaventure* in early 1957 and operated the aircraft from November 1960 until September 1962.

Production: Banshee production totaled 894 aircraft, the last an F2H-4 delivered in August 1953. The production of additional F2H-3 and F2H-3P (photo) aircraft was cancelled in favor of more advanced jet-propelled aircraft.

Operational:	F2H-2B	1951–1958 in nuclear strike role
	F2H-3	1952–1955 in nuclear strike role
Design:	McDonnell	
Aircraft data:	F2H-2B	
Crew:	1	
Engines:	2 Westinghouse J34-WE-34 turbojet; 3,250 lbst each	
Weights:	empty	11,268 lb
	combat	18,126 lb
	max takeoff	23,200 lb
Dimensions:	length	40 ft ½ in
	wingspan	41 ft 9 in
		45 ft with wingtip tanks
	wing area	294 ft^2
	height	14 ft 5½ in
Speed:	maximum	586 mph at sea level
	cruise	501 mph
Ceiling:	48,500 ft	
Range:	1,200 miles normal	
	1,475 miles maximum	
Radar:	F2H-2B	AN/APG-30 fire control
	F2H-3	AN/APQ-41 search
Guns:	4 20-mm M-3 cannon (150 rounds per barrel)	
Ordnance:	1 nuclear bomb Mk 7, Mk 8	
or	3,250 lb conventional bombs	

F3H-2N Demon (later F-3C)

The F3H Demon was developed as a platform for the radar-guided Sparrow air-to-air missile and its related search and guidance radar. The Demon differed in many ways from its McDonnell-produced predecessors—the FH Phantom and F2H Demon—most notably being a swept-wing, single-engine aircraft.

The F3H was a second-rate aircraft. Its immediate predecessor was the F2H Banshee, an excellent fighter that saw combat from carriers in Navy squadrons and was flown by Marines from shore bases during the Korean War. The subsequent McDonnell fighter was the F4H Phantom, unquestionably the West's best multipurpose fighter aircraft—land- and carrier-based—during the 1960s and 1970s.

Early F3Hs suffered from being underpowered. Many of the early aircraft never flew but were used for static ground training. Once a suitable engine became available, the F3H entered squadron service with the U.S. Navy. Naval aviator and writer Jerry O'Rourke, observing the aircraft's poor maneuverability and acceleration, wrote, "It was often heard in those days that 'Demon pilots were the only thinking fighter pilots.' In truth, Demon pilots who were not advanced planners and thinkers didn't last very long. A good effect of the F3H-2 was to make all naval aviators into better planners and thinkers."[12]

Design: The F3H was a large, attractive, swept-wing aircraft with a high-set tail structure, above the engine exhaust.

There were numerous accidents with the flying aircraft, some attributed to engine problems.

McDonnell F3H-2N (F-3C) Demon. (U.S. Navy)

12 Capt. Jerry O'Rourke, USN, "Fighters That Never Got to the Fight: Part II," U.S. Naval Institute *Proceedings* (April 1962), p. 79.

All F3H models had four 20-mm cannon, and the F3H-1N and -2/2N variants could carry four AIM-9 Sidewinder air-to-air missiles; the F3H-2M variant could carry four AIM-7 Sparrow radar-guided missiles. These were among the most potent fighter aircraft of the period.

The F3H-3 could carry 6,000 pounds of conventional bombs.

Some squadrons removed the two upper 20-mm guns, saving about 500 pounds for the guns and another 50 pounds for their ammunition. Removing all guns made the aircraft "nose light." The F3H-2N also could carry bombs in place of missiles, with eight wing and fuselage pylons available.

Fitted with the low-altitude bombing system (LABS).

The aircraft had folding wings for carrier stowage.

Designation: The surviving F3Hs were redesignated F-3 on 1 October 1962.

Operational: The first XF3H-1 flight was on 7 August 1951, F3H-1N first flight was on 24 December 1953, and F3H-2N first flight was on 23 April 1955. The F3H-1N did not enter service. The F3H-2N entered service in December 1955.

Production: With two XF3H-1 prototypes, McDonnell produced a total of 519 Demons including 167 nuclear-capable aircraft—27 F3H-1N and 140 F3H-2N models. Another 31 F3H-1N aircraft never entered service because of underpowered engines. Production ended in November 1959. An F3H-2P variant was proposed but not built. Similarly, an improved F3H-3 was ordered but none was built.

Operational:	Dec 1955–Aug 1964	
Design:	McDonnell	
Aircraft data:	F3H-2N	
Crew:	1	
Engines:	1 Allison J71-A-2 turbojet; 14,500 lbst	
Weights:	empty	20,645 lb
	max takeoff	35,378 lb
Dimensions:	length	59 ft
	wingspan	35 ft 4 in
	wing area	519 ft^2
	height	14 ft 7 in
Speed:	maximum	715 mph at sea level
		645 mph at 35,000 ft
	cruise	555 mph
Ceiling:	39,000 ft	
Range:	1,020 miles with nuclear bomb	
Radar:	AN/APG-51A fire control	
Guns:	4 20-mm Mk 12 cannon (180 rounds per barrel)	
Ordnance:	1 nuclear bomb Mk 7, Mk 12	
or	4 Sidewinder air-to-air missiles	

North American FJ-4 (AF-1E) Fury. (U.S. Navy)

FJ-4B Fury (later AF-1E)

The FJ Fury was one of three all-jet carrier fighters ordered by the U.S. Navy during World War II (the others were the McDonnell FD Phantom—later redesignated FH—and the Chance-Vought F6U Pirate). The North American XFJ-1 Fury was ordered in January 1945, the first carrier-based fighter produced by that firm. The FJ-2 and later variants were swept-wing aircraft.

The FJ series had a considerable external store capability with the FJ-4B variant configured to carry nuclear weapons. This aircraft was widely flown by U.S. Navy and Marine Corps squadrons in the conventional and nuclear attack roles.

The FJ Fury was the progenitor of the F-86 Sabre fighter.

Design: The initial, straight-wing FJ-1 series was followed by the swept-wing FJ-2, which was identical to the U.S. Air Force F-86E Sabre with the naval aircraft having carrier launch/arresting hook features (although the F-86E had .50-caliber machine guns in contrast to the 20-mm cannon in the naval aircraft).[13] The improved FJ-3 and FJ-4 aircraft followed.

13 The FJ-1 Fury did mount six .50-caliber machine guns. That aircraft could not carry external stores.

The FJ-4 Fury featured six wing store points, additional armor, the LABS, and an improved control system for high-speed flight at low altitude.

The nuclear-capable FJ-4B had a maximum external payload of 6,000 pounds on six wing pylons. It could lift up to five 540-pound Bullpup air-to-surface missiles or up to four Sidewinder air-to-air missiles or a nuclear weapon.[14]

Designation: The FJ series was changed to F-1 in the 1 October 1962 redesignation of U.S. military aircraft; the nuclear-capable FJ-4B became the AF-1E.

Two FJ-4B aircraft experimentally fitted with rocket motors became FJ-4F.

Operational: The first XFJ-1 flight was on 11 September 1946, the first swept-wing XFJ-2 flight was on 14 February 1952, and the first FJ-4B flight was on 4 December 1956.

The swept-wing FJ-2 entered Marine Corps service in January 1954 (with all 200 FJ-2s being flown by Marine units). The FJ-4B entered service with Marine and Navy squadrons in June 1957. (All ten navy squadrons and three Marine squadrons that flew the FJ-4B variant served in the Pacific Fleet.)

Production: North American built 1,148 FJ-series aircraft, of which 1,112 were swept-wing FJ-2/3/4 variants. Of those, 222 were FJ-4B variants. Production ended in May 1958.

Operational:	June 1957–Sep 1962	
Design:	North American	
Aircraft data:	FJ-4B	
Crew:	1	
Engines:	1 Wright J65-W-16A turbojet; 7,700 lbst	
Weights:	empty	13,778 lb
	max takeoff	28,000 lb
Dimensions:	length	36 ft 4 in
	wingspan	39 ft 1 in
	wing area	338.66 ft^2
	height	13 ft 11 in
Speed:		max 655 mph at sea level with external stores
		635 mph at 10,000 feet with external stores
	cruise	520 mph
Range:	1,940 miles with nuclear bomb, 1 200-gallon tank, 2 150-gallon tanks	
Ceiling:	40,400 ft	
Radar:	AN/APG-30A fire control	
Guns:	4 20-mm Mk 12 cannon (144 rounds per barrel)	
Ordnance:	1 nuclear bomb Mk 7, Mk 12, Mk 28	
or	6,000 lb missiles, conventional bombs	

F7U-3 Cutlass

The Chance-Vought F7U Cutlass was one of the most aerodynamically advanced aircraft of its time. A swept-wing, tailless fighter, its career as a U.S. Navy carrier-based fighter was brief but dramatic. The F7U-3 variant was nuclear capable.

The F7U was underpowered and suffered from a myriad of mechanical problems, especially with its hydraulic system. From mid-1952 to mid-1956 there were 78 mishaps and accidents involving the Cutlass, of which more than one-fourth were fatal. Aviation historian Lindsay Peacock observed that "the Cutlass was an inherently more dangerous beast than any of its contemporaries."[15] The F7U was withdrawn from fleet squadrons in November 1957, only two years after the last aircraft delivery.

Although its service career was short and it never saw combat, the F7U Cutlass was an important addition to the Navy's arsenal, especially in the F7U-3M missile version. The F7U must be counted as a pioneer in the development of advanced post–World War II aircraft designs.

Design: Development of the Cutlass began in 1945 as Allied intelligence revealed German advances in high-performance, tailless aircraft. This intelligence led the U.S. Navy to sponsor several advanced carrier-based fighters. Chance-Vought's recent fighter experience included the highly successful F4U Corsair, the radical F5U "flying pancake," and the F6U Pirate, one of the Navy's first turbojet aircraft and the first to be fitted with an afterburner (i.e., the auxiliary combustion chamber to provide extra thrust for limited periods).

The firm's advanced F7U design promised high speed and a rapid rate of climb. The aircraft had a swept-wing (38°) without horizontal tail surfaces and with fins and

14 When Bullpup missiles were carried, one pylon was used for the missile guidance pod.

15 Lindsay Peacock, "Vought's Chancy Cutlass," *Air International* (November 1992), p. 28.

Chance-Vought F7U-3 Cutlass. (U.S. Navy)

rudders located on the wings; pitch and roll control were combined in elevons on the wing. The outer wing panels folded upward for carrier stowage.

The first XF7U-1 flew in 1948, but the plane was underpowered and the production of improved engines lagged. The plane's radical design also encountered previously unknown aerodynamic problems (which pioneered technology for later aircraft of more advanced design).

The definitive F7U-3 was powered by two improved J46 turbojet engines and the plane could exceed Mach 1. The engines provided for single-engine flight for the large fighter, even in the critical period after a wave-off during a carrier approach, with the working engine in afterburner. In air-to-air combat the F7U-3 also could outperform its contemporaries—the FJ-2 Fury and F9F-6 Cougar.

All F7Us were single-seat aircraft and the standard armament was four 20-millimeter cannon (deleted from the F7U-3P). Only the F7U-3 variant had provisions for external weapons.

The F7U-3M could carry four Sparrow air-to-air missiles. All F7U-3s could carry up to 7,000 pounds of conventional bombs and drop tanks on four wing pylons. A rocket package with 32 2.75-inch folding-fin rockets or a 220-gallon fuel tank could be carried on the fuselage centerline.

Operational: The first flight of XF7U-1 was on 29 September 1948. The 20 F7U-1s did not enter operational service but were flown by the naval training establishment. The first flight of the F7U-3 was on 20 December 1951, and IOC was in January 1953.

Production: Cutlass production totaled only 307 aircraft, of which 290 were the definitive F7U-3 variants; of those, 180 were nuclear-capable F7U-3s, 98 were F7U-3M missile carriers, and 12 were F7U-3P photo planes. No F7U-2 variants were produced. There were three X7U-1 prototypes and 20 F7U-1 aircraft.

Operational:	Jan 1953–Nov 1957	
Design:	Chance-Vought (United Aircraft)	
Aircraft data:	F7U-3	
Crew:	1	
Engines:	2 Westinghouse J46-WE-8B turbojet; 5,725 lbst each in afterburner	
Weights:	empty	18,210 lb
	max takeoff	approx 35,000 lb
Dimensions:	length	44 ft 3 in
	wingspan	39 ft 9 in
	wing area	535 ft²
	height	14 ft 7 in
Speed:	maximum	677 mph at sea level
	593 mph at 15,000 ft	
	cruise	517 mph
Ceiling:	36,100 ft	
Range:	885 miles with 2 150-gallon drop tanks	
	600 miles with Mk 12 nuclear bomb	
Radar:	AN/APG-30 fire control	
Guns:	4 20-mm cannon (180 rounds per barrel)	
Ordnance:	1 nuclear bomb Mk 7, Mk 8, Mk 12	
or	7,000 lb conventional bombs	

Lockheed P-7A LRAACA. (Lockheed)

MARITIME PATROL/ASW AIRCRAFT

The P6M Seamaster flying boat was a strategic attack/mining/reconnaissance aircraft; it is described in Chapter 3 of this volume.

Aircraft[a]		Service[b]	IOC	Type
P-7	LRAACA	N	cancelled	land
P5M P-5	Marlin	N	1952	seaplane
P2V-5F P-2E	Neptune	N	1957	land
P3V P-3	Orion	N	1962	land
S-3	Viking	N	1974	carrier
S2F S-2	Tracker	N	1954	carrier

a Second column indicates 1962 redesignation
b N = Navy

P-2 Neptune (see P2V Neptune)

P-3 Orion (see P3V Orion)

P-5 Marlin (see P5M Marlin)

P-7A LRAACA

The P-7 long-range air anti-submarine warfare capable aircraft (LRAACA) was intended as a replacement for the P-3 Orion in the maritime patrol/ASW roles. Development and procurement of the P-7 was cancelled by the Navy on 20 July 1990 because the contractor, Lockheed, "failed to make adequate progress toward completion of all contract phases, which were to have resulted in the delivery of two prototype aircraft in April and December 1992."

The aircraft was to have a nuclear depth bomb capability. This feature is not being provided in the subsequent, post–Cold War P-8 Poseidon maritime patrol aircraft.

In the mid-1980s the Navy held a competition for the P-3 replacement between modifications of the Lockheed Boeing 757 and McDonnell Douglas MD-90.[16] The modified P-3 design submitted by Lockheed Aeronautical Systems of Burbank, California, was selected in October 1988 on the basis of acquisition and life-cycle costs, according to the Navy. Lockheed subsequently experienced major cost problems with the P-7 program as well as technical difficulties. In March 1989—shortly before cancellation—the estimated acquisition cost for 125 P-7A aircraft was about $7.9 billion with an estimated production cost of $56.7 million per aircraft. The original cost estimate for the program was $3.5 billion.

Design: The LRAACA bore a strong resemblance to the P-3 Orion; there was to be a large internal weapons bay plus 12 wing hardpoints for weapons. The four GE-38 engines were estimated to have a 25 percent lower fuel consumption over the aircraft operating envelope compared with the P-3's T56 engines. The improved performance was due in part to five-blade composite propellers.

Production: Initial Navy planning was for 125 aircraft plus two prototypes, but more aircraft were anticipated in addition to eventual foreign orders. The German government in April 1988 announced that it planned to select the P-7 to replace its Dassault-Breguet Atlantic ASW aircraft for long-range maritime patrol duties.

Operational:	cancelled	
Design:	Lockheed	
Aircraft data:	P-7	
Crew:	10	
Engines:	4 General electric GE-38 turboprop; 5,150 shp each	
Weights:	empty	105,000 lb
	max takeoff	171,350 lb
Dimensions:	length	112 ft 8 in
	wingspan	106 ft 7 in
	wing area	1,438 ft^2
	height	32 ft 11 in
Speed:	maximum	
	cruise	
Ceiling:		
Range:	radius	1,840 miles with 6 hours on station
Radar:	search	
Guns:	none	
Ordnance:	3 nuclear depth bombs B57	
or	23,660 lb conventional bombs, torpedoes (internal)	
or	Harpoon anti-ship missiles (external)	

P5M Marlin (later P-5)

The Martin P5M Marlin was the last operational flying boat patrol aircraft to be flown by the U.S. Navy and the last flying boat of any nation to participate in a major conflict. P5Ms were employed extensively during the Vietnam

16 Boeing proposed a modified version of the 757 twin-engine commercial transport, and McDonnell Douglas a modified MD-87 twin-engine transport, the latter with ultrahigh-bypass-ratio engines. Gulfstream Aerospace had also proposed a twin-engine, wide-body Gulfstream 4 for the LRAACA program, but it was not accepted as a viable competitor.

War in Operation Market Time, the U.S. effort to interdict arms being brought into South Vietnam by sea.

The P5M evolved from the Martin PBM Mariner, which was flown in large numbers by the U.S. Navy during World War II. The single P5M prototype was ordered as a standard PBM-5; it was converted on the production line to an extensively modified configuration and flew in 1948 as the XP5M-1.

Design: The P5M had a distinctive gull-wing that mounted twin piston engines and fixed stabilizing floats. The major changes of the P5M over the PBM were improved engines, a lengthened and narrower fuselage with a full-length planing hull, and a tall, single tail fin. Internal bomb bays were fitted in the two engine nacelles to accommodate a total of 12,000 pounds of bombs, mines, or two anti-submarine torpedoes, or nuclear depth bombs—the Mk 90 Betty and, later, the Mk 101 Lulu. In addition, eight 500-pound rockets could be carried under the wings.

The prototype XP5M-1 mounted six 20-mm cannon in twin remote-controlled, radar-directed turrets in the nose, dorsal, and tail positions. In the P5M-1 the nose guns were replaced by a large AN/APS-44 search radar, and in 80 aircraft the tail guns were deleted during an upgrade to provide an AN/ASQ-8 magnetic anomaly detection (MAD) antenna. The dorsal turret subsequently was removed, deleting all defensive armament. The aircraft also were fitted to drop and monitor sonobuoys.

The P5M-2 introduced an improved hull, more powerful engines, and a T-tail configuration with the AN/ASQ-8 MAD antenna fitted in a pod atop the vertical stabilizer. The tail 20-mm cannon initially were retained in the P5M-2 but were deleted in the 1960s to provide space for additional ASW equipment.

A single SP-5B was fitted with a Pratt & Whitney J60 turbojet engine producing 3,000 pounds static thrust. The 1964 installation provided significant improvements in performance but was not pursued because of the short remaining service life of the surviving flying boats in the U.S. Navy. A four-engine P5M-3 was developed; however, it did not progress past the mock-up stage.

Designation: The original U.S. Navy P5M designation was modified by the suffix *S* as upgraded anti-submarine equipment was provided, that is, P5M-1S and P5M-2S. These became SP-5A and SP-5B, respectively, in the 1 October 1962 redesignation of U.S. military aircraft.

Operational: The first flight of the single XP5M-1 was on 30 May 1948; first flight of the P5M-1 was on 21 June 1951. The production P5M-1 entered service in April 1952.

During Operation Market Time, the flying boats operated from seaplane tenders off the coast of South Vietnam and from shore bases in the Philippines from February 1965 to May 1967.

An SP-5B aircraft from Patrol Squadron 40 made the last U.S operational flight of a flying boat on 6 November 1967. This marked the end of flying boat operations by the U.S. Navy.

Production: A total of 239 aircraft were produced for the U.S. Navy:

Aircraft	Built	First flight
XP5M-1	1	1948
P5M-1	107	1951
P5M-2	131	1953

In addition, the U.S. Coast Guard took delivery of 7 P5M-1G and 4 P5M-2G aircraft for search-and-rescue

The Martin P5M Marlin was the U.S. Navy's last flying boat to see combat service, being used in offshore patrol during the Vietnam War. These aircraft—like this P5M-2S (SP-5B)—carried bombs, torpedoes, or nuclear depth bombs in their large engine nacelles. The design originally provided a heavy defensive gun armament, discarded as its principal target became Soviet submarines. (U.S. Navy)

(unarmed), and the French Navy took delivery of 10 P5M-2 aircraft. (The Coast Guard P5M-1G Marlins later were employed by the Navy as trainers and changed to P5M-1T.)

The last U.S. Navy P5M-2S was delivered in December 1960.

Operational:	Apr 1952–Nov 1967	
Design:	Martin	
Aircraft data:	P5M-2S	
Crew:	12	
Engines:	2 General Electric J47-GE-25 turbojet; 7,200 lbst each	
Weights:	empty	49,218 lb
	max takeoff	78,000 lb
Dimensions:	length	100 ft 2 in
	wingspan	118 ft 2 in
	wing area	1,406 ft^2
	height	30 ft 11 in
Speed:	maximum	270 mph at 17,300 ft 250 mph at sea level
	cruise	160 mph
Ceiling:	21,900 ft	
Range:	2,470 miles with 2 torpedoes, 8 5-inch rockets	
Radar:	AN/APS-80 search	
Guns:	deleted	
Ordnance:	2 nuclear depth charges Mk 90 Betty or Mk 101 Lulu	
or	8,000 lb conventional bombs, mines, ASW torpedoes (internal)	
or	4,000 lb conventional bombs, rockets (wing pylons)	

P2V-5F Neptune (later P-2E)

The P2V Neptune, which first entered U.S. Navy squadrons in 1947, was the most capable maritime patrol/antisubmarine aircraft of its era. Although produced in large numbers, only 12 aircraft were converted for carrier takeoff—P2V-3C variants—and the production P2V-5F variants were configured for carrying nuclear weapons. (The P2V-3C aircraft is described on pages 108–10)

The U.S. Air Force flew nine Neptunes with the designation RB-69A in the ELINT role for the Central Intelligence Agency. Delivered in 1955–1957, these were given the Lockheed designation P2V-7U. During the Vietnam War the U.S. Navy also flew the AP-2H and OP-2E variants in the ground sensor monitoring and night interdiction roles, respectively.

Neptunes were also flown by nine allied navies and air forces. Twenty-five P2V-7s delivered to the Royal Canadian Air Force carried U.S. nuclear depth bombs (Mk 101 Lulu), probably from February 1968 to June 1970. The Netherlands Navy acquired 12 P2V-5s beginning in 1967 that eventually carried the B57 nuclear depth bomb; they were stored at naval aviation weapons facilities at Machrihanish and St. Mawgan, United Kingdom, under U.S. custody.

Design: The P2V was the U.S. Navy's only land-based maritime patrol aircraft to be designed from the outset for that role, all others being adaptations of bomber or commercial aircraft.

The aircraft had a conventional straight-wing configuration, twin radial engines, and a tall tail fin. Early variants had a solid nose, subsequently replaced in different variants with a glazed bombardier nose, solid nose, six fixed 20-mm cannon in the nose, or two 20-mm cannon in a

This Lockheed P2V-5 (later P-2E) Neptune, showing the lines of highly successful aircraft, was the first land-based plane produced specifically for the maritime patrol/ASW role. Note the tail "stinger" MAD antenna, the turbojet pods outboard of the piston engines, and the searchlight fitted in the right wingtip fuel tank. Neptunes were configured for many specialized roles, some flown by the Air Force and CIA (RB-69A). (U.S. Navy)

powered nose turret. Other armament modifications included a twin 20-mm dorsal turret, and a tail turret with twin .50-caliber machine guns or two 20-mm cannon (with six 20-mm guns in twin turrets being the maximum armament).

The P2V-5 model initially differed primarily from earlier aircraft in having a 20-mm power-operated nose turret; during production, improved wingtip tanks were provided, with radar and searchlights as well as fuel cells fitted in the tanks. The nose turret was subsequently deleted in favor of an observer position, and the 20-mm tail turret was replaced by a MAD "stinger." Later the .50-caliber dorsal gun turret was also deleted and two turbojet engines were fitted under the wings for improved takeoff performance and improved "dash-over-target" performance.

Weapons were carried primarily in the internal bomb bay; eight 5-inch rockets could be carried on eight wing pylons (maximum capacity 500 pounds each). For normal ASW patrol missions, the aircraft would carry two homing torpedoes plus rockets, the latter for attacking surfaced submarines.

Most Neptune variants had 400-gallon wingtip tanks.

Some variants were optimized for aerial minelaying with the P2V-6M (later MP-2F) fitted to carry the Petrel air-to-surface missile. The P2V-3W and later aircraft had the large AN/APS-20 search radar. The P2V-3Z was an armored VIP transport. The small number of P2V-7LP (LP-2J) aircraft were fitted for Arctic photo reconnaissance.

Lockheed proposed a lighter variant of the Neptune, designated P2V-8, but it was not pursued because of the advent of the P3V Orion.

Designation: The surviving P2Vs were redesignated in the P-2 series on 1 October 1962.

Operational: The first XP2V-1 flight was on 15 May 1945. First flight of the definitive P2V-5 was on 29 December 1950. The P2V-5 entered squadron service in 1957.

The last U.S. Navy Neptune retired from the maritime patrol role was an SP-2H that last flew in February 1970. The last Navy Neptune was a DP-2H drone-control aircraft, retired on 7 July 1980.

A P2V-1 named *Truculent Turtle* established an unrefueled distance record of 11,235.6 miles on 29 September–1 October 1946, flying from Perth, Australia, to Ohio. The record stood for 16 years when it was broken on 11 January 1962 by a B-52H flying from Kadena, Okinawa, to Torrejon, Spain, a distance of 12,532.28 miles.

Production: A total of 1,051 aircraft were built by Lockheed including two XP2V-1 prototypes. More P2V-5s were produced than any other variant, with 424 aircraft delivered from September 1949 to March 1951. The last P2V-7 was delivered in April 1952. In addition, Kawasaki in Japan produced 130 P-2J aircraft (in addition to 16 P2V-7s built by Lockheed for Japanese service).

Operational:	1957–1970	
Design:	Lockheed	
Aircraft data:	P2V-5F	
Crew:	9 or 10	
Engines:	2 Wright R-3350-36 piston; 3,250 hp each	
	2 Westinghouse J34-WE-34 turbojet; 3,250 lbst each	
Weights:	empty	48,500 lb
	max takeoff	80,000 lb
Dimensions:	length	89 ft 4 in
	wingspan	101 ft 4 in
	wing area	1,000 ft^2
	height	28 ft 1 in
Speed:	max	400 mph at 19,000 ft
		370 mph at sea level
	cruise	185 mph
Ceiling:	23,000 ft	
Range:	2,900 miles with two ASW torpedoes	
Radar:	AN/APS-20B search	
Guns:	deleted	
Ordnance:	1 nuclear depth bomb B57, Mk 90 Betty, Mk 101 Lulu	
or	12,000 lb conventional bombs, torpedoes, rockets	

P3V Orion (later P-3)

The Orion is a long-range maritime reconnaissance and ASW aircraft with the Harpoon anti-ship missile providing a surface attack capability. The aircraft replaced the P2V/P-2 Neptune and P5M/P-5 Marlin in U.S. Navy patrol squadrons.

The U.S. aircraft could carry nuclear depth bombs before those weapons were removed from the U.S. arsenal in 1992.

Several specialized electronic reconnaissance variants of the aircraft have been produced with the EP-3E currently in U.S. Navy service.

The Orion has been extensively updated with improved ASW equipment and non-ASW reconnaissance and communications equipment. The latter support the aircraft's

A Lockheed P-3B Orion shows the fine lines of the aircraft, adapted from the commercial Electra turbo-prop airliner. The weapons bay is fitted forward of the wing; the after fuselage has sonobuoy chutes; a MAD antenna is fitted in the tail; and a searchlight is fitted under the right wing. Fifteen other navies and air forces have flown the Orion, which will remain in U.S. naval service until at least 2015. (U.S. Navy)

over-land operations that have been prevalent since U.S. combat operations in Afghanistan and Iraq began in 2001.

Design: The Orion was adopted from the commercial Electra transport with a lengthened fuselage and other design modifications. It is powered by four turboprop engines. Up to 15,000 pounds of rockets, missiles, mines, ASW torpedoes, or nuclear depth bombs can be carried in the internal weapons bay and on ten wing pylons. The P-3C aircraft are fitted to carry the Harpoon. (In the early 1970s some P-3B variants were fitted to carry the AGM-12 Bullpup missile.) The plane's ASW equipment includes radar, tail-mounted ASQ-81 MAD, and 48 external (fuselage) sonobuoy chutes plus four in-flight reloadable (internal) chutes; a total of 84 buoys are normally carried.

The P-3C variants have undergone a series of modernizations: Update I of the mid-1970s included a computer upgrade, Omega navigation system, additional tactical displays, and a new tactical program for computer-aided analysis of incoming data. Update II introduced in 1977 had additional navigation capabilities and provision for the Harpoon anti-ship missile. The latest improvement series is Update III, introduced in 1984; this provides an IBM Proteus signal processor system and new avionics. An Update IV implemented in the 1990s primarily provided improvements in signal processing; the AN/APS-137(V) surveillance radar was also fitted.

Plans to provide an in-flight refueling capability have not been pursued.

A planned P-3G maritime patrol/ASW aircraft for the U.S. Navy was to have had new engines and updated avionics. That aircraft was incorrectly identified as a P-3F in some official documents (six P-3F variants were produced for Iran). The procurement of 125 P-3G aircraft was envisioned; however, the Navy did not pursue the P-3G option because only Lockheed responded to the Navy's request for a proposal, and the P-7 LRAACA was planned in its place. Proposals were also being considered to re-engine and upgrade the avionics of the P-3C force, with a possible enlargement of the weapons bay to carry extended-range Harpoon missiles internally; this upgrade would have been labeled P-3H.

Former P-3 ASW aircraft fly as VIP transports and in various Navy and non-Navy research roles.

Designation: The Orion was originally designated P3V; it was changed to P-3 series on 1 October 1962.

Operational: The first flight (aerodynamic airframe) was on 19 August 1958, YP-3A on 25 November 1959, P-3A on 15 April 1961, and YP-3C on 18 September 1968. P-3A IOC was in August 1962, P-3B IOC in January 1966, and P-3C IOC in July 1970.

Orions—new and ex–U.S. Navy—have been flown by 15 other countries: The Netherlands Navy flew 13 NP-3C aircraft that could employ U.S. B57 nuclear depth bombs stored at Machrihanish, United Kingdom. The first Dutch Orions were delivered in November 1981; the nuclear mission ended in 1992.

Production: Lockheed delivered a total of 551 Orions to the U.S. Navy. The last U.S. Navy Orion to be produced was a P-3C delivered on 17 April 1990.

More than 700 Orions were produced in the United States by Lockheed; the last were eight P-3C aircraft for South Korea delivered in 1995. Kawasaki in Japan assembled five airframes produced by Lockheed and built more than 100 P-3C, EP-3, UP-3C, and UP-3D aircraft for Japanese service.

Operational:	1962–1992 with nuclear capability	
Design:	Lockheed	
Aircraft data:	P-3C	
Crew:	10 (command pilot, 2 pilots, flight engineer, navigator/communications officer, tactical coordinator, 3 systems operators; 1 technician; provision for 2 additional observers)	
Engines:	4 Allison T56-A-14 turboprop; 4,910 hp each	
Weights:	empty	61,491 lb
	max takeoff	142,000 lb
Dimensions:	length	116 ft 10 in
	wingspan	99 ft 8 in
	wing area	1,300 ft^2
	height	33 ft 8½ in
Speed:	maximum	475 mph at 15,000 ft
	cruise	380 mph at 25,000 ft
	loiter	230 mph with two engines shut down
Ceiling:	28,300 ft	
Range:		radius 1,550 miles with 13 hours on station
Radar:	AN/APS-115 or AN/APS-137(V) search	
Guns:	none	
Ordnance:	3 nuclear depth bombs B57	
or	weapons bay + wing points 8 Mk 46/50 torpedoes + 4 Mk 4 6/50 torpedoes	
or	2 2,000-lb mines + 4 Mk 46/50 torpedoes	
or	4 1,000-lb mines + 4 Mk 46/50 torpedoes	
or	8 Mk 46/50 torpedoes + 16 5-inch rockets	
or	HARM, Maverick missiles	

S-3 Viking

The S-3 Viking was the U.S. Navy's last fixed-wing antisubmarine aircraft that operated from aircraft carriers. Its ASW equipment was removed in the mid-1990s and the aircraft subsequently served in the carrier-based surveillance and tanker roles until phased out of service in 2009. In those roles—in the S-3B configuration—the aircraft

A Lockheed S-3A Viking carrier-based ASW aircraft shows its extended MAD antenna. This aircraft has drop tanks under the wings; weapons could be carried in the bomb bay and under the wings—including nuclear depth bombs. From the mid-1990s until 2008 the aircraft were employed in the surveillance and tanker roles, their ASW gear having been removed. (U.S. Navy)

could carry the Harpoon and Maverick air-to-surface missiles. Aerial tanking was undertaken with external drogue pod and fuel tanks.

In the ASW role the aircraft could carry a nuclear depth bomb.

The Lockheed S-3 Viking was built to replace the Grumman S2F/S-2 Tracker as the Navy's ship-based, fixed-wing ASW aircraft.

Design: The S-3 was designed to be within the approximate dimensions of the piston-engine Tracker but to be faster and to carry more advanced ASW equipment. The internal weapons bay, sized to hold four lightweight ASW torpedoes, could carry 2,400 pounds of weapons or auxiliary fuel tanks. There were also two wing pylons, which in the S-3B were upgraded to carry the Harpoon anti-ship missile. The S-3B also had an improved acoustic processor and the improved AN/APS-137 inverse synthetic aperture radar (ISAR) with a major avionics upgrade program for 124 S-3B aircraft begun in 1995. ASW systems included the AN/ASQ-81 MAD, forward-looking infrared (FLIR,) and 60 sonobuoys in fuselage chutes.

The wings and tail fin folded for carrier stowage.

Five preproduction S-3A aircraft were modified to a US-3A cargo configuration for operation from carriers in the Western Pacific and Indian Ocean; those planes were taken out of service in the early 1990s. A KS-3 tanker variant was proposed as well as pod tanks and drogue system for the US-3A variant; however, no development of a specialized tanker was undertaken.

Sixteen S-3A aircraft were converted to the ELINT role and were designated ES-3A.

Operational: The first S-3A flight was on 21 January 1972. S-3A IOC was in February 1974.

S-3A/B aircraft served as ground attack aircraft in low-threat areas during Operation Desert Storm in 1991. The last S-3B Viking was retired in December 2008.

Production: The last of 187 S-3A aircraft was completed in 1978. About 119 were upgraded to S-3B by 1994. Proposals for additional S-3 production did not come to fruition.

Operational:	1974–1992 in nuclear configuration	
Design:	Lockheed	
Aircraft data:	S-3B	
Crew:	4 (pilot, copilot, 2 ASW systems operators)	
Engines:	2 General Electric TF34-GE-400 turbofan; 9,275 lbst each	
Weights:	empty	26,783 lb
	max takeoff	
Dimensions:	length	53 ft 4 in
	wingspan	68 ft 8 in
	wing area	598 ft^2
	height	22 ft 9 in
Speed:	maximum	505 mph at sea level
	cruise	400+ mph
	loiter	240 mph at 20,000 ft
Ceiling:	40,000 ft	
Range:	patrol 2,650 miles	
Radar:	AN/APS-137 ISAR search	
Guns:	none	
Ordnance:	1 nuclear depth bomb or weapons bay + wing points	
	4 Mk 46/50 torpedoes + 6 500-lb bombs	
or	4 500-lb bombs + 6 500-lb bombs	

S2F Tracker (later S-2)

The Tracker was a long-serving, widely used anti-submarine aircraft developed by the U.S. Navy specifically for carrier operation. The aircraft was one of the first to combine advanced search and attack capabilities in a single airframe, resulting in a relatively large carrier aircraft for its time.

The Tracker was replaced in U.S. Navy ASW squadrons by the S-3 Viking. However, the Tracker has survived for many years in other air forces, flown from both carriers and land bases. Trackers—some with upgraded engines—were flown by 14 other air forces.

In U.S. service, the Tracker could carry nuclear depth bombs.

De Havilland Canada built 100 Trackers; a portion of these, probably some 15 to 18 aircraft, could carry U.S. nuclear depth charges (Mk 101 Lulu and B57) in the period 1961–1964 from land bases in Canada. The Grumman-built IS-2F in Italian service could also carry the B57 beginning in 1970.

Design: The Tracker had a high-wing, twin-engine configuration, giving a compact appearance with the pilot's cockpit relatively far forward. Features included an internal weapons bay, retractable radome under the rear fuselage, a retractable MAD boom in the tail, and a searchlight mounted in the right wing. Eight (later ten) expendable sonobuoys were housed in the rear of each engine nacelle. The wings folded for carrier stowage.

A Grumman S2F (S-2) Tracker reveals its weapons load with four lightweight ASW torpedoes on wing pylons, two air-to-surface rockets between the torpedoes, and a Mk 101 Lulu nuclear depth bomb in the aircraft's open bomb bay. The large fuselage AN/APS-88A radome behind the bomb bay is lowered and the MAD tail "stinger" is extended. These aircraft served on U.S. and other nations' aircraft carriers. (U.S. Navy)

The S2F-2 (S-2C) was basically an S2F-1 with the left side of the fuselage extended to carry a Mk 90 Betty nuclear depth bomb (and tail surfaces slightly enlarged). The larger S2F-3 (S-2D) and S2F-3S (S-2E) could carry the Mk 57 and Mk 101 Lulu nuclear depth bombs; the S-2D/E were virtually identical except for avionics equipment.

In a "maximum" ASW armament loadout, the S-2D/E could carry one Mk 101 Lulu nuclear depth bomb, four Mk 46 ASW torpedoes, and two 5-inch rockets.

A planned turboprop version was not pursued.

Tracker derivatives included the WF/E-1 Tracer airborne early warning aircraft (AEW), TF-1Q electronic countermeasures (ECM) aircraft, and the TF/C-1 Trader carrier on-board delivery (COD) aircraft.

Designation: The S2F series was redesignated S-2 on 1 October 1962.

The aircraft was invariably referred to as the "Stoof."

Operational: The first flight of XS2F-1 was on 4 December 1952. The S2F-1 entered squadron service in February 1954.

The last ASW aircraft were retired by the U.S. Navy in 1975. Specialized variants flew from shore bases until 1986.

Production: Grumman produced 1,169 S2Fs from 1952 through December 1967, including more than 100 supplied directly to other countries:

Aircraft[a]		Built	First flight
XS2F-1		2	1952
YS2F-1		15	1953
S2F-1	S-2A	740	1952
S2F-2	S-2C[b]	60	1954
S2F-3	S-2D[b]	100	1959
S2F-3S	S-2E[b]	252	

a Second column indicates 1962 redesignation.
b Nuclear-capable aircraft

Another 100 Trackers were built by de Havilland Canada—43 CS2F-1 and 57 CS2F-2. Other variants were upgrades or modifications of the above aircraft. These included training, utility, target tow, reconnaissance, and avionics development aircraft.

Operational:	1955–1975 in nuclear role	
Design:	Grumman	
Aircraft data:	S2F-3	
Crew:	4 (pilot, copilot, 2 ASW systems operators)	
Engines:	2 Curtiss-Wright R-1820-82A piston; 1,525 hp each	
Weights:	empty	19,726 lb
	max takeoff	29,150 lb
Dimensions:	length	43 ft 6 in
	wingspan	72 ft 7 in
	wing area	496 ft^2
	height	16 ft 7½ in
Speed:	maximum	240 mph at 5,000 feet[17]
	cruise	150 mph
Ceiling:	20,100 ft	
Range:	460 miles with 7¾ hour endurance with internal stores only[18]	
	360 miles with 6 hour endurance with internal nuclear depth bomb + external 4 torpedoes	
Radar:	AN/APS-88A search	
Guns:	none	
Ordnance:	1 nuclear depth bomb B57, Mk 101 Lulu (internal) plus 4 ASW torpedoes (external)	

ASW HELICOPTERS

Aircraft[a]			Service	IOC	Type
SH-60F		Seahawk	Navy	1989	carrier
DSN	QH-50	DASH	Navy	1962	ship[b]
HSS-1	SH-34	Seabat	Navy	1955	carrier
HSS-2	SH-3	Sea King	Navy	1958	carrier

a Second column is post-1962 designation.
b Embarked in cruisers, destroyers, frigates.

SH-3 Sea King (see HSS-2 Sea King)

SH-34 Seabat (see HSS-1 Seabat)

QH-50 Dash (see DSN Dash)

SH-60F Seahawk

The SH-60F is the U.S. Navy's carrier-based ASW variant of the ubiquitous H-60 Hawk-series helicopter. First flown by the U.S. Army as the UH-60 Black Hawk troop carrier—succeeding the widely used Bell UH-1 Huey (Iroquois)—the U.S. Navy adopted the H-60 for the light airborne multipurpose system (LAMPS) to provide an anti-submarine helicopter for use aboard cruisers, destroyers, and frigates. Large numbers of the SH-60B Seahawk were procured for the LAMPS role, and, subsequently, the helicopter was adopted for carrier operations as the SH-60F. In the latter role the Seahawk replaced the SH-3 Sea King.

Only the carrier-based SH-60F was fitted to deliver nuclear depth bombs because the surface warships did not have facilities for handling them.

The SH-60B and SH-60F are both being replaced in of the shipboard ASW roles by the multipurpose MH-60R Seahawk.

Design: The Seahawk is adopted from the UH-60A Black Hawk, the U.S. Army's basic transport/utility helicopter. That selection was made in large part to reduce LAMPS program costs.

The H-60 series has a conventional configuration with twin turboshaft engines mounted above the spacious cabin that, in the troop variants, can accommodate 11 fully equipped troops. The UH-60A can hook-lift up to 8,000 pounds. The Army variants are armored and have special survivability features, minimal avionics, and normally mount two 7.62-mm machine guns in door positions. All H-60s have a tail-wheel landing gear.

A Navy YSH-60B was modified in 1984 to test the AN/AQS-13F active dipping sonar and automated flight control system for the carrier-based SH-60F. That helicopter has dipping sonar with a 1,500-foot cable. The radar, MAD, sonobuoys, and some other equipment of the SH-60B has been deleted in the SH-60F.

17 Without wing stores but with radome and MAD boom extended, retraction of the radome and MAD boom increased speed by approximately 12 mph.
18 With radome and MAD boom extended.

Sikorsky SH-60F Seahawk. (U.S. Navy)

SH-60Fs can mount 7.62-mm machine guns in their doors.

The U.S. Navy also acquired the HH-60H combat search and rescue (SAR) and special operations variant, the Coast Guard the HH-60J Jayhawk for SAR, and the Marine Corps VH-60A for presidential-VIP support.

SH-60F endurance is 4¼ hours.

Designation: During development the SH-60F was designated as the CV Inner Zone ASW helicopter or CV-Helo.

Status: The first Army YUH-60 test bed flight was on 17 October 1974, prototype SH-60B on 12 December 1979, and first SH-60F on 19 March 1987. SH-60B IOC was in September 1983 and SH-60F IOC in June 1989.

Seventeen foreign nations fly military variants of H-60/S-70 helicopters.

Production: The U.S. Navy procured 82 SH-60F helicopters for carrier operation, the last delivered in December 1994. The second production aircraft was redesignated YSH-60F for research duties.

Mitsubishi produced the SH-60J LAMPS variant for Japanese service.

Operational:	1989–1992 in nuclear role	
Design:	Sikorsky	
Aircraft data:	SH-60F	
Crew:	4 (pilot, tactical officer/copilot, 2 ASW sensor operators)	
Engines:	2 General Electric T700-GE-401C turboshaft; 1,900 shp each	
Weights:	empty	13,648 lb
	max takeoff	21,800 lb
Dimensions:	length	64 ft 10 in (overall)
		50 ft ¾ in (fuselage)
	rotor diam	53 ft 8 in
	height	17 ft
Speed:	maximum 155 mph at sea level cruise	
Ceiling:	19,000 ft	
Range:	radius 60 miles with 3 hours loiter	
	radius 175 miles with 1 hour loiter	
Radar:	none	
Guns:	(see text)	
Ordnance:	1 nuclear depth bomb Mk 101 Lulu	
or	2 ASW torpedoes	

DSN DASH (later QH-50)

The U.S. Navy's drone anti-submarine helicopter (DASH) was an early unmanned aerial vehicle (UAV), probably the only one designed to carry a nuclear weapon. The DASH program was both a success and a failure.

By the late 1950s the Navy was installing the large AN/SQS-26 sonar in ASW ships with an effective detection range beyond ship-launched torpedoes and ASROC (anti-submarine rocket), the latter being able to deliver a torpedo—or nuclear depth bomb—some seven nautical miles. The British and Soviet Navies developed manned helicopters to exploit shipboard sonar detections. The U.S. Navy decided to develop an unmanned helicopter. It was hoped that DASH could fly from smaller ships than could a manned helicopter, operate in bad weather that would keep a manned helicopter on deck, and have a higher readiness. Its operational range was approximately 30 miles, based on the ship's radar range.

DASH was developed from the Gyrodyne firm's one-man "Rotorcycle," intended to carry a single Marine and his equipment. The Marine Corps had evaluated two

Gyrodyne DSN-3 (QH-50C) DASH with MK 46 torpedo. (U.S. Navy)

XRON-1 and three YRON-1 prototypes manned Rotorcycles, the first XRON-1 flying in 1956. The DASH was to be a modification of the "off-the-shelf" RON-1 design.

Design: The DASH fuselage was an aluminum tubular structure to which was attached the engine, landing gear, and other equipment. Contrarotating rotors alleviated the need for a tail boom and anti-torque tail rotor.

The DSN-1 and DSN-2 were similar to the manned RON with provisions for a safety pilot during shipboard trials. The DSN-3 (QH-50C) and QH-50D were production variants, the latter with fiberglass rotor blades. The more capable QH-53E was under development when the program ended (the YQH-53E having flown in September 1969).

During a DASH mission, the drone was "piloted" during takeoff and landing by the ship's DASH officer from a console adjacent to the ship's flight deck. During the mission, the drone was controlled by an officer in the ship's combat information center who would "fly" the drone radio control to the target area and release the weapon.

The drone carried either one Mark 46 or two Mark 44 acoustic homing torpedoes. Although the Navy planned to use DASH to deliver a Mk 57 nuclear depth bomb, it was not deployed because of problems with logistic support for the DASH/Mark 57 combination and problems with DASH.

No sensors were carried by DASH.

Flight endurance was just over two hours.

Designation: The original DSN designation was changed to QH-50 on 1 October 1962.

Operational: Gyrodyne's unmanned DSN-1 first flew on 31 December 1958, and a DSN-1 made the first shipboard landing—with a pilot on board—on 1 July 1960 on the frigate *Mitscher* (DL 2); the first unmanned landing occurred on 7 December 1960 aboard the destroyer *Hazelwood* (DD 531).

The definitive DSN-3 variant became operational in November 1962 and was first deployed aboard ship, the destroyer *Wallace L. Lind* (DD 703), in January 1963. The improved QH-50D entered service in 1966.

Drone availability lagged behind the availability of DASH-capable ships; for example, in December 1961 there were 41 ships but no operational DASHs; by mid-1964 there were 128 ships with 50 operational DASHs. In all, more than 200 large frigates (DL/DLG/ DLGN), destroyers (DD), and escort ships (DE/DEG) were modified with a small flight deck and hangar, plus aviation support facilities for two DASH "birds" on each ship.

The large number of DASH crashes were due in large part to the Navy's operational policies. Destroyers fitted with DASH that deployed with aircraft carriers would not fly their planes en route to overseas areas to conserve the helicopters. In forward areas they could not fly their drones while the carriers were operating aircraft because of the fear of electronic interference. Thus, during six-month deployments to forward areas, the ships rarely operated the drones. Furthermore, there was a desire to not risk the drones and retain them in a high state of readiness for combat. DASH operators had little experience, and when they were flown the loss rate was high.

More than 400 drones were lost, leading to the DASH program being halted in January 1971. A few continued to serve on U.S. warships during the Vietnam War for gunnery spotting (fitted with television cameras), and they were flown for several years by the Japanese Maritime Self-Defense Force.

Production: Following a dozen development aircraft, from 1960 to 1969 the Gyrodyne produced 732 operational drones for the Navy plus one DSN-3 delivered to the Army.

Aircraft		Built
DSN-1	QH-50A	9
DSN-2	QH-50B	3
DSN-3	QH-50C	357
	QH-50D	375

Operational:	Nov 1962–Jan 1971	
Design:	Gyrodyne	
Aircraft data:	DSN-3	
Crew:	unmanned	
Engines:	1 Boeing T50-BO-8A gas turbine; 300 hp	
Weights:	empty	1,172 lb
	max takeoff	2,303 lb with 2 Mk 44 torpedoes
		2,181 pounds with Mk 57 bomb
Dimensions:	length	20 ft
	rotor diam	20 ft
	height	9 ft 8½ in
Speed:	maximum	90 mph at sea level
	cruise	65 mph at sea level
Ceiling:	17,470 ft	
Range:	83 miles	
Radar:	none	

Guns:	none
Ordnance:	1 nuclear depth bomb Mk 57
or	2 ASW homing torpedoes

HSS-1 Seabat (later SH-34)

The Sikorsky HSS-1 was a widely used anti-submarine helicopter of a basic type that was also flown in large numbers by the U.S. Marine Corps (HUS) and U.S. Army (CH-34) as troop carriers.

The H-34 series was also produced in Britain and France and was flown by the military services of 27 other nations. For two decades, from 1955 into the 1970s, the H-34 series was the backbone of several nations' military operations, ashore and afloat.

The HSS-1 was the first U.S. Navy helicopter that could carry a nuclear depth bomb.

Design: Evolving from the HO4S (H-19) design, the later helicopter had a cockpit above and slightly forward of the cabin, with the radial engine mounted in the nose, providing easy access for maintenance. The tail boom was an extension of the fuselage, terminating in a vertical stabilizer with the four-blade tail rotor. The main rotor—56 feet in diameter—had four blades, which folded for shipboard stowage, as did the tail pylon. (This reduced the helicopter's maximum length from 65½ feet to 37 feet.)

All models had a fixed, three-wheel landing gear except for 40 Marine HUS-1A and 6 Coast Guard HUS-1G helicopters, which were fitted with pontoons. The cabin could seat 12 to 18 troops, depending upon the amount of equipment carried, or eight stretchers, or could lift 5,000 pounds of cargo on a sling.

For the ASW mission the HSS-1 was fitted with AN/AQS-4 dipping sonar and could carry 2 Mk 24, 43, or 44 homing torpedoes, or a nuclear depth bomb. The HSS-1 and HSS-1N were virtually identical, the principal difference being the automatic flight stabilization and night-flying capabilities of the latter helicopter.

A Marine HUS-1 was employed in June 1960 to test the feasibility of launching the Bullpup air-to-surface missile from a helicopter. That concept was not pursued.

Designation: The HSS-1 and HSS-1N were redesignated SH-34G and SH-34J, respectively, on 1 October 1962.

The U.S. Marine Corps procured the helicopter as the HUS-1 (UH-34D) and HUS-1A (UH-34E) using the name Seahorse.

Operational: The first prototype XHSS-1 flight was on 8 March 1954, first flight of HSS-1 was in September 1954, and first flight of HSS-1N was in July 1957. The HSS-1 entered squadron service in August 1955 and the HSS-1N in May 1958.

Sikorsky HSS-1 (SH-3) Seabat hovering with dipping sonar lowered. (U.S. Navy)

The last H-34 in U.S. military service is believed to have been the UH-34D, in the Navy's inventory until March 1974.

Production: Beginning in 1954 the U.S. Navy took delivery of 3 XHSS-1 prototype, 215 HSS-1, and 156 HSS-1N helicopters; the production helicopters were redesignated SH-34G and SH-34J, respectively, in 1962.

The U.S. Marine Corps bought 549 of the aircraft in the HUS-1 series.

The U.S. Army accepted its first H-34 Choctaw in 1955, and it became the Army's principal transport helicopter. The Army bought 437 CH-34s, and at least 26 were transferred from the Marine Corps. A few aircraft were also flown by the Air Force. Eight HSS-1Z (later VH-34D) variants had VIP accommodations and were assigned to a joint Army–Marine Corps detachment flying the president and other government executives. Small numbers of these helicopters were also flown by the U.S. Air Force and U.S. Coast Guard.

Westland in England produced some 400 helicopters called Wessex, and Sud-Aviation in France built 166 airframes for French military service. With Sikorsky, Westland, and Sud-Aviation also producing the helicopter for other nations and civilian sales, the production total was just over 2,300 helicopters.

Operational:	1958–1974 in nuclear role
Design:	Sikorsky
Aircraft data:	HSS-1N
Crew:	4 (pilot, copilot, 2 ASW systems operators)
Engines:	1 Wright R-1820-84B piston; 1,525 hp

Weights:	empty	8,648 lb
	max takeoff	13,300 lb
Dimensions:	length	65 ft 6 in
	rotor diam	56 ft
	height	15 ft 8 in
Speed:	maximum	145 mph at sea level
	cruise	97 mph at sea level
Ceiling:	16,200 ft	
Range:	260 miles	
	radius	95 miles with 2 ASW torpedoes
Radar:	none	
Guns:	none	
Ordnance:	1 nuclear depth bomb Mk 101 Lulu	
or	2 ASW torpedoes	

HSS-2 Sea King (later SH-3)

The Sea King was the standard ship-based ASW helicopter of the U.S. and several other navies from the early 1960s until the early 1990s. It was flown from aircraft carriers in the U.S. Navy until replaced by the SH-60F Seahawk. (The Royal Navy flew the helicopter from destroyers as well as carriers.)[19] It was also flown by other U.S. and foreign services in a variety of roles and configurations.

All Sea Kings have been retired from U.S. Navy service.

Design: The Sea King had a "boat" hull but did not normally alight on the water. The tail pylon and main rotor blades folded for shipboard stowage. The SH-3 series had a fully articulated, five-blade main rotor and a five-blade anti-torque rotor mounted at the top of the tail boom.

Designed specifically for naval use, the large SH-3A served as the basis for a large number of Sikorsky military and commercial models. Variants were flown by the U.S. Army, Air Force, and Marine Corps, the last being the VH-3A/D VIP employed as "Marine One" to transport the president and other VIPs. Several Sea King variants were fitted with rear ramps.

Navy ASW variants had AN/AQS-13 dipping sonar, AN/AQS-81 MAD, and other anti-submarine equipment. The dipping sonar had a 450-foot operating depth. They could carry four ASW torpedoes or a nuclear depth bomb. The SH-3H was fitted with the LN-66 search radar; it could carry only two ASW torpedoes or the Mk 101 Lulu nuclear depth bomb.

Sikorsky HSS-2A (SH-3D) Sea King. (U.S. Navy)

During Operations Desert Shield/Desert Storm in 1990–1991, SH-3H helicopters aboard participating aircraft carriers were fitted with a flexible 7.62-mm M60 machine gun mounted in the (starboard) door opening.

The original production HSS-2/SH-3A and HSS-2A/SH-3D variants were upgraded to the SH-3H ASW variant. Others became SH-3G utility helicopters; these were unarmed but in theory could be reconverted to an ASW role. The SH-3G could carry up to 15 passengers.

Two Sea Kings designated YSH-3J were used to test sensors for the SH-60B Seahawk. Other U.S. Navy variants were the HH-3A rescue and the RH-3A mine countermeasures versions.

In addition to ASW and troop-carrying Sea King variants, the Royal Navy developed an airborne early-warning AEW configuration during the Falklands War (1982).

Main rotor blades and tail rotor section folded for shipboard stowage.

Designation: Originally designated HSS-2, it became the SH-3 series on 1 October 1962. (The HSS-1 was redesignated SH-34.)

Naval variants (and some others) are named Sea King; the U.S. Air Force HH-3 rescue variants were named Jolly Green Giant, and U.S. Coast HH-3F rescue variants were named Pelican.

Operational: The first XHSS-2 flight was on 11 March 1959; first flight of a converted SH-3H in April 1972, Navy HSS-1 IOC in June 1961, and SH-3D in 1966. The last SH-3H helicopters were retired from U.S. Navy service in May 1996; the last Navy SH-3G utility helicopters were retired in 2007.

The Sea King has been flown in military configurations by 24 other countries.

19 The Royal Navy continues to operate anti-submarine HAS-series Sea Kings from aircraft carriers. The Royal Navy also had nuclear depth bombs available for their helicopters.

In February 1962 an HSS-1 became the first helicopter to break the 200-mph barrier by registering 210.6 mph over a 19-kilometer course. Two U.S. Air Force HH-3Es made the first nonstop trans-Atlantic helicopter flight on 31 May–1 June 1967; each helicopter was refueled nine times in flight by C-130 Hercules tankers.

Production: U.S. Navy production of Sea King ASW variants totaled 326 helicopters:

Aircraft		Built
XHSS-2		1
YHSS-2		7
HSS-2	SH-3A[a]	245
HSS-2A	SH-3D	73

a Second column indicates 1962 redesignation.

Subsequently, 145 SH-3A/D aircraft were converted to the SH-3H anti-submarine configuration. Another 105 SH-3A/D aircraft were converted to the SH-3G utility configuration.

Additional Sea Kings have been built by United Aircraft Canada, Westland in Britain, Augusta in Italy, and Mitsubishi in Japan. In total, more than 1,100 Sea Kings have been produced for military service.

Operational:	June 1961–1992 in nuclear ASW role	
Design:	Sikorsky	
Aircraft data:	HSS-1	
Crew:	4 (pilot, copilot, 2 ASW systems operators)	
Engines:	2 General Electric T58-GE-6 turboshaft; 1,050 shp each	
Weights:	empty	10,814 lb
	max takeoff	19,000 lb
Dimensions:	length	65 ft 6 in (overall)
		54 ft 8 in (fuselage)
	rotor diam	62 ft
	height	15 ft 4 in
Speed:	maximum	155 mph at sea level with 2 torpedoes
	cruise	140 mph at sea level
Ceiling:	12,100 ft	
Range:	575 miles with 2 torpedoes	
	555 miles with 4 torpedoes	
Radar:	none	
Guns:	(see text)	
Ordnance:	1 nuclear depth bomb Mk 101 Lulu	
or	4 ASW torpedoes	

CHAPTER FIVE

Strategic Missiles

Strategic missiles are those intended to strike an enemy's homeland. Missiles are listed by name rather than designation because of the multiple designations that were applied.

Missile designations: Early missile designations employed a common scheme for Army, Air Force, and Navy missiles; the first two services were assigned odd sequential numbers and the Navy even numbers, as AAM-A-1 and ASM-N-2, respectively.

These early designations were:

AAM	air-to-air missile
ASM	air-to-surface missile
AUM	air-to-underwater missile
SAM	surface-to-air missile
SUM	surface-to-underwater missile
UAM	underwater-to-air missile
USM	underwater-to-surface missile

Subsequently, the Air Force adopted the following designations:

GAM	guided aircraft missile
GAR	guided aircraft rocket
IM	interceptor missile
SM	strategic missile
TM	tactical missile

The Army and Navy retained the older scheme.

The Air Force also assigned these designations in the "standard" aircraft sequence. Thus, the Convair B-60 bomber was followed by the TM-61 Matador missile, SM-62 Snark, GAM-63 Rascal, etc. Many official publications listed these designations with the "B" prefix, as B-61, B-62, and B-63, respectively. (The next aircraft in the bomber series was the Douglas B-66 Destroyer.)

Finally, in 1962 the Department of Defense established a joint designation system for guided missiles and rockets.

Launch Environment	Mission	Type
A air	D decoy	M guided missile
B multiple	G surface attack	R rocket
C coffin	I aerial intercept	
H silo stored		
L silo launched		
M mobile		
P soft pad		
R ship		
U underwater (submarine)		

LAND-BASED MISSILES

Missile	Type[a]	Service[b]	IOC	Warhead
Atlas D	ICBM	AF	1959	W49
Atlas E	ICBM	AF	1961	W38
Atlas F	ICBM	AF	1962	W38
Boojum	SCM	AF	cancelled	—
Gryphon	GLCM	AF	1984	W84
Jupiter	IRBM	USA, AF	1961	W49
Midgetman	ICBM	AF	cancelled	W87
Minuteman I	ICBM	AF	1962	W56/W59
Minuteman III	CBM	AF	1966	W56
Minuteman IIII	CBM	AF	1970	W62/W78
Navaho	SCM	AF	cancelled	W13
Peacekeeper	ICBM	AF	1986	W87
Pershing II	MRBM	USA	1983	W85
Pluto	SCM	AF	cancelled	—
Snark	SCM	AF	1960	W39

Thor	IRBM	AF	1959	W49
Titan I	ICBM	AF	1962	W38
Titan II	ICBM	AF	1963	W53

a GLCM = ground-launched cruise missile; ICBM = intercontental ballistic missile; IRBM = intermediate-range ballistic missile; MRBM = medium-range ballistic missile; SCM = strategic cruise missile.
b AF = Air Force; USA = Army.

Atlas D SM-65D/CGM-16D

The Atlas D was the first operational U.S. intercontinental ballistic missile (ICBM). The Atlas was initiated in response to the Army Air Forces requirement of August 1945 for a 5,000-mile ballistic missile.

The program initially progressed rapidly under a General Dynamics team headed by Belgian-born Karel J. Bossart. However, in 1947 the government was forced to cut funding for the program; General Dynamics put its own money into the Atlas and three test vehicles were successfully flown in 1948, testing a variety of advanced features for a liquid-propellant missile.

The Atlas effort was again government supported from 1951 but at a low funding level; from 1951 to 1954 the Atlas program received $26.2 million while the Snark and Navaho strategic cruise missiles were provided a total of $450 million.

Production of the Atlas D was authorized in January 1955.

Design: Early designs provided for a missile weight of some 220 tons. The development of smaller warheads permitted a smaller missile to carry the 1,500-pound W49 warhead (estimated 1-MT yield).

The Atlas and Titan I missiles used cryogenic—extremely cold—fuels. They burned kerosene and liquid oxygen, which were carried in separate tanks and stored at a temperature of minus-280°F.

The Atlas A through C were development missile programs that did not become operational.

Designation: During development the Atlas was designated weapon system MX-774.

Operational: Flight testing began on 11 June 1957, using the Atlas A missiles with only booster engines and dummy nose cones for short-distance flights. A total of eight Atlas A missiles were launched—all on the Atlantic Missile Range—through June 1958.

The Atlas B series provided a test missile with a propulsion system that was close to operational capability. The Atlas 4-B, the first successful launch and the second in the series B test flights, was launched on 2 August 1958. The eighth missile in the series, Atlas 10-B, attained orbit with the Project Score payload on 18 December 1958, becoming the world's first communications satellite in the first successful use of the Atlas as a space launch vehicle. Nine B missiles were launched from July 1958 to February 1959

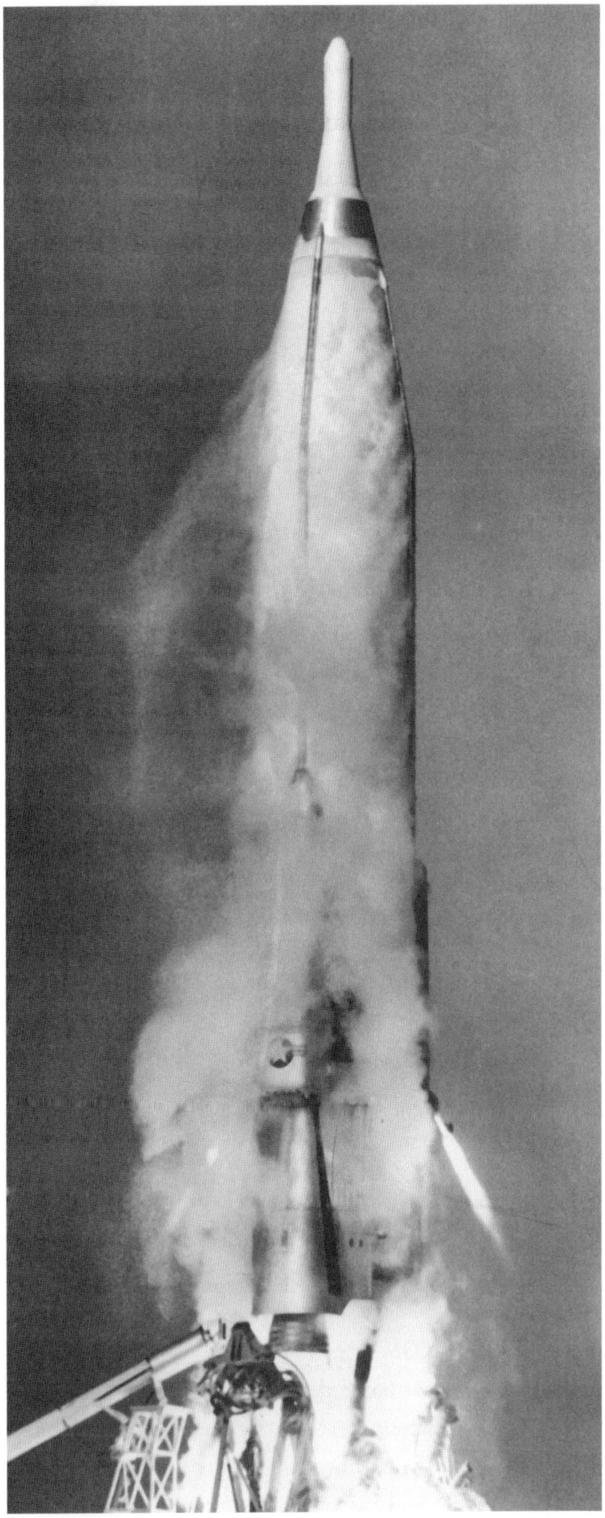

Convair Atlas D ICBM. (U.S. Air Force)

with testing of both sustainer engine and separable nose cones beginning in mid-1958 and the first full-range flight of 6,000 miles being achieved on 28 November 1958.

Six Atlas C missiles were launched from December 1958 to August 1959.

The Atlas D became operational on 9 September 1959, at Vandenberg AFB, California. The missile attained alert status on 31 October 1959. The last of the four Atlas D squadrons went on alert in May 1961.

At peak strength there were 30 Atlas D missiles deployed: 6 at Vandenberg AFB; 15 at Warren AFB, Wyoming; and 9 at Offutt AFB, Nebraska. The missiles were in service for five years, with the last being removed from alert status on 1 October 1964.

On 20 May 1960, the Air Force launched an Atlas D from Cape Canaveral (later Kennedy Space Center) carrying a 1½-ton payload a distance of 9,040 miles to impact into the Indian Ocean; this was the longest distance ever flown by a U.S. ICBM.

Operational:	Sep 1959–Sep 1964
Design:	General Dynamics/Convair
Weight:	267,136 lb
Length:	81 ft 7 in
Diameter:	10 ft
Wingspan:	(ballistic)
Propulsion:	1½-stage liquid-propellant rocket motor; 360,000 lbst
Range:	6,325 miles
Guidance:	radio-inertial
Warhead:	1 W49Y2 nuclear (1.4 MT)
Launch mode:	horizontal in "soft" above ground launcher; raised to a vertical position for firing

Atlas E SM-65E/CGM-16E

The Atlas E was virtually identical to the Atlas D but was deployed in a different configuration than the unprotected Atlas D. A heavily reinforced concrete roof covered the Atlas E launcher and provided protection against nuclear blast (hardened to 25 psi).

To launch the missile, the concrete roof had to be retracted and the missile raised to vertical launch position.

Design: The W38 warhead used in this missile was also fitted to the Atlas F and Titan I ICBMs.

Operational: At the peak Atlas E deployment there were 27 missiles at three air force bases—Fairchild AFB, Washington; Forbes AFB, Kansas; and Warren AFB, Wyoming.

On 4 January 1965 the first Atlas E ICBMs were taken off alert at Forbes AFB. On 31 March 1965, the last Atlas E ICBMs were removed from alert, completing operational phaseout of this missile.

Operational:	Sep 1961–Mar 1965
Design:	General Dynamics/Convair
Weight:	267,136 lb
Length:	81 ft 7 in
Diameter:	10 ft
Wingspan:	(ballistic)
Propulsion:	1½-stage liquid-propellant rocket motor; 360,000 lbst
Range:	7,800 miles
Guidance:	all inertial
Warhead:	1 W38Y1 nuclear (4.5 MT)
Launch mode:	horizontal in a semihard concrete coffin-type launcher buried in the ground

Atlas F SM-65F/HGM-16F

The Atlas F ICBM was deployed in a vertical silo—174 feet deep and 52 feet in diameter—with an underground launch control center adjacent to the silo.

At the peak of Atlas ICBM deployment there were 13 Atlas squadrons (four D, three E, and six F). They operated a total of 129 missiles.

The Atlas F was phased out of service after only 29 months of being on alert because of its vulnerability, slow reaction time, and cost and maintenance compared to the new Minuteman ICBM. The average ratio of men per missile for each Atlas was about 80 compared to 12 for the Minuteman. (The earlier Atlas D and E missiles were phased out for similar reasons.)

Design: The Avco reentry vehicle weighed 3,825 pounds and was fitted with penetration aids to help overcome possible ballistic missile defenses. The Atlas F was designed to deliver a warhead at intercontinental range with a 2.3-mile circular error probability (CEP).

Operational: The first successful Atlas F test flight was in July 1961. The first operational Atlas F was deployed in August 1962. At peak Atlas F deployment there were 72 launchers at six air force bases: Schilling AFB, Kansas; Lincoln AFB, Nebraska; Altus AFB, Oklahoma; Dyess AFB, Texas; Walker AFB, New Mexico; and Plattsburgh AFB,

New York. All six squadrons were operational by the end of 1962.

On 1 December 1964, the first Atlas F was removed from alert at Dyess AFB; the last was removed from service on 12 April 1965.

Operational:	Aug 1962–Apr 1965
Design:	General Dynamics/Convair
Weight:	267,136 lb
Length:	81 ft 7 in
Diameter:	10 ft
Wingspan:	(ballistic)
Propulsion:	1½-stage liquid-propellant rocket motor; 389,000 lbst
Range:	7,800 miles
Guidance:	all-inertial
Warhead:	1 W38Y1 nuclear (4.5 MT)
Launch mode:	hardened silo

Boojum

In March 1946 the U.S. Army Air Forces awarded Northrop a study contract to develop a supersonic missile with a range of 1,500 to 5,000 miles carrying a 5,000-pound nuclear warhead. Called the Boojum, the missile was to have been a supersonic follow-on to the Snark strategic cruise missile (also developed by Northrop).

The Boojum project was cancelled in 1948.

Design: The Boojum was to be a two-stage missile with a delta-wing, tailless configuration. The turbojet engines were to be mounted far out on the wings. After flying some 1,500 miles, the conformal, slipper-shaped fuel tank would be jettisoned from under the fuselage.

Designation: This missile was assigned the AAF project designation MX-775B.

The names "Boojum" and "Snark" were taken from the writings of Lewis Carroll.

Operational:	cancelled
Design:	Northrop
Weight:	112,131 lb
Length:	85 ft 4 in 104 ft 10 in with external conformal fuel tank
Wingspan:	50 ft 11 in
Diameter:	
Propulsion:	2 General Electric J47 turbojet with afterburners
Speed:	Mach 2
Range:	5,070 miles at Mach 1.8 4,975 miles at Mach 2
Guidance:	
Warhead:	nuclear
Launch mode:	land based

Gryphon BGM-109G

The Gryphon ground-launched cruise missile (GLCM) was a long-range, subsonic, strategic cruise missile developed for use in the European theater. The GLCM was a land-launched version of the widely used Navy Tomahawk-series cruise missile.

On 12 January 1977, the U.S. government made the decision to develop a ground-launched variant of the Tomahawk, in part to counter Soviet deployment of the SS-20 Saber (Soviet RSD-10 Pioneer) intermediate-range ballistic missile beginning in 1977. The NATO leadership agreed on 12 December 1979 to the deployment of 464 U.S. Air Force–operated GLCMs in western Europe. The first GLCMs arrived in Britain in December 1983 with deployments rapidly following in three other NATO nations.

The original Gryphon/GLCM program called for 565 GLCMs with 137 launch vehicles. In the end, only 464 missiles and 116 launch vehicles were scheduled to be deployed at six bases in five NATO countries; none was deployed in The Netherlands, and only 309 missiles were in Europe when the deployment was halted. The planned deployments were:

General Dynamics/McDonnell Douglas Gryphon GLCM. (U.S. Air Force)

USAF Tactical Missile Wing	Location	Missiles	Launchers
38th	Wuescheim Air Station, West Germany	96	24
485th	Florennes Air Base, Belgium	48	12
486th	Woensdrecht Air Base, The Netherlands	48	12
487th	Comiso Air Station, Italy	112	28
501st	RAF Greenham Common, England	96	24
550th	RAF Molesworth, England	64	16

The 868th Tactical Missile Training Group at Davis-Monthan AFB, Arizona, provided GLCM training.

The Gryphon/GLCM's operational service was brief because the Intermediate-range Nuclear Forces (INF) treaty was signed on 8 December 1987 and entered into force on 1 June 1988. Banned under the INF agreement, the missiles were withdrawn from service between 1988 and 1991 and destroyed. On 1 May 1991 the final GLCM missile and launcher elimination was conducted at Davis-Monthan AFB. In all, the United States eliminated 443 missiles and 123 launchers.

Design: The GLCM was virtually identical to contemporary Navy's Tomahawk land-attack missiles (see Tomahawk entry).

Mission reliability was estimated to be between 74 and 79 percent.

Designation: The name Gryphon was officially adopted for the GLCM but was not widely used.

Operational: The first flight of prototype GLCM was in December 1979.

The GLCM firing unit ("flight") was composed of four transporter-erector-launcher (TEL) vehicles, each of which was fitted with four missiles. Thus, a flight of four TELs had 16 missiles plus 2 launch control vehicles (one primary, one backup), 16 support vehicles, and 69 personnel. The vehicles all were air transportable in C-5, C-130, and C-141 cargo aircraft. Each TEL with four missiles and mated nuclear warheads weighed 77,900 pounds, was 55⅔ feet long, and had self-contained electrical power.

The first GLCMs to deploy arrived at RAF Greenham Common on 14 November 1983; they attained IOC on 1 January 1984. The other NATO deployments followed.

The first GLCMs were withdrawn from western Europe on 8 September 1988 (RAF Molesworth). The last deployed GLCMs were shipped from Comiso, Italy, on 26 March 1991.

Operational:	Jan 1984–1991
Design:	General Dynamics/McDonnell Douglas
Weight:	3,770 lb fueled 3,240 lb without fuel
Length:	21 ft
Diameter:	20¾ in
Wingspan:	8 ft 2 in
Propulsion:	1 Williams F107-WR-400 turbofan
Range:	1,550 miles
Guidance:	inertial navigation with TERCOM; radar altimeter
Warhead:	1 W84 nuclear (0.2-150 KT)
Launch mode:	TEL vehicle

Jupiter SM-78/PGM-19A

The Jupiter was the U.S. Army's entry into the intermediate range ballistic missile (IRBM) field. It was developed in competition with the Air Force's Thor missile; both missiles were developed and deployed.

The IRBM program was the result of a group of leading scientists convened in 1955 by President Eisenhower to look into the potential of the Soviet Union for carrying out a surprise attack—"a nuclear Pearl Harbor"—against the United States. The report, issued on 13 September 1955, recommended the high-priority development of both land- and sea-based IRBMs. The Army's response was the Redstone-derived Jupiter.

On 8 November 1955, Secretary of Defense Charles E. Wilson directed that the Jupiter should be a collaborative effort with the Navy for launching from surface ships. Navy participation in the Jupiter program led to the Polaris program because of Navy opposition to liquid-propellant missiles on board ships and submarines (see entry for Jupiter-S.)

The separate Army–Air Force IRBM efforts raised questions about service responsibilities, and on 26 November 1956, Secretary Wilson issued a directive stating that "operational employment of the land-based Intermediate Range Ballistic Missile system will be the sole responsibility of the U.S. Air Force," and "the U.S. Army will not plan at this time for the operational employment of the Intermediate Range Ballistic Missile or for any other missiles with ranges beyond 200 miles." Accordingly, the Army proceeded with the Jupiter IRBM for use by the Air Force.

There is a consensus that both Jupiter and Thor were obsolete by the time they were deployed in part because of the availability by 1960 of the solid-propellant Polaris and other advanced strategic nuclear weapons.

Design: The Jupiter's warhead weight was 1,600 pounds.

Operational: The first Jupiter was launched on 1 March 1957. The first full-range flight occurred on 31 May 1957.

The missile became operational in 1961 with 15 missiles deployed in Gioia Del Colle, Italy, for joint Air Force–local operation within the NATO structure. Another 30 missiles were installed in Cigli Air Base, Turkey; however, the missiles in Turkey were never fully operational.

A squadron of 15 missiles was organized into five "flights" of three missiles each, located several miles apart to reduce their vulnerability.

The Jupiter IRBMs were in service only from 1961 until 1963, when they were dismantled at the specific direction of President Kennedy in the wake of the 1962 Cuban missile crisis.

A Jupiter C launch vehicle was used to orbit the first American satellite—*Explorer I*—on 31 January 1958. A month earlier the attempt to launch a satellite by the nonmilitary Vanguard booster had been a sensational failure.

Operational:	Italy	14 Apr 1961–1963
	Turkey	6 Nov 1962–1963
Design:	Army Redstone Arsenal and Chrysler	
Weight:	109,000 lb	
Length:	60 ft 1 in	
Diameter:	8 ft 9 in	
Wingspan:	(ballistic)	
Propulsion:	1-stage liquid-propellant rocket motor; 150,000 lbst	
Range:	1,750 miles	
Guidance:	inertial	
Warhead:	1 W49Y2 nuclear (1.4 MT)	
Launch mode:	vertical position from an above-ground launch site	

Redstone Army Arsenal/Chrysler Jupiter IRBM. (U.S. Air Force)

Midgetman MGM-134

The small ICBM (SICBM)—popularly called "Midgetman"—came into existence in 1983 for two primary reasons. The first was an ongoing desire by the Strategic Air Command to have a single-warhead missile to implement nuclear warfighting plans. While the overriding tendency has been to provide multiple warheads on ICBMs, nuclear war planners and targeters desired single-warhead missiles for more flexibility. The Air Force has always had single-warhead ICBMs, and the SICBM was seen as a replacement for the Minuteman II.

A second reason for the SICBM program was the political controversies surrounding the MX/Peacekeeper missile. The SICBM's development was hastened in part because of the inability to find an acceptable basing mode to alleviate the MX's hypothetical or potential vulnerability to Soviet preemptive missile attack. Congressional concerns led to formation of a Presidential Commission on Strategic Forces to resolve the controversy. Among the so-called Scowcroft Commission's report of April 1983 was that development of a single-warhead ICBM weighing about 15 tons should proceed and that the missile should be deployed in hardened silos or shelters and hardened mobile launchers in the early 1990s.[1]

Design: The mobile launcher concept had been considered in the United States for more than 20 years as a method of survivability against Soviet missile attack. Among the mobility methods considered were using rail launch cars, having multiple (covered) launch positions for each missile, and hiding actual missiles in dummy missile fields. The SAC conducted a series of tests as early as 1960 to determine the feasibility of deployed Minuteman ICBMs on mobile railroad car launchers (Operation Big Star). In 1969–1970, in a preview of the debate over the MX/Peacekeeper missile, studies were conducted about deploying Minuteman missiles on special trucks to move upon notice of strategic warning.

The SICBM system was initially discussed as a force of 500 missiles on 500 hard mobile launchers (HML). The initial deployment would be 150 to 200 HMLs at Malmstrom AFB, Montana. Equivalent numbers would then be based at Warren AFB, Wyoming and, lastly, at Ellsworth AFB, South Dakota. Two HMLs would be collocated with an existing silo-based Minuteman ICBM.

The HML consisted of a manned tractor and an unmanned missile launcher with off-road capability. The launch vehicles could keep the missile on alert status and could receive and execute retargeting and launch commands. The two-member crew operating the HML would not participate in the launch of the missile, which would be remotely controlled.

The missile was "cold launched," that is, ejected by gas pressure with the rocket engine igniting when it was free of the HML.

During peacetime the HMLs would be parked inside manned alert shelters near the Minuteman launch facilities. To protect against theft and unauthorized use, the security system for the SICBM would rely on a combination of access delay systems and quick response security teams. The HMLs would be "hardened" to withstand approximately 30 psi overpressure from nearby nuclear explosions.

Operational: The first prototype Midgetman was launched in 1989 but was unsuccessful. The first successful launch occurred on 18 April 1991.

Early in 1988 Secretary of Defense Frank Carlucci recommended terminating the SICBM, principally for cost reasons. In December 1991, President George H. W. Bush announced that the mobile elements of the SICBM program would be cancelled. In the aftermath of the collapse of the Soviet Union in late 1991 the entire program was cancelled.

Operational:	cancelled
Design:	Martin Marietta
Weight:	37,000 lb
Length:	53 ft
Diameter:	3 ft 10 in
Wingspan:	(ballistic)
Propulsion:	3-stage solid-propellant rocket motors
Range:	6,900 miles
Guidance:	lightweight advanced inertial reference sphere (AIRS)
Warhead:	1 W87 nuclear (300 KT)
Launch mode:	hard mobile launcher

1 Lt. Gen. Brent Scowcroft, USAF, served as National Security Advisor to presidents Gerald R. Ford and George H. W. Bush, and as Military Assistant to President Richard M. Nixon.

A Boeing Minuteman I was launched from a C-5A Galaxy transport on 24 October 1974, over the Western Test Range off the California coast. The missile was pulled from the aircraft by drogue parachutes and then, held upright by the parachutes, it fell to about 5,000 feet before the rocket motor ignited. A Navy A3D Skywarrior served as a chase plane. (U.S. Air Force)

Minuteman I HSM-80/LGM-30A/LGM-30B

The Minuteman ICBM was the most numerous U.S. strategic missile with a force of 1,000 missiles deployed at the height of the Cold War. The Minuteman was the second U.S. solid-propellant strategic missile to become operational (following the Polaris A-1 submarine-launched ballistic missile).

On 27 February 1958, Secretary of Defense Neil McElroy approved development of a solid-propellant ICBM with a range of 6,300 miles to be launched from hardened underground silos. The decision was greatly influenced by Soviet ICBM developments and the orbiting of the *Sputnik 1* satellite in October 1957.

On 9 October 1958, Boeing was chosen as the Minuteman missile contractor.

Design: The LGM-30A had a single target capability while the LGM-30B had two pre-stored targets.

Each flight of ten Minuteman missiles was controlled from a launch control center (LCC) buried 50 feet below ground. The LCC was equipped with multiple communications systems, including high frequency, single sideband, ultra-high frequency, low frequency, satellite communications, and dedicated telephone and data systems. Missiles could also be launched from the LCC and from an EC-135 Looking Glass airborne aircraft.

Designation: The Minuteman—originally called Weapon System Q—was named Minuteman in September 1957.

Operational: The first Minuteman IA was placed on alert at Malmstrom AFB, Montana, on 27 October 1962, in the midst of the Cuban missile crisis. In April 1963 the first Minuteman IB was emplaced in silo at Ellsworth AFB, South Dakota.

Boeing Minuteman I ICBM. (U.S. Air Force)

Secretary of Defense Robert McNamara, on 20 November 1964, set the final Minuteman force level at 1,000. Accordingly, Minuteman I deployments were halted with 800 missiles. The total number of Minuteman I weapons procured was 925: 125 for research and development, and 800 for deployment.

The 800 missiles were deployed in 16 squadrons: 150 A models at Malmstrom, and 650 B models with 150 each at Ellsworth; Minot AFB, North Dakota; and Whiteman AFB, Missouri; and 200 at Warren AFB.

The last Minuteman I missile was retired on 3 September 1974.

On 1 March 1965, the first and only successful U.S. ICBM launch from an operational silo occurred with a Minuteman IB. The firing, nicknamed Project Long Life, was conducted from Ellsworth AFB with the warhead removed. The flight lasted seven seconds.

Boeing MM II ICBM in its silo (note technician at right). (U.S. Air Force)

Operational:	Minuteman IA	
	11 Dec 1962–1974	
	Minuteman IB	
	30 Sep 1963–1974	
Design:	Boeing	
Weight:	69,920 lb	
Length:	IA 53 ft 8 in	
	IB 55 ft 9 in	
Diameter:	6 ft 1 in	
Wingspan:	(ballistic)	
Propulsion:	3-stage solid-propellant rocket motors	
	1st stage	167,000 lbst
	2nd stage	49,200 lbst
	3rd stage	18,900 lbst
Range:	7,250 miles	
Guidance:	inertial gimballed	
Warhead:	1 W59Y1 nuclear (1 MT)	
or	1 W56 nuclear (1.2 MT)	
Launch mode:	hardened silo	

Minuteman II LGM-30F

The Minuteman II was an improved ICBM with a titanium second stage and increased range.

Design: In addition to the 450 Minuteman II missiles deployed with nuclear warheads, several additional missiles were fitted with the emergency rocket communications system (ERCS) to broadcast emergency action messages (EAM) for nuclear weapons release on UHF frequencies for about 30 minutes after launch.

The missile had a CEP estimated at 1,200 to 2,000 feet.

Operational: At peak deployment there were 450 Minuteman II ICBMs deployed equally at Malmstrom, Ellsworth, and Whiteman Air Force bases.

The first Minuteman II squadron became operational on 25 April 1966 at Grand Forks AFB, North Dakota. The 450th missile became operational on 21 April 1967.

Two attempts in 1966 and one in 1968 to launch a Minuteman II missile from an operational silo at Grand Forks AFB were unsuccessful.

On 13 December 1966, the first ERCS research and development test was conducted using a Minuteman II missile.

President George H. W. Bush ordered the deactivation of the Minuteman II force on 27 September 1991. The following day they were taken off alert. The Minuteman II silos at Malmstrom were provided with Minuteman III missiles.

Production: A total of 668 Minuteman II missiles were produced.

Operational:	25 Apr 1966–28 Sep 1991
Design:	Boeing
Weight:	72,090 lb
Length:	55 ft 9 in
Diameter:	5 ft 6 in
Propulsion:	3-stage solid-propellant rocket motors
Range:	7,750 miles

Guidance:	inertial gimballed
Warhead:	1 W56 Mod 4 nuclear (1.2 MT)
Launch mode:	hardened silo

Minuteman III LGM-30G

The Minuteman III ICBM introduced multiple warheads to U.S. land-based strategic missiles.

Development of the Minuteman III with a multiple independently targeted reentry vehicle (MIRV) was approved on 8 December 1965.

Design: The Minuteman IIIs were housed in underground hardened and dispersed silos (88½ feet deep with a diameter of 12⅙ feet) with missile suspension, shock isolated floor, debris collection system, and electromagnetic pulse (EMP) protection. The silos are hardened to approximately 2,000 psi.

Minuteman was originally fielded with the capability to store 100 employment options in the missile's onboard computer. The Minuteman Expanded Execution Plan program increased the number of stored attack options from 100 to 2,400.

The missile had a CEP estimated at 600 to 700 feet. The throw weight was 1,760 pounds.

A Pave Pepper flight test on 16 May 1975, demonstrated a seven-MIRV configuration for the Minuteman III using a lighter reentry vehicle.

Operational: The first launch of a Minuteman III from Cape Kennedy occurred on 16 August 1968.

The first MIRVed Minuteman III was emplaced in a silo at Minot AFB on 17 April 1970 and went on alert on 19 August 1970. The last Minuteman III was deployed at Malmstrom AFB on 11 July 1975, bringing the deployed force to 450 Minuteman IIs and 550 Minuteman IIIs. These 550 Minuteman IIIs were deployed at four bases—50 at Malmstrom, 150 at Minot, 150 at Grand Forks, and 200 at Warren. Between 6 January 1986, and mid-1988 50 Minuteman III missiles with Mk 12 RVs were removed from the Warren silos and replaced with 50 MX/Peacekeeper missiles. Between December 1979 and February 1983, the 150 missiles at Minot and the 150 at Grand Forks were refitted with the Mk 12A reentry vehicle carrying a higher yield W78 warhead.

In the late 1990s the 500 Minuteman IIIs were consolidated at the Minot, Warren, and Malmstrom bases. In July 2007 the first of 50 ICBMs of the 564th Missile Squadron at Malmstrom AFB was deactivated. The Air Force planned to reduce the ICBM force from 500 to 450 by mid-2008. Beginning in October 2006 the Air Force began replacing W62 warheads from a portion of the Minuteman III force with W87 warheads from retired MX/Peacekeeper ICBMs. With the abandonment of the START Treaty and its ban on MIRVs, the Air Force planned for 450 ICBMs to carry 500 warheads by the end of 2012.

Production: A total of 840 Minuteman III missiles were procured, including 46 for research and development.

Operational:	29 Dec 1970–current	
Design:	Boeing	
Weight:	77,160 lb	
Length:	59 ft 8 in	
Diameter:	5 ft 6 in	
Wingspan:	(ballistic)	
Propulsion:	3-stage solid-propellant rocket motors	
	1st stage	210,000 lbst
	2nd stage	60,300 lbst
	3rd stage	34,400 lbst
Range:	with Mk 12 RV	9,200 miles
	with Mk 12A RV	8,000 miles
Guidance:	inertial gimballed	
Warhead:	200 missiles with 3 MIRV W62 nuclear (170 KT); warheads to be retired by 2009 and replaced with W87 nuclear (300 KT)	
	300 missiles with 3 MIRV W78 nuclear (335 KT)	
Launch mode:	hardened silo	

Navaho SM-64

The Air Force sponsored development of the North American–developed Navaho—a supersonic strategic cruise missile—simultaneous with the Snark missile (see Snark entry). The Navaho's performance goals were eventually set at a 5,500-mile range carrying a nuclear warhead. A proposal to air-launch the weapon from a B-36 bomber was briefly considered for the Navaho.

Although a ramjet-powered cruise missile, in some respects the Navaho was to be a hybrid cruise-ballistic missile. With a designed launch weight of 290,000 pounds, as large as a contemporary bomber, the Navaho was to be launched pointing skyward at 90° with boost thrust provided by a piggyback booster consisting of three rocket engines. After climbing vertically, the missile would arch over, the massive booster assembly would fall away, and the two internal ramjets would propel the missile toward its target at a speed of Mach 3 and an altitude of more than 60,000 feet.

By 1950 the Air Force specified that a strategic cruise missile would be developed in three stages: the X-10 turbojet test vehicle followed by the SM-64, 4,100-mile interim missile, and finally a 6,325-mile operational weapon. By 1955 the Air Force assigned a high priority to the strategic cruise missile program—second only to ICBM programs—with an operational date of October 1960.

Operational: The first XSM-64 launch attempt in November 1956 ended in failure, as did most of those that followed in the first half of 1957.

The Air Force cancelled the program on 8 July 1957, but authorized additional test flights that were conducted in 1957 and 1958. Most ended in failure. The program cost $700 million with a total flight time for Navaho missiles less than 1½ hours.

Operational:	cancelled
Design:	North American
Weight:	missile 120,500 lb
	boosters 169,500 lb
Length:	87 ft 5 in (missile)
Wingspan:	40 ft 2 in
Wing area:	751 ft^2
Height:	14 ft 11 in
Propulsion:	2 Wright RJ47-W-7 turbojet; 10,035 lbst each
	1 LR83-NA-1 rocket booster; 415,500 lbst
Range:	6,325 miles
Guidance:	inertial
Warhead:	1 W13 nuclear
Launch mode:	launched vertically from a concrete pad using an erector vehicle

Peacekeeper (MX) LGM-118A

The Air Force began a study of the "improved capability missile" (WS-120A) in 1963 to succeed the Minuteman ICBMs. The original Air Force plan was to deploy the new MIRVed missile in fixed Minuteman silos. This basing mode was soon questioned as the controversy increased over strategic missile basing (see Midgetman). The search for more survivable basing modes for the MX would last more than a decade and lead three U.S. presidents to propose six different modes for MX basing. (More than 30 modes were studied and rejected for technical, cost, political, and public acceptability reasons.)

Martin Marietta Peacekeeper ICBM. (U.S. Air Force)

Design: Advanced development of the MX missile began in May 1974. Eleven days before signing the SALT II Treaty on 7 June 1979, President Carter announced his decision to begin full-scale engineering development of a 192,000-pound missile, the largest design option under consideration. And on 7 September 1979, he announced plans to deploy MX in the "racetrack" multiple protective shelters (MPS) basing mode in Utah and Nevada. The scheme envisioned 200 missiles hidden among 4,600 shelters. Each missile would be deployed on a closed-loop roadway approximately 25 miles in circumference, hiding in one of 23 garage-like shelters. With 200 racetracks, the plan would have taken up approximately 30,000 square miles and was expected to have a substantial impact on the states' land, water, environment, and population. Congressional hearings were held in 1979–1980 to examine the impact MX would have on Utah and Nevada.

Upon entering office in January 1981 the Reagan administration announced its support for continuing development of the MX but also decided to reexamine the basing issue. On 2 October 1981, President Reagan, in announcing a comprehensive strategic modernization program, cancelled the MX/MPS basing scheme. As an interim plan, he approved placing 50 MX missiles in superhardened Minuteman silos with continued research on three long-term basing possibilities: continuous patrol aircraft (each carrying an air-launched MX missile), deep underground basing, and ballistic missile defense of MX silos.

In April 1982 the Senate Armed Services Committee rejected the interim plan outright and requested the secretary of defense to deliver, not later than 1 December 1982, a report on basing mode selection. Throughout the spring and summer of 1982 attention shifted to a new basing scheme called closely spaced basing or "dense pack." Finally on 22 November 1982, President Reagan announced his recommendation to place 100 MX missiles in the closely spaced basing mode at Warren AFB, Wyoming. (This concept placed the MX missiles in superhard capsules that were spaced close together; this close spacing theoretically forces attacking weapons to arrive in close proximity to each other, creating "fratricide" when early arriving Soviet warheads detonate, disabling nearby incoming warheads. According to the theory, this phenomenon would limit the size and pace of a Soviet attack so that a significant number of MX missiles would survive.)

This plan was not well received by the public or Congress. Congress reacted by denying procurement money for MX in the fiscal year 1983 budget. In December 1982 Congress requested a report on MX alternatives, and President Reagan established the Commission on Strategic Forces to resolve the controversy. On 6 April 1983, the commission, chaired by Lieutenant General Brent Scowcroft, submitted its report to the president. Its major recommendations were

- immediate deployment of 100 MX missiles in existing Minuteman silos, replacing older Minuteman and Titan II missiles,
- research to resolve uncertainties regarding increased silo hardness,
- investigation of different types of land-based vehicles and launchers, including hardened vehicles, and
- engineering design of a new single-warhead ICBM.

On April 19 President Reagan transmitted to the Congress his recommendation approving the Scowcroft Report. In the following weeks numerous congressional hearings were held to review the report and recommendation.

Throughout 1984 the MX controversy continued and became entwined in election year politics and heightened concerns over large military expenditures. The Reagan administration's MX program suffered a serious setback when, in the summer of 1985, Congress voted to limit the number of MX missiles to 50 deployed in Minuteman silos. Additional missiles would be deployed only if the president proposed and the Congress approved a more survivable basing mode.

From the summer of 1985 to the fall of 1986 the Reagan administration reviewed eight candidate basing schemes for the MX. On 19 December 1986 President Reagan endorsed his fourth basing scheme for MX. Called rail garrison, it would have deployed 50 missiles on 25 trains at seven or more secure garrisons within existing Air Force bases. The main base was Warren AFB. On 11 February 1987, the Air Force identified ten more candidate installations for possible rail garrison basing. Each garrison would consist of about 45 to 50 acres. Each train would have seven cars—a locomotive, two missile cars, two security cars, a launch command and control car, and a maintenance car. The crew of each ICBM garrison train would be 29, consisting of 3 civilian railroad personnel, 1 train commander, a combat missile crew of 4, a maintenance team of 6, and a security team of 15.

The specially designed missile launch cars would weigh in excess of 135,000 pounds and would be 89 feet long. Three or four trains at each site would be parked in shelters constructed of earthen berms and corrugated steel. During normal day-to-day operations, the trains would be on strategic alert in their garrisons. Upon strategic warning the trains would be dispersed onto approximately 120,000 usable miles of track of the U.S. civilian rail system.

One train per garrison would be capable of dispersing within minutes, and the remaining trains within several hours. The trains would be capable of operating autonomously for several weeks if necessary. The missiles could be launched from virtually anywhere and would not need to be at pre-designated points. The missiles could also be launched directly from their shelters. The rail garrison trains would closely resemble existing commercial trains and would have few unique or distinguishable features, complicating Soviet missile targeting procedures.

The trains would not be hardened against blast effects, although the launch control car would be radiation hardened.

The Air Force estimated that the basing program would cost $7.4 billion plus another $3.2 billion for the missiles. A contract was awarded to Boeing in September 1987 for basing, test, and system support for the rail garrison program.

Preliminary findings indicated that approximately 125,000 miles of U.S. railroad track would be available for use by rail-garrison trains. The operational date was set for June 1992, defined as one train on alert with two missiles, plus one spare train at Warren AFB. Full operational capability was scheduled for June 1994.

The Peacekeeper throw weight was 8,690 pounds. The warhead carried a penetration aids package that included decoys, advanced anti-radar chaff, and the penetration aids deployment system (PADS).

Operational: The first MX flight test was conducted on 17 June 1983.

On 3 January 1986, the first Minuteman III was removed from its silo at Warren AFB, and on 17 October 1986, the first MX missile went on alert. On 22 December 1986, IOC was achieved with ten missiles on alert. In December 1988 MX achieved full operational capability with 50 silo-based missiles.

Subsequently, in accord with the START II agreements, the Peacekeeper missiles were retired: 4 were withdrawn from alert in 2002, 17 in 2003, 19 in 2004, and 10 in 2005, the last one on 19 September 2005.

Production: A total of 114 missiles were produced with the program costs estimated at $20 billion through 1998.

Operational:	22 Dec 1986–19 Sep 2005	
Design:	Martin Marietta	
Weight:	195,000 lb	
Length:	71 ft	
Diameter:	7 ft 7 in	
Wingspan:	(ballistic)	
Propulsion:	4-stage solid-propellant rocket motors	
	1st stage	570,000 lbst
	2nd stage	335,000 lbst
	3rd stage	77,000 lbst
	4th stage	reentry vehicles
Range:	6,700 miles	
Guidance:	inertial floating ball	
Warhead:	10 Mk 21 MIRV W87-1 nuclear (300 KT each)	
Launch mode:	hardened silo	

Pershing II MGM-31C

The Pershing II was a U.S. Army medium-range ballistic missile deployed in West Germany on the eve of the U.S.-Soviet agreement to ban such intermediate-range nuclear forces (INF). The 400-KT warhead of the U.S. Army's Pershing Ia missile became a problem in the early 1970s because its size precluded its effective use as an Army battlefield support weapon. Accordingly, the decision was made to develop a new "front end" section for the Pershing, with a lower yield warhead, increased accuracy, and a maneuvering reentry vehicle (MaRV) with active radar terminal guidance. The new missile was named Pershing II.

Martin Marietta Pershing II MRBM. (U.S. Army)

As flight tests of the new RV on existing Pershing Ia rocket stages took place in May 1978, the Pershing II requirements changed in response to the Soviet Union's deployment of the SS-20 Saber (Soviet RSD-10 Pioneer) missile in eastern Europe. That missile had a range in excess of 1,000 miles with a MIRV capability.

Accordingly, the decision was made to counter the Soviet move with U.S.-developed weapons that could be based in western Europe and threaten Soviet forces throughout eastern Europe. These would be both a cruise missile (see the Gryphon GLCM entry) and a ballistic missile—the Pershing II.

To achieve the goals set for the Pershing II new rocket motors employing high-energy fuels and lightweight casings made of Kevlar would be used. This change coupled with the warhead change effectively made the Pershing II a completely new missile. However, the Pershing II would be fully compatible with existing Pershing Ia ground equipment, including launchers (which required minor modifications).

The Pershing II was subject to the INF treaty that went into effect on 1 June 1988. At the time of the treaty signing six months earlier there were 247 accountable Pershing II missiles and 167 launchers.

Design: The MGM-31C reentry vehicle housed a W85 warhead with four variable yield options of 0.3, 5, 10, and 80 KT. With the Singer Kearfott inertial guidance system and the Goodyear Aerospace active radar terminal guidance unit in the warhead, the MGM-31C achieved an accuracy of about 100 feet CEP at ranges of up to 1,200 miles.

Operational: The first flight test of an MGM-31C missile in July 1982 was a failure; the second launch, in November 1982, was a success. On 18 September 1983 the 18th and final flight test of Pershing II occurred; there were nine launches at the White Sands Missile Range, New Mexico, and nine at Cape Canaveral with two failures and seven successes at each site.

The first Pershing II battery with nine missiles became operational in December 1983 in West Germany. By December 1985 all 108 U.S. Army Pershing Ia missiles in western Europe had been replaced by Pershing II. All were assigned to battalions of the U.S. Army's 9th Field Artillery Regiment in West Germany:

Battalion	Location	Launchers
1st Bn	New Ulm	36
2nd Bn	Schwaebisch Gmuend	36
4th Bn	Neckers Ulm	18
4th Bn	Heilbronn	18

In addition, the 3rd Battalion, the training unit, was located at Fort Sill, Oklahoma.

The Pershing II units in West Germany could launch their missiles within 25 minutes after a release order issued by the Army's European Headquarters.

Under terms of the INF treaty, the first nine Pershing IIs stood down at their base near Heilbronn, West Germany, on 1 September 1988; the last missile was shipped from West Germany on 13 March 1991. The last Pershing II battery at Fort Sill stood down on 1 November 1990.[2]

Operational:	Dec 1983–1991
Design:	Martin Marietta
Weight:	16,476 lb
Length:	34 ft 10 in
Diameter:	3 ft 4 in
Wingspan:	(ballistic)
Propulsion:	2-stage Hercules solid-propellant rocket motors
Range:	1,120 miles
Guidance:	inertial with active radar-guided MaRV
Warhead:	1 W85 nuclear (0.3–80 KT)
Launch mode:	mobile erector-launcher

Pluto

The Pluto supersonic low-altitude missile (SLAM) was an ambitious program to develop an "unmanned bomber" powered by a nuclear ramjet engine. SLAM has been described as a "locomotive-size missile that would travel at near treetop level at three times the speed of sound, tossing out hydrogen bombs as it roared overhead."[3]

The name "Pluto" was originally assigned to the missile's nuclear ramjet engine; however, in practice the name was used for the entire program.

The critical component of the program was the missile's nuclear ramjet engine. Ramjet technology offered the promise of high speeds for a large vehicle, while nuclear

2 One Pershing II missile and one SS-20 missile are preserved and on display at the National Air and Space Museum on the National Mall in Washington, D.C., and a pair are similarly preserved at the Military Museum in Moscow.

3 Gregg Herken, "The Flying Crowbar," *Air & Space Magazine* (April/May 1990), p. 28.

technology could provide the "fuel" necessary to power the vehicle for intercontinental ranges. The ramjet was to propel the aircraft at Mach 3 speeds, with air temperature at entry to the engine almost 1,000°F with an exit (exhaust) temperature of 2,200°F.

Engine size and weight as well as nuclear-radiation issues were major constraints on the project.

On 1 January 1957, the Lawrence Livermore National Laboratory was selected to develop the nuclear engine. On 14 May 1961, after extensive development efforts, the world's first nuclear ramjet engine—Tory-IIA—mounted on a railroad car, was tested for 45 seconds at Jackass Flats, Nevada. This was a scale model. The rail-car engine mounting enabled remote-controlled retrieval into the protective shed where the engine could be fueled and maintained by remote control because of residual radioactivity.

Three years later, on 20 May 1964, the Tory-IIC nuclear ramjet was run for five minutes at full power, producing 513 megawatts and the equivalent of over 35,000 pounds of thrust.

Despite these successful tests, the rapid development of ICBMs and the operational problems associated with what was essentially an unmanned nuclear-powered bomber led to the decision on 1 July 1964, to cancel the Pluto program. The entire effort, costing some $320 million by the end of fiscal year 1964, was related to engine development. The engine weighed 14,410 pounds.

The airframe was not developed.

Operational:	cancelled
Design:	Livermore (engine)
Propulsion:	nuclear ramjet; approx 35,000 lbst
Range:	intercontinental
Guidance:	preset and terrain contour matching (TERCOM)
Warhead:	5,500 lb nuclear bombs; tentative nuclear loadout options:
	1 — 26 MT
	5 — 1.3 MT
	9 — 1.1 MT
	14 — 750 KT
	16 — 200 KT
	36 — 50 KT
	42 — 5 KT
Launch mode:	land takeoff

Snark SSM-A-3/SM-62

Snark was the first nuclear-strike missile developed under the auspices of the Army Air Forces. It had a very brief operational service life.

In August 1945 the AAF established a requirement for a missile capable of 600-mph speeds with a 5,000-mile range carrying a 2,000-pound conventional warhead. Early in 1946, Northrop received a study contract to develop a missile—to be named Snark—and a supersonic missile with a range of 1,500 to 5,000 miles carrying a 5,000-pound warhead, the Boojum. While the Boojum missile was quickly aborted, the Snark was pursued, as was the more ambitious Navaho missile from North American Aviation.

Even before the first successful flight of a Snark test vehicle in 1951, the Air Force increased the Snark's performance requirements to include a supersonic dash at the end of a 6,350-mile flight and to carry a warhead of 7,000 pounds (later reduced to 6,250 pounds) with a CEP of 1,500 feet. The Snark thus became a nuclear system, to be fitted with the W39 warhead. Northrop immediately scaled up its existing missile to meet the new requirements, and the redesigned missile began test flights in November 1952.

Numerous problems beset the Snark program, and flight tests were not completed until May 1955. This was almost two years past the originally planned operational date of the missile.

The Snark became operational 1960 and was retired in 1961. Even after becoming operational with the Strate-

Northrop Snark strategic cruise missile. (U.S. Air Force)

gic Air Command, the missile fell far short of expectations in flight tests. Historian Kenneth Werrell summarized the Snark's problems as "the weapon's low reliability . . . , inability to penetrate, lack of positive control, and vulnerable, unprotected launch sites."[4]

The United States did not deploy another land-based strategic cruise missile for 25 years. That weapon was the Gryphon (neé Tomahawk), which was briefly deployed in Europe in the mid-1980s.

Design: The Snark was essentially a pilotless bomber (its sequence number 62 was in the bomber series). The missile had a swept-wing configuration without horizontal tail surfaces. It was land-launched from a zero-length launcher with two large, solid-propellant boosters. At one point a nuclear reactor was proposed for a Snark ramjet propulsion (see Pluto entry).

The missile's guidance was self-contained, being preset with inertial updating, and thus was not jammable. The remarkable CEP was 1.6 miles at the end of an intercontinental flight.

As the Snark approached the target, the warhead was ejected from the airframe and the warhead achieved an ultimate velocity of about Mach 1. (The airframe would then begin to tumble and break up.)

Designation: The Snark was initiated under the weapons system designation MX-775A. The name "Snark" was derived from the writings of Lewis Carroll.

Operational: The first successful Snark test flight was on 31 October 1957, and the missile traveled 4,400 miles. This flight was six weeks ahead of the first successful Atlas missile test on 17 December 1957. Thus, the Snark qualified as the first U.S. intercontinental missile.

The final Snark test flight was on 5 December 1960.

On 18 March 1960—ten years after the first test launch of a Snark test vehicle—the first operational missile was placed on alert at Presque Isle, Maine. On 28 February 1961, SAC headquarters declared the 702d Strategic Missile Wing at Presque Isle AFB to be operational.

The Snark program survived just over a year, until June 1961, when the missiles were taken off alert and retired. This was less than four months after the last of the 30 SM-62A production missiles had been declared combat ready (February 1961). In directing the deactivation of the Snark force, President Kennedy labeled the Snark "obsolete and of marginal utility value."

The number of Snark crashes near the Florida missile test center at Cape Canaveral led to the term "Snark infested waters." At the other extreme, in December 1956 a Snark flew "too far"—it failed to respond to control signals and was last seen flying toward the jungles of Brazil. In 1982 a Brazilian farmer found the remains of the errant missile.

Operational:	18 March 1960–June 1961	
Design:	Northrop	
Weight:	missile	49,603 lb
	boosters	11,365 lb
Length:	69 ft 11 in	
Wingspan:	42 ft 4 in	
Wing area:	326 ft^2	
Height:	14 ft 10 in	
Propulsion:	1 Pratt & Whitney J57-P-17 turbojet; 10,500 lbst	
	2 solid-fuel rocket boosters; 65,000 lbst each	
Speed:	615 mph	
Range:	5,700 miles	
Ceiling:	49,000 ft	
Guidance:	inertial aided by stellar monitoring	
Warhead:	1 W39Y1 Mod 1 nuclear (3.8 MT)	
Launch mode:	mobile short-rail with rocket boosters	

Thor SM-75/PGM-17

The Thor was one of two IRBMs developed by the United States; it was developed for the Air Force. The development of the Thor was initiated in November 1955, after the start of the Army's Jupiter IRBM, because the Air Force was institutionally incapable of accepting an Army-designed missile. The Thor became operational shortly before the Jupiter. Air Force leaders envisioned that Thor IRBMs would be emplaced around the periphery of the Soviet Union pending the availability of ICBMs, similar to the basing of B-47 bombers overseas pending a buildup of the B-52 bomber force.[5]

The Air Force issued a contract to Douglas Aircraft and directed that, to the extent possible, Atlas ICBM technology was to be employed in the Thor. The Thor was deployed in 1959–1960 in Great Britain under Royal

4 Kenneth P. Werrell, *The Evolution of the Cruise Missile* (Maxwell Air Force Base, Ala.: Air University Press, September 1985), p. 97.
5 Thomas C. Reed, *At the Abyss: An Insider's History of the Cold War* (New York: Random House, 2004), p. 79.

Douglas Thor IRBM. (U.S. Air Force)

Air Force (RAF) operational control with a joint U.S. Air Force–Royal Air Force "two-key" procedure for launching the missiles; that is, specified individuals in both air forces had to certify launch orders simultaneously to initiate the missile firing sequence. Historian Julian Hartt, in his book *The Mighty Thor* (1961), explained:

> On the consoles there were two locks with similar nomenclature stamped above them. The RAF officer's key, on decision of the Prime Minister, can initiate the countdown for combat launching. But it cannot release the nuclear warhead. The USAF officer's key, on decision of the President of the United States, can bring to life that nuclear warhead. But it cannot launch a missile.[6]

The Thor provided Britain with a nuclear missile strike capability five years earlier than would have been possible with a British weapons system. (In fact, Britain did not achieve a post-Thor strategic missile capability until the Polaris SLBM went to sea in HMS *Resolution* in June 1968.)

Design: The Thor was a single-stage, liquid-fuel IRBM designed to deliver a nuclear warhead at ranges between 345 and 1,740 miles.

It was emplaced in open, vulnerable, aboveground launch sites.

Operational: The first successful Thor flight test was in September 1957. The first launch by an RAF crew was from Cape Canaveral on 16 April 1959.

The Thor was declared operational in September 1958, although it would not be deployed for another year. Sixty missiles were deployed in Great Britain beginning in December 1959 where they were operated by the Royal Air Force. At the time of the Cuban missile crisis in October 1962, the RAF had 59 missiles available for launch with the 60th used for training.

The Thor's rapid development and deployment led aviation writer Bill Gunston to observe, "This timescale has never been equalled for any major weapon system."[7] After a massive trans-Atlantic airlift of Thor missiles, support equipment, and personnel, the first Thor IRBMs became operational under the U.S. Strategic Air Command in Britain in December 1959. The entire Thor IRBM force was operational by mid-1960 under British operational control:

RAF Strategic Missile Squadron	Location	SAC Turnover	Inactivated
7th	RAF Hemswell	11 Sep 1959	May 1963
8th	RAF Driffield	22 Dec 1959	Apr 1963
44th	RAF North Luffenham	29 Apr 1960	Sep 1963
77th	RAF Feltwell	22 June 1959	July 1963

Each squadron had five geographically dispersed launch positions, each containing three launchers. The missiles were stored horizontally on transporter-erector trailers and covered by a retractable 100-foot-long garage-like shelter. To fire, the shelter was rolled back and the missile was lifted to a vertical position and fueled. Fifteen missiles and about 500 personnel were assigned to each squadron.

6 Quoted in Humphrey Wynn, *The RAF Strategic Nuclear Deterrent Forces: Their Origins, Roles and Deployment 1946–1969* (London: Her Majesty's Stationery Office, 1994), p. 345, fn 2.

7 Bill Gunston, *The Illustrated Encyclopedia of the World's Missiles & Rockets* (London: Salamander Books, 1979), p. 60.

The Thor and Jupiter IRBMs had relatively short service lives; both missile systems were deactivated and discarded in 1963 at the direction of President Kennedy following the Cuban missile crisis. There is a consensus that both missiles were obsolete by the time that they were deployed, in part because of the availability by 1960 of the solid-propellant Polaris and other advanced nuclear-capable weapons.

The Thor squadrons were inactivated in 1963 and all missiles and warheads were returned to the United States for disposal.

A Thor missile served as the first-stage booster for the U.S. Pioneer moon rocket launched on 11 October 1958.

Operational:	9 Dec 1959–Sep 1963
Design:	Douglas
Weight:	110,000 lb
Length:	64 ft 10 in
Diameter:	8 ft
Wingspan:	(ballistic)
Propulsion:	1-stage liquid-propellant rocket motor; 150,000 lbst 2 liquid-propellant rocket boosters; 1,000 lbst each
Range:	1,740 miles
Guidance:	all-inertial
Warhead:	1 W49Y2 nuclear (1.4 MT)
Launch mode:	unprotected launch pad

Titan I SM-68/LGM-25/HGM-25A

With the Atlas D/E/F, the Titan I was a first-generation U.S. ICBM. All of these missiles had liquid propellants and were capable of reaching key targets in the Soviet Union from launching sites in the north-central United States. These early missiles were initially installed in "soft," aboveground installations, highly vulnerable to bomber or missile attacks. Atlas deployments peaked at 133 missiles in 1962; the Titan I program peaked at 54 missiles the following year. Six squadrons of nine Titan I missiles were placed at five Air Force bases: Lowry, Colorado; Mountain Home, Idaho; Beale, California; Larson, Washington; and Ellsworth, South Dakota.

Subsequently the Titan I missiles were placed in hardened, underground silos. Elevators lifted the missiles to ground level for launching (thus making them vulnerable to attack, albeit for brief periods). All Titan I missiles were dismantled by 1964; they were replaced by the improved Titan II.

Design: When "hardened," a Titan I complex consisted of three reinforced concrete launch silos, each 160 feet deep and 40 feet in diameter, together with connecting tunnels and an underground power source and launch control center. The Titan I would be fueled while in the silo and then raised to the surface by elevator.

The missile's Mark 4 reentry vehicle weighed 3,956 pounds. The Atlas E/F and Titan I used the same W38 nuclear warhead.

Operational: The first Titan I ICBM was emplaced on 6 October 1961, at Lowry AFB; the first Titan I squadron became operational at Lowry AFB on 20 April 1962. All six Titan I squadrons were operational by 16 August 1962 and reached full missile strength in 1963.

On 19 November 1964, Secretary McNamara announced that all Titan I (and Atlas) ICBMs would be

Martin Titan I ICBM. (U.S. Air Force)

phased out by the end of June 1965. The Air Force code-name for the Titan I phaseout was Deep Trouble. The first Titan Is were taken off alert on 4 January 1965 and the last on 1 April 1965.

Production: Titan I production totaled 155 missiles—47 research and development and 108 production.

Operational:	20 Apr 1962–1 Apr 1965	
Design:	Martin	
Weight:	226,000 lb	
Length:	98 ft 3 in	
Diameter:	10 ft	
Wingspan:	(ballistic)	
Propulsion:	2-stage liquid-propellant rocket motors	
	1st stage	300,000 lbst
	2nd stage	80,000 lbst
Range:	6,400 miles	
Guidance:	radio-inertial	
Warhead:	1 W38 Mod 0 nuclear (4.5 MT)	
Launch mode:	hardened silo, elevated for launch	

Titan II SM-68B/LGM-25C

The Titan II was the largest U.S. ICBM to be developed and deployed, weighing almost five times as much as the Minuteman and carrying a Mk 53 warhead of about nine megatons to the same distance. This was the largest nuclear warhead in the U.S. missile arsenal, either land-based or sea-based.

Martin received a contract for development of the Titan II in June 1960.

Air Force planning in the early 1960s provided for 1,200 Minuteman missiles and 120 Titan II weapons as the second-generation SAC missile force. At peak deployment there were 54 missiles in six squadrons at three sites: Davis-Monthan AFB, Arizona; McConnell AFB, Kansas; and Little Rock AFB, Arkansas.

Design: The Titan II weighed 50 percent more than the Titan I and was longer and had a greater diameter. The later missile used storable liquid fuels consisting of Aerozine 50 and dinitrogen tetroxide. The missile was launched from within the silo.

Operational: The first Titan II flight was in December 1961. The missile attained IOC in October 1963.

The Titan II retirement program began in October 1982 with one missile per month being dismantled; the last Titan II was deactivated on 5 May 1987 at Little Rock AFB.

Martin Marietta Titan II ICBM. (U.S. Air Force)

Twelve Titan II boosters were used by NASA in the Gemini manned space program in the 1960s. Titan IIs with an Agena third-stage rocket have been used to launch U.S. reconnaissance satellites.

There were long-standing safety concerns associated with the Titan II ICBM, and it had a history of fires, accidents, and fuel leaks. On 8 August 1965, 53 contract workers died in a flash fire while working on a silo near Searcy, Arkansas. Another accident occurred near Damascus, Arkansas, on 19 September 1980, when an Air Force repairman dropped a heavy wrench socket while working in a Titan II silo. The socket punctured the missile's first stage causing the pressurized fuel tank to leak. About 8½ hours later fuel vapors within the silo ignited and exploded, causing 1 death and injuries to 21 others. The tremendous explosion blew open the 740-ton silo door and catapulted the nuclear warhead 600 feet into nearby woods. There was no radiation leakage.

Production: The number of Titan IIs procured was 127: 19 research and development and 108 production missiles.

Operational:	Oct 1963–May 1987	
Design:	Martin Marietta	
Weight:	327,000 lb	
Length:	103 ft	
Diameter:	10 ft	
Wingspan:	(ballistic)	
Propulsion:	2-stage (storable) liquid-propellant rocket motors	
	1st stage	430,000 lbst
	2nd stage	100,000 lbst
Range:	8,400 miles	
Guidance:	all-inertial	
Warhead:	1 W53Y1 nuclear (9 MT)	
Launch mode:	hardened silo	

SUBMARINE-LAUNCHED MISSILES

Missile	Type[a]	Service	IOC	Warhead
EXPO	SLBM	Navy	—	became Trident I
Jupiter-S	SLBM	Navy	cancelled	W49
Loon	SLCM	Navy	cancelled	Mk 8
Polaris A-1	SLBM	Navy	1960	W47
Polaris A-2	SLBM	Navy	1962	W47
Polaris A-3	SLBM	Navy	1964	W58
Poseidon C-3	SLBM	Navy	1971	W68
Regulus I	SLCM	Navy	1959	Mk 5/W27
Regulus II	SLCM	Navy	cancelled	W27
Rigel	SLCM	Navy	cancelled	Mk 5
Tomahawk	SLCM	Navy	1987	W80
Trident I	SLBM	Navy	1979	W76
Trident II	SLBM	Navy	1990	W76/W88
Tritino	SLCM	Navy	cancelled	—
Triton	SLCM	Navy	cancelled	Mk 5/W27

a SLBM = submarine-launched ballistic missile; SLCM = submarine-launched cruise missile

TABLE 5-1. Submarine-Launched Missile Deployments

Missile	First patrol begun	Last patrol completed	Total Patrols
Regulus I	*Tunny* 23 Oct 1959	*Halibut* 14 July 1964	41
Polaris A-1	*George Washington* 15 Nov 1960	*Abraham Lincoln* 14 Oct 1965	83
Polaris A-2	*Ethan Allen* 26 June 1962	*John Marshall* 9 June 1974	338
Polaris A-3	*Daniel Webster* 28 Sep 1964	*Robert E. Lee* 1 Oct 1981[a]	844
Poseidon C-3	*James Madison* 31 Mar 1971	*Kamehameha* and *Ulysses S. Grant*[b] 1 Oct 1991	1,182
Trident C-4	*Francis Scott Key* 20 Oct 1979	*Michigan* 15 Dec 2003	887
Trident D-5	*Tennessee* 29 Mar 1990	(missile in service)	495[c]

a This is the date that the *Robert E. Lee* was taken off alert; she was still at sea at the time, but Polaris patrols officially ended on 1 October 1981.
b The *Kamehameha* and *Ulysses S. Grant* were taken off alert on this date. The submarines were still at sea, returning to port on 15 and 16 October 1981, respectively; their deterrent patrols officially ended on 1 October 1991.
c Through 30 June 2008.

The Jupiter-S (Seaborne) was to have been a sea-based adaption of the Army's Jupiter IRBM. This artist's concept shows how the four missiles would have been fitted in the sail and hull of the submarine—an arrangement similar to early Soviet missile submarines. The Jupiter-S would have been launched from the surface. The U.S. Navy eschewed the liquid-propellant Jupiter and shifted to the solid-fuel Polaris. (U.S. Navy)

EXPO

A final version of the Polaris/Poseidon SLBM series was the so-called extended-range Poseidon (EXPO) missile, which entered advanced development in the early 1970s. This missile was soon renamed Trident I (C-4).

Jupiter-S

In the fall of 1955, as a result of anticipated Soviet advances in strategic missiles, Secretary of Defense Charles E. Wilson directed the U.S. Navy to join the Army in development of an IRBM that could be launched from surface ships. The Army was developing the Jupiter IRBM, based on its earlier Redstone battlefield ballistic missile. The seaborne version was designated Jupiter-S.

The proposed naval version of the Jupiter IRBM was intended for launching from surface ships and possibly submarines. The joint Army–Navy IRBM program was established on 8 November 1955. Initial plans were for merchant ships to be converted to carry the missile with a planned IOC of 1959. Studies were also undertaken to address the feasibility of fitting the missile in submarines, with three missiles envisioned for each IRBM-armed submarine. The submarines would operate at the near surface, with the top of their sail structure exposed, to launch the missiles.

The Navy objected strenuously to the joint program because the Navy considered liquid propellants too dangerous to handle at sea, and the 60-foot missile would be difficult to install in ships or submarines. The naval Jupiter-S was to be shorter but wider than the land-launched Jupiter.

The joint effort ended in December 1956 when the U.S. Navy withdrew from the Jupiter program and initiated the solid-propellant Polaris SLBM program.

Operational:	cancelled
Design:	Army Redstone Arsenal and Navy
Weight:	approx 110,000 lbs
Length:	50 ft
Diameter:	10 ft in
Wingspan:	(ballistic)
Propulsion:	liquid-propellant rocket motors; 150,000 lbst
Range:	1,725 miles
Guidance:	inertial
Warhead:	1 W49 nuclear (1.4 MT)
Platforms:	merchant ships submarines

Loon LTV-N-2[8]

The Loon was the U.S. Navy's version of the German-developed V-1 "buzz bomb" (see Historical comments). The U.S. Navy briefly considered a nuclear-armed version of the Loon for submarine launching as an interim land-attack weapon.

8 LTV = launching test vehicle.

V-1 components were gathered by the British soon after the first missiles struck England in June 1944. A shipment of V-1 parts were at Wright Field, Ohio, by late July, and within two weeks Army engineers were able to assemble an operable V-1.

An American copy of the V-1—dubbed JB-1 for Jet Bomb No. 2 by the Army—was rushed into production. One thousand were initially ordered from Republic Aviation, with early plans indicating that some 75,000 would be required for the campaign against Japan. Launching JB-2s from bombers was not thought feasible, and the Army looked to Navy landing ships as launch platforms. However, the end of the war in August 1945 led to termination of the JB-2 effort with 1,391 missiles having been produced.

The Navy took delivery of 351 missiles from the Army order. The Navy, beginning in September 1944, had considered launching the JB-2 from escort aircraft carriers, but that plan was also stillborn. Instead, the Navy fitted the submarine *Cusk* (SS 348) in early 1947 with a 50-foot-long launching ramp, small hangar, and equipment to launch the missile in support of the development of the submarine-launched Regulus missile.[9] The missile was called Loon by the Navy and designated as a launch test vehicle—LTV-N-2.

The *Cusk* became the world's first submarine to launch a guided missile—a Loon—on 18 February 1947. (That missile crashed after traveling 6,000 yards, apparently due to a control malfunction.)

The potential of the Loon as a land-attack weapon was immediately considered, especially since the first American submarine-launched missile, the Regulus, would not be available with a nuclear warhead before 1954.[10] The warhead proposed was the so-called Elsie. This was the Mk 8 nuclear weapon but with a yield of only 3 KT. It was developed by Los Alamos specifically for use by naval aircraft against hardened land targets—such as submarine "pens" (bunkers).

Providing a nuclear warhead in the Loon would have produced the world's first nuclear-armed missile. However, the project died in 1950, primarily because of the limited range and accuracy of the missile. The latter was

Willys-Aero JB-2 cruise missile. (U.S. Army Air Forces)

estimated as a 40 percent probability of striking within a mile of the target and 70 percent of hitting within five miles. With a 3-KT warhead, this was considered too poor for practical use.

The submarine *Carbonero* (SS/SSG 337) was also employed in Loon launch trials.

Designation: The Army developed the JB-2 as weapon system MX-544. The Navy's initial designation as a test vehicle was KGW-1, changed to KUW-1 in 1946, and changed to LTV-N-2 in 1947.

Historical: The V-1 was the first guided missile in history to be employed in large numbers against an enemy. The German Air Force began launching the V-1 (for *Vergeltungswaffe* or reprisal weapon) against London on 13 June 1944. German records indicate that 8,564 of the missiles were launched against England and the Belgium port of Antwerp; about 43 percent failed or were diverted or destroyed by fighters, anti-aircraft fire, or barrage balloons. Over England, fighters—especially the jet-propelled Meteor—could intercept the missiles; more than a thousand anti-aircraft guns were set up in belts beneath the missiles' path.

In the first two weeks of the V-1 campaign, the missiles fell on London at an average of nearly 100 every 24 hours. During an 80-day period the V-1s wrecked or damaged more than 1,000,000 buildings in England, killed 6,184 persons, and seriously injured 17,981 others.

The V-1 was 25 feet, 11 inches long, with a wingspan of 17 feet, 5 inches, and weighed 4,806 pounds at launch.

9 The *Cusk*, a fleet submarine of the *Balao* (SS 285) class, was originally commissioned on 5 February 1946. She was modified for the Loon-launching role in 1947 and redesignated a guided missile submarine (SSG 348) on 20 January 1948. The hangar fitted to the *Cusk* in her SSGN role could hold a single Loon with wings detached.

10 Deputy Chief of Naval Operations (Air) memo to Chief of Naval Operations, via Op-03, subj: "LOON with Atomic Warhead: Operational use of," 15 May 1950.

A Loon cruise missile is launched from the submarine *Carbonero* (SS 337) off Point Mugu, California. The Loon—navalized version of the Army's JB-2 and the German V-1 missiles—was considered for a nuclear warhead and a land-attack role when it appeared that the Regulus cruise missile would be delayed. Note the booster rocket under the missile (about to fall away) and the pulse-jet engine atop the missile. (U.S. Navy)

The standard range was 150 miles with a speed of 400 mph. It carried a 1,870-pound warhead. At least 29,000 V-1s were produced

Most V-1s were launched from ground ramps, but from July 1944 some 1,200 of the missiles were air-launched from He 111H twin-engine bombers. Also, a scheme was developed for the Ar 234B Blitz jet-propelled bomber to tow a V-1 for air launching. A proposed piloted version of the V-1 was flight tested but never used operationally.

Operational: The nuclear-armed version of the Loon was cancelled. The following data is for the standard LVT-N-2 test vehicle.

Operational:	cancelled as nuclear weapon	
Design:	Willys-Aero	
Weight:	empty	2,245 lb
	payload	1,875 lb
	gross	5,200 lb
Length:	27 ft 1 in	
Wingspan:	17 ft 8½ in	
Wing area:	60.7 ft^2	
Height:	5 ft (bottom of fuselage to top of engine)	
Propulsion:	1 Ford PJ-31-1 pulse-jet engine; 900 lbst	
	1 solid-propellant rocket booster	
Speed:	440 mph	
Range:	150 miles	
Guidance:	preset	
Warhead:	1,875-lb conventional	
Launch mode:	ramp	

Polaris A-1 UGM-27A

This was the first SLBM developed and deployed by the United States and the world's first long-range ballistic missile with solid-propellant rocket motors. The first submarine to carry the Polaris A-1 missile was the *George Washington* (SSBN 598), which deployed on the first Polaris missile patrol in November 1960. The Soviet SS-N-4 Sark (Soviet R-13) went to sea earlier, but the Polaris SLBM was more advanced and truly a revolutionary weapon system.

After the efforts of a joint Army–Navy missile program from November 1955 to December 1956 to develop an IRBM that could be launched from ships as well as from land sites, the Navy embarked on the fleet ballistic missile (FBM) program (see entry for Jupiter-S).

The development of smaller nuclear warheads and higher specific impulse solid-fuel propellants permitted (1) a break from the Army's Jupiter program in December 1956, (2) formal initiation of the Polaris SLBM program with a solid-propellant missile, and (3) a shift from surface ships to submarines as the principal launch platform.

On 8 February 1957, Admiral Arleigh Burke, Chief of Naval Operations from 1955 to 1961 and a strong backer of the SLBM program, issued a requirement for a 1,500-nautical-mile missile launched from a submarine to be operational by 1965. A range of some 1,500 n.miles was stipulated to permit a submarine in the Norwegian Sea to target the Soviet capital of Moscow—some 1,100 n.miles inland. The February 1957 schedule with a goal of 1965 was soon followed by a series of revisions and accelerations in the Polaris program. On 4 October 1957, the Soviets orbited the world's first artificial satellite; on 23 October the Secretary of the Navy proposed acceleration of Polaris to provide a 1,200-n.mile missile by December 1959, with three SLBM submarines available by mid-1962, and a 1,500-n.mile missile available by mid-1963.

A month later the program was further accelerated to provide the 1,200-n.mile missile by October 1960, and in December 1957 the plan provided for the first submarine in December 1959 and the second in March 1960.

To produce the submarines in so short a time, on the last day of 1957 the Navy reordered a recently begun nuclear-propelled torpedo-attack submarine (SSN) and a second not-yet-started unit as ballistic missile submarines. Given the highest national priorities, the first Polaris submarine, renamed *George Washington*, was rushed to completion (see Operational comments) and was armed with the 1,200-n.mile A-1 missile. There was little remarkable in the SSBN design—basically the Navy "stretched" the highly successful *Skipjack* (SSN 585) design. The redesign added 130 feet in length to permit the installation of special navigation, missile control, and other mission support equipment as well as 16 launch tubes for the Polaris missile. This resulted in the largest submarine built by any nation to that time, except for the nuclear-propelled, radar picket *Triton* (SSRN 586). That submarine, however, was a one-of-a-kind craft, intended primarily to test a two-reactor propulsion plant.

Polaris missiles were also proposed for installation in Navy surface ships. Space and weight for Polaris missile tubes were provided in the nuclear cruiser *Long Beach* (CGN 9) as well as several conventional missile cruisers; in addition, the Italian missile cruiser *Giuseppe Garibaldi*, rebuilt as a missile cruiser in 1957–1962, was fitted with four Polaris missile tubes. But no Polaris SLBMs were ever installed in surface ships except for the U.S. test ship *Observation Island* (T-AG 154).

During 1964 several NATO nations were involved in a program to develop the Multilateral Force (MLF) of multination manned surface ships armed with Polaris missiles. This proposal also was cancelled.

Design: The A-1 was the first SLBM developed and deployed in the West. Among the most important SLBM submarine components were the fire control and navigation systems. The latter was particularly critical in view of the ranges of the missile and the need for the submarine to

Lockheed Polaris A-1 SLBM; the Polaris A-2 was similar in appearance. (U.S. Navy)

remain submerged (except for masts and antennas raised on a periodic basis). Recalling that this was in the period before the development of navigation satellites, the ships inertial navigation system (SINS) developed for the Polaris program was also a remarkable technological achievement. SINS had to provide accurate position information regardless of surface and atmospheric conditions.

While the Polaris circular error of probability (CEP) remains classified, a declassified secretary of the Navy memorandum of 30 January 1958, credited the A-1 missile with a CEP of three to four miles and the A-2 missile with a CEP of two miles. These CEPs were probably bettered in service.[11]

The missiles were launched from the submarine at a depth of approximately 50 to 60 feet from the top of the sail structure to the surface. A steam generator launch system ejected the missile from the launch tube, propelling it through the water column to the surface. After the

11 Secretary of the Navy, Memorandum "Augmentation and Acceleration of the Fleet Ballistic Missile (POLARIS) Program," 30 January 1958.

missile attained sufficient velocity and distance from the submarine, the first-stage motor ignited. This procedure has been followed for subsequent submarine-launched ballistic missiles.

Operational: The first successful flight test of a Polaris test vehicle from Cape Canaveral occurred on 20 April 1959. The first seaborne launch of a Polaris test missile was from the USS *Observation Island* at sea off Cape Canaveral on 27 August 1959. The *Washington* successfully launched two A-1 missiles on 20 July 1960, off Cape Canaveral, Florida, marking the first time that a missile was fired from a submerged U.S. submarine. The missiles were fired at an interval of 2 hours, 53 minutes; they streaked more than 1,150 miles down the Atlantic test range.

On 15 November 1960, the *Washington* sailed from Charleston, South Carolina, on the Navy's first Polaris deterrent patrol carrying 16 Polaris A-1 missiles. The submarine returned to New London, Connecticut, on 21 January 1961, after a submerged cruise of 66 days, 10 hours. The voyage was the longest U.S. submerged submarine operation made to that date.

On 6 May 1962, the submarine *Ethan Allen* (SSBN 608), operating near Christmas Island in the Pacific, fired a Polaris A-1 missile with a nuclear warhead that detonated. This was the only full-systems test of a U.S. strategic missile—land- or sea-based—from launch tube through detonation. The test had the code name Operation Frigate Bird.

The Polaris A-1 was fitted in the five nuclear-propelled submarines of the *George Washington* class, which had an IOC of 1960–1961; each submarine carried 16 missiles. All were later rearmed with the Polaris A-3 missile.

The A-1 was operational from 15 November 1960 until 14 October 1965, when the last submarine carrying the missile, the *Abraham Lincoln* (SSBN 602), returned to a U.S. port.

Operational:	15 Nov 1960–14 Oct 1965
Design:	Lockheed
Weight:	28,000 lb
Length:	28 ft
Diameter:	4 ft 6 in
Wingspan:	(ballistic)
Propulsion:	2-stage solid-propellant rocket motors
Range:	1,380 miles
Guidance:	inertial
Warhead:	1 W47Y1 nuclear (600 KT)
Launch mode:	5 nuclear-propelled submarines of the *George Washington* class (16 tubes)

Polaris A-2 UGM-27B

The Polaris A-2 was an interim SLBM, developed to the original Polaris requirement for a 1,500-n.mile missile. It was rapidly succeeded in service by the Polaris A-3 missile. (See Polaris A-1 entry for development history.)

Operational: The Polaris A-2 was carried in 13 submarines:

Submarine Class	Built	IOC	Missiles
Ethan Allen	5	1961–1963	16
Lafayette	8	1963–1964	16

On 23 October 1961, the *Ethan Allen* launched the first Polaris A-2 missile fired from a submarine.

The five *Ethan Allen*-class submarines were rearmed with the A-3 missile and the eight submarines of the *Lafayette* (SSBN 616) class with the Poseidon C-3 missile.

All Polaris submarine launches have been made submerged with two exceptions—the surface launches were from the *Henry Clay* (SSBN 625) firing an A-2 missile on 20 April 1964, and the *Nathaniel Green* (SSBN 636) firing an A-3 missile on 15 March 1965.

Operational:	26 June 1962–9 June 1974
Design:	Lockheed
Weight:	32,500 lb
Length:	31 ft
Diameter:	4 ft 6 in
Wingspan:	(ballistic)
Propulsion:	2-stage solid-propellant rocket motors
Range:	1,730 miles
Guidance:	inertial
Warhead:	1 W47Y2 nuclear (1.2 MT)
Launch mode:	see Operational comments

Polaris A-3 UGM-27C

The A-3 was the ultimate Polaris SLBM design, its successor as an operational missile being the much larger Poseidon C-3 missile. The A-3 differed from earlier Polaris missiles by having a greater range and a multiple reentry vehicle (MRV) warhead that "shotgunned" three nuclear warheads onto a single target to overcome the limitations of missile accuracy, providing a wider area of damage

The USS *Sam Rayburn* (SSBN 635) shows the innovative arrangement for 16 missile tubes of U.S. Polaris-Poseidon missile submarines. Earlier Soviet ballistic missile submarines had carried two or three SLBMs. Also, the U.S. Polaris missile could be fired while the submarine remained fully submerged, another innovative feature of Polaris. (U.S. Navy)

against soft targets. This was the only U.S. strategic missile to be deployed with MRV warheads.

The Polaris A-3 replaced the manned bomber in British service as the nation's strategic deterrence force. The British A-3 missiles were produced in the United States by Lockheed and fitted with a British-developed warhead. The original British A-3 missile was replaced by the upgraded Chavelin A-3TK variant.

Design: The A-3 payload was 1,347 pounds at 2,590 miles. The payload could consist of three Mk 2 warheads (1,100 pounds) or three Mk 12 warheads plus penetration aides; the latter configuration was never deployed.

Operational: The A-3 was operational in 18 U.S. and 4 British submarines.

Submarine Class	Built	IOC	Missiles
George Washington	5	1966–1967	16
Ethan Allen	5	1974–1976	16
Lafayette	8	1964–1967	16
Resolution	4	1967–1969	16

The USS *Andrew Jackson* (SSBN 619) fired the first A-3 missile on 26 October 1963.

The eight *Lafayette*-class submarines were rearmed with the Poseidon missile; the ten earlier A-3 submarines (converted from the A-1 and A-2 missiles) were retired.

The first British SLBM patrol began on 22 June 1968, with HMS *Resolution* carrying 16 A-3 missiles; the last of four British Polaris submarines, the *Repulse*, completed the

Lockheed Polaris A-3 SLBM. (U.S. Navy)

last British A-3 patrol on 13 May 1996. (The Polaris was succeeded by the Trident D-5 missile in the British strategic deterrent force.)

Operational:	28 Sep 1964–1 Oct 1981
Design:	Lockheed
Weight:	35,700 lb
Length:	32 ft
Diameter:	4 ft 6 in
Wingspan:	(ballistic)
Propulsion:	2-stage solid-propellant rocket motors
Range:	2,880 miles
Guidance:	inertial
Warhead:	3 MRV W58 nuclear (200 KT each)
Launch mode:	see Operational notes

Poseidon C-3 UGM-73

The Poseidon was the world's first strategic missile to carry multiple independently targeted reentry vehicles (MIRV). Up to 14 warheads could be fitted, albeit at the reduction of range from its predecessor, the Polaris A-3 missile. The MIRV program—adopted for subsequent U.S. ICBMs and SLBMs—was intended primarily to overcome expected Soviet anti-ballistic missile (ABM) systems and to provide coverage of an expanding number of targets in the communist world.

The Poseidon missiles were normally loaded with only ten warheads to provide approximately the same range as the A-3. Coupled with the MIRV warhead, the Poseidon also introduced a new level of accuracy to SLBMs, approaching that of land-based ICBMs.

The December 1987 U.S.-Soviet summit agreement to negotiate reductions in strategic missile forces stated that each Poseidon missile would be counted as carrying ten MIRVs. The Reagan administration's decision to delay the retirement of *Lafayette*-class submarines armed with Poseidon SLBMs allowed the number of MIRVed missile launchers to exceed the limits established under the unratified (but informally observed) SALT II agreement of 1979 between the United States and the Soviet Union.

Design: By reducing the relatively severe shock mitigation requirement imposed for Polaris, additional diam-

Lockheed Poseidon C-3 SLBM. (U.S. Navy)

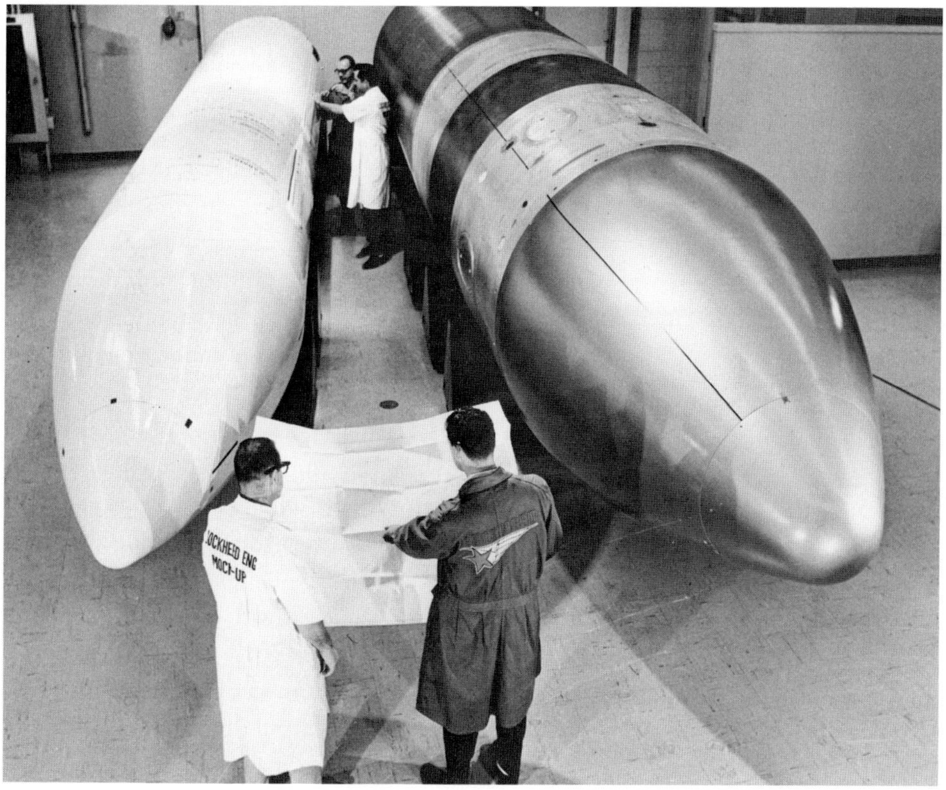

Polaris A-3 (left) and Poseidon C-3 SLBMs. (Lockheed)

eter could be obtained using the existing missile tubes in U.S. Polaris submarines. Thus missile diameters as large as 6 feet 2 inches could be obtained in the existing submarines, which provided the opportunity for large warheads. This capability matured into the B-3 concept, a missile carrying multiple Mk 12 reentry bodies or even the larger Mk 17, or large numbers of the smaller Mk 3 MIRVs, which evolved into the Poseidon C-3. This weapon was significantly longer, had a greater diameter, and was almost twice the weight of the Polaris.

As deployed, the Poseidon's post-boost "bus" could carry up to 14 Mk 3 reentry vehicles with a relatively large attack "footprint." The missile's throw weight was approximately 3,300 pounds. Estimated CEP was approximately 1,500 feet.

Operational: The Poseidon was fitted in 31 nuclear-propelled submarines of the *Lafayette* class that were originally built to carry the Polaris A-2 missile (first 8 units) or A-3 missile (23 units). All 31 submarines were rearmed with the Poseidon missile by September 1974. The "conversion" of the submarines to launch the Poseidon missile required relatively minor work. (Twelve of these submarines were subsequently rearmed with the Trident C-4 missile.)

The first flight test model of the Poseidon missile (designated C3X-1) was successfully launched from Cape Kennedy on 16 August 1968.

The first submarine to launch a Poseidon missile was the *James Madison* (SSBN 627) on 3 August 1970.

Operational:	31 Mar 1971–1 Oct 1991
Design:	Lockheed
Weight:	64,000 lb
Length:	34 ft
Diameter:	6 ft 2 in
Wingspan:	(ballistic)
Propulsion:	2-stage solid-propellant rocket motors
Range:	2,800 miles with 10 RVs
Guidance:	inertial
Warhead:	10 to 14 MIRV W68 nuclear (50 KT each)
Launch mode:	31 submarines of the *Lafayette* class (16 tubes)

Regulus I SSM-N-8/RGM-6

The Regulus I was the U.S. Navy's first missile to carry a nuclear warhead and was its first operational submarine-

Chance-Vought Regulus I SLCM. (U.S. Navy)

launched missile. Known as the Regulus attack missile (RAM), the weapon was fitted in five U.S. Navy submarines that conducted deterrent patrols in the Western Pacific.[12] It was also carried briefly in aircraft carriers and heavy cruisers.

The missile was designed primarily for the land-attack role from surfaced submarines and surface warships. A secondary anti-ship capability was provided.

An initial Navy contract with Chance-Vought Aircraft on 3 October 1945 initiated the Regulus program. Fabrication of the Regulus test vehicles began in 1948, and their success led to the Navy issuing an operational requirement for the missile in May 1950.

The Regulus I was to have been succeeded by the more-capable Regulus II. However, the Regulus I was retained in service after the later weapon was cancelled. When the Regulus I was retired in 1964 it marked the end of long-range cruise missiles in the U.S. Navy until introduction of the Tomahawk in 1982.

Design: The Regulus concept was based on previous tests of the American versions of the German V-1 pulse-jet rocket from surfaced submarines (U.S. Navy LTV-N-2 Loon).

The missile was carried in a submarine hangar, then extracted and wings spread, engine started, and launched via a short-length ramp. (The Soviet surface-launched missile systems were much more efficient with the missile container also providing the launch mechanism. The Soviet Navy operated 62 submarines—conventional and nuclear—that carried the similar SS-N-3 Shaddock (Soviet P-5 Pyatyorka.[13])

The Regulus missile resembled a contemporary, swept-wing fighter aircraft with the air intake in the nose, a small, swept-back vertical tail surface (there was no horizontal tail), and—in training models—a fully retractable landing gear. The missile was launched with two solid-propellant rocket boosters.

On a regular basis, controllers were trained on unarmed Regulus missiles that could be launched with retractable landing gear, flown on a simulated mission, and then landed at an airfield. (More than 1,000 such practice flights were made by the Navy with both operational and training missiles.)

Originally the Regulus was to be armed with a conventional warhead (4,000 pounds), in part because of the limited data on nuclear warheads made available to the Navy. In 1949 a nuclear warhead was proposed for the missile, and operational Regulus missiles were only fitted with nuclear warheads, initially the Mk 5 and, from 1957, the W27.

Regulus guidance consisted of an autopilot with radio commands from the launching submarine, or other submarines, surface ships, or aircraft. Radio commands were transmitted using the Trounce system. In the first joint cruiser-submarine Regulus operation on 19 November 1957, the cruiser *Helena* (CA 75) launched the missile and

12 Two definitive works on the Regulus missile are David K. Stumpf, *Regulus: The Forgotten Weapon* (Paducah, Ky.: Turner Publishing, 1996), and Berend Derk Bruins, "U.S. Naval Bombardment Missiles, 1940–1958: A Study of the Weapons Innovation Process," PhD dissertation (New York: Columbia University, 1981).

13 The Soviet submarines were:

Number	Type		Project	Missiles
6	SSG	Whiskey-Twin Cylinder	644	2
16	SSG	Juliett	651	4
6	SSG	Whiskey-Long Bin	665	4
5	SSGN	Echo I	659	6
29	SSGN	Echo II	675	8

guided it for 129 miles; the submarine *Cusk* then assumed guidance control for 81 miles; the guidance was then given over to the submarine *Carbonero*, which guided the missile for the last 104 miles to target. The missile struck about 150 yards of the target point. The CEP for the Regulus I was reported to be 300 yards in a 1957 Navy evaluation.

Several attack routes could be flown, with approaches to the target from up to 40,000 feet down to treetop level. Diving on its target, Regulus attained a speed of Mach 1.1.

The missile had folding wings and tail fin for shipboard stowage.

Designation: The Regulus I was originally designated SSM-N-8 in the multiservice designation system; it was changed to RGM-6 in 1963. The recoverable training versions were designated KDU-1 and, after 1963, BQM-6C.

Operational: The first XSSM-N-8 test vehicle flew on 29 March 1951 under the control of an airborne controller. Fitted with wheels, the Regulus took off under its own power, circled the airfield, and landed safely.

The first shipboard launch was made from the missile test ship *Norton Sound* (AVM 1, ex–AV 11) on 3 November 1952, and the aircraft carrier *Princeton* (CVA 37) launched a Regulus missile on 16 December 1952. The first submarine launch of a Regulus I occurred on 15 July 1953 from the surfaced submarine *Tunny* (SSG 282).

The Regulus I became operational in May 1954 on board the *Tunny* in the Atlantic Fleet. In the Pacific the first operational missiles went to sea on the cruiser *Los Angeles* (CA 135) in 1955. Several heavy cruisers as well as six *Essex*-class carriers are believed to have deployed with Regulus I missiles, with other carriers having also launched the missile in trials.

During the Lebanon crisis of 1958, the *Tunny*—carrying two Regulus missiles—was ordered into the North Pacific to relieve a nuclear-capable aircraft carrier, and the missile submarine took up station above the Arctic Circle. This was more than two years before the first U.S. ballistic missile submarine, the *George Washington*, began the first Polaris missile patrol.

From 23 October 1959 to 14 July 1964, submarines carrying Regulus I missiles were continuously on patrol in the North Pacific area. During this period Regulus missiles and warheads were stored in Chichi Jima, an island in the Japanese Bonin group. A total of 41 Regulus deterrent patrols were conducted in that period, with one or two submarines with a total of four or five missiles normally "on station," their missiles aimed at targets in the Soviet Far East.

A total of five Regulus submarines were in service:

Submarine		IOC	Missiles
SSG 282	*Tunny*	1953	2 Regulus I
SSG 317	*Barbero*	1955	2 Regulus I
SSG 574	*Grayback*	1958	4 Regulus I or 2 Regulus II
SSG 577	*Growler*	1958	4 Regulus I or 2 Regulus II
SSGN 587	*Halibut*	1960	5 Regulus I or 2 Regulus II

Additional nuclear submarines were planned, but with cancellation of the Regulus II those submarines were reordered as torpedo-attack submarines of the *Thresher* (SSN 593) class.

Submarine launching of the Regulus was a complex evolution: the submarine surfaced, crewmen came on deck, the missile hangar was opened, final preparations were made to the Regulus (including unfolding the wings and tail), it was moved onto a launch ramp, the crew withdrew, and the missile was launched. It was a difficult procedure on a submarine deck, impossible in rough seas.

Zero-length launchers for the Regulus were provided to four heavy cruisers and to eight aircraft carriers of the *Essex* (CV 9) class, with each ship carrying several missiles. (The missiles were removed from the carriers by 1956 as the A3D Skywarrior and A4D Skyhawk attack aircraft became available.) The cruisers could accommodate three Regulus I or two Regulus II missiles; additional missiles could be carried if pre-loading "marriage" of the warhead and missile were accepted (as was done in submarines).

There were also proposals for employing former aircraft carriers to carry and launch hundreds of Regulus missiles.[14]

The first operational flight of the KDU-1 drone version of the Regulus I occurred on 31 March 1955. More than 900 Regulus/KDU-1 launches were recorded, with one reusable training missile having been flown 18 times.

Production: Chance Vought produced 514 Regulus missiles—228 tactical missiles, 209 fleet training missiles, and 77 drones. The last was delivered in December 1958.

14 See George Fielding Eliot, *Victory without War 1958–1961* (Annapolis, Md.: U.S. Naval Institute, 1958).

The diesel-electric submarine *Grayback* (SSG 574) entering port with a Regulus I cruise missile in launching position. Twin hangars for four Regulus I missiles are faired into the bow section. Alternatively, the similar *Grayback* and *Growler* (SSG 577) could have carried two Regulus II missiles. The first underwater-launch cruise missiles went to sea in 1967—in Soviet submarines. (U.S. Navy)

Operational:	23 Oct 1959–14 July 1964
Design:	Chance Vought
Weight:	13,485 lb
Length:	34 ft 3¾ in
Diameter:	4 ft 1¾ in
Wingspan:	21 ft
Height:	8 ft 10 in
Propulsion:	1 Allison J33-A-14 turbojet engine; 4,600 lbst
	2 Aerojet 2KS-33000 jet-assisted takeoff boosters; 16,500 lbst each
Speed:	high subsonic (Mach 0.85–0.95)
Range:	575 miles
Guidance:	preset or radio control (Trounce)
Warhead:	1 Mk 5 nuclear (10–45 KT)
or	1 W27 nuclear (2 MT)
Launch mode:	zero-length ramp on submarine and surface ships

Regulus II SSM-N-9/RGM-15A

The Regulus II was an advanced land-attack missile for launching from surfaced submarines and surface warships. Early successes of Regulus I test vehicles led to the development of an improved missile, with the Regulus II being initiated in July 1953. Another factor in the decision to proceed with Regulus II was the difficulties being encountered with the development of the Rigel missile. The Regulus II missile was a supersonic (Mach 2) missile with increased range and accuracy over the Regulus I.

Despite a highly successful development and test program, on 12 December 1958, then–Secretary of the Navy Thomas S. Gates ordered the termination of the Regulus II program because of the decision to accelerate development and deployment of the Polaris SLBM.

At the time the project ended 12 cruisers and 12 nuclear-propelled submarines were planned for Regulus II installations, in addition to two diesel-electric submarines already in service.[15] As part of the termination, the four new construction submarines of the *Permit* (SSGN 594) class intended to carry the Regulus II (four missiles per submarine) were reordered as torpedo-attack submarines (SSN) of the *Thresher* class; two diesel submarines already completed and one nuclear submarine under construction went to sea with the Regulus I.

Perhaps less grandiose, but still significant with respect to potential roles for the Regulus II, was the interest expressed by the U.S. Air Force in employing the weapon as an air-to-surface missile carried by B-52s (informally

15 USS *Grayback* and USS *Growler* were commissioned earlier in 1958. They continued to serve as Regulus I submarines until 1963. Subsequently, the *Grayback* was converted to a troop transport in 1967–1968 (redesignated APSS 574 and then LPSS 574); she served in that role until 1984.

Chance-Vought Regulus II SLCM. (U.S. Navy)

discussed in 1956) and by the French Navy in using the missile from its two aircraft carriers (discussed in 1958). Both the German and Italian governments also expressed some interest in the missile.

Design: A more-streamlined and efficient design than the Regulus I, this missile had a pointed nose, small canards, and swept-back wings. The turbojet engine was mounted under the after portion of the fuselage; it was launched with a solid-propellant rocket booster fitted under the fuselage. Speed was approximately Mach 2. The missile could reach an altitude of more than 58,000 feet or could fly at treetop level to avoid radar detection.

There were proposals to extend the range beyond that indicated by employing external wing tanks.

The Regulus II was to carry the W27 warhead (initial planning provided for the Mk 8). No conventional warhead was envisioned.

Like the Regulus I, the missile could be fitted with a fully retractable landing gear for training flights. The missile had folding wings and tail fin for shipboard stowage.

Designation: The Regulus II, originally designated SSM-N-9, was given the designation RGM-15A in the post-1963 scheme. The training/target version was originally designated KD2U.

Operational: The first Regulus II test flight was conducted on 29 May 1956. The first submarine launch occurred on 16 September 1958 from the surfaced submarine *Grayback*; this was the only submarine launching of the missile. One additional shipboard launching was made on 11 December 1958 from the Regulus test ship *King County* (AG 157, ex–LST 857).

A total of 48 missile test flights were conducted prior to cancellation of the project with 30 being considered successful—that is, achieving all primary objectives—14 partially successful, and 4 failures.

At the time of cancellation the missile was to have become operational in January 1960.

Production: The missile production run was from March 1956 through 1958. More than 100 were used for training or expended as targets.

Operational:	cancelled
Design:	Chance Vought
Weight:	22,564 lb missile
	4,653 lb rocket boosters
Length:	57 ft + 10 ft 2 in nose boom
Diameter:	
Wingspan:	20 ft ½ in
Wing area:	145 feet^2
Height:	12 ft 6 in
Propulsion:	1 General Electric J-79-GE-3 turbojet; 15,000 lbst in afterburner
	1 Aerojet General rocket booster; 115,000 lbst for 4 seconds
Speed:	maximum 1,320 mph (Mach 2)
	cruise 630 mph (Mach 0.94)
Range:	655 miles with 515 miles at Mach 2
	920 miles with 115 miles at Mach 2
	1,200 miles with no distance at Mach 2
Guidance:	inertial
Warhead:	1 W27Y1 nuclear (2 MT)
Launch mode:	zero-length ramp on submarines and surface ships

Rigel SSM-N-6

The Rigel was a highly advanced concept for its time, providing for a high-speed (Mach 2) land-attack missile launched from surfaced submarines. The program was initiated by the Navy in 1946 and in 1948 was designated as the potential follow-on to the Regulus I/II missiles.

General Dynamics/Convair Tomahawk TLAM. (Raytheon)

Difficulties with the catapult-launch requirements—with a large ramp required because of the ramjet propulsion—led to cancellation of the project on 5 August 1953.

Design: The Rigel had an aircraft-type configuration with canards forward, twin ramjets mounted on small tail wings, and four solid-propellant rocket boosters mounted between the four tail surfaces. Despite the power of its ramjet and boosters, the Rigel would have required a catapult some 400 feet long, making it impossible to operate from a submarine. Subsequent efforts to develop a scheme for launching the missile directly from its stowage canister were not successful prior to cancellation.

Early planning called for a Mach 2 missile and a 3,000-pound warhead with a range of 575 miles, a cruise altitude of 50,000 feet, and a CEP of 600 yards.

Operational: The first Rigel test vehicle flight on 4 May 1950 reached a speed of Mach 1.9 and a range of 20.7 miles; a later vehicle attained Mach 2 and travelled 73.7 miles. No tactical missiles were produced.

All flight tests employed a $\frac{6}{10}$th model of the proposed tactical missile. No full-scale missile was produced because of the limitations of ramjet engines during the test phase.

Operational:	cancelled
Design:	Grumman
Weight:	approx 23,800 lb
Length:	
Wingspan:	
Height	
Propulsion:	2 ramjet engines + 4 rocket boosters
Range:	approx 690 miles
Guidance:	radio control + preprogram or terminal homing
Warhead:	1 Mk 5 nuclear
Launch mode:	submarines

Tomahawk BGM-109/UGM-109

The Tomahawk land-attack missile (TLAM) became prominent as the ship-launched cruise missile (SLCM) in the 1974–1975 negotiations between the United States and Soviet Union over strategic arms limitations. The deployment of 375 SLCMs on 25 surface warships was proposed by the United States to balance 300 Tu-22M Backfire bombers in the Soviet strategic force.

The Chief of Naval Operations, Admiral James L. Holloway, opposed the SLCM proposal, in part because U.S. naval aviation officers feared the SLCM as a possible competitor for carrier-based strike aircraft.[16] Pushed into developing the SLCM, in 1979 the Navy began using the terms Tomahawk land-attack missile (TLAM) and Tomahawk anti-ship missile (TASM) to distinguish the principal variants. The TLAM weapons were developed with nuclear and several conventional warhead options, with the TSAM variant have a conventional, 1,000-pound warhead.

The Tomahawk was deployed in armored box launchers (ABL) on 4 battleships, 5 cruisers, and 7 destroyers of the *Spruance* (DD 963) class; it was fitted in the vertical launchers (Mk 41) of 22 cruisers of the *Ticonderoga* (CG 47) class, 62 destroyers of the *Arleigh Burke* (DDG 51) class, and 24 *Spruance*-class destroyers. It also can be fired from standard (21-inch) submarine torpedo tubes and, in the Improved *Los Angeles* (SSN 688) and the *Virginia* (SSN 774) classes, from 12 vertical launch tubes. In 2003–2008 the Navy converted four early Trident submarines of the

16 See Henry Kissinger, *Years of Renewal* (New York: Simon & Schuster, 1999), pp. 853–61.

Ohio (SSBN 724) class to combination transport/TLAM submarines with the capability to carry 154 TLAMs in 22 ex-Trident tubes. Thus, the "holes" available in the fleet for Tomahawk missiles have reached several thousand by 2009, although far fewer missiles were available.

The TASM missiles were phased out of service; in 1992 the TLAM-nuclear weapons were taken off ships and, when this book went to press, were stored at the Trident submarine bases at Bangor, Washington, and Kings Bay, Georgia. Today only conventional TLAM versions of the Tomahawk are carried in U.S. surface ships and submarines, and in British torpedo-attack submarines (SSN).

Design: Development of the SLCM began in June 1972, and the Navy selected a General Dynamics design over a competitive Ling-Temco-Vought (LTV) design in January 1974.

The missile was designed from the outset to fit into a 21-inch-diameter, 21-foot-long torpedo tubes, in large part because of the efforts of Admiral Elmo R. Zumwalt, Chief of Naval Operations from 1970 to 1974, who wanted the weapon to be suitable for all types of surface warships and submarines.

Upon exiting its container/launcher, the Tomahawk's stub wings and air intake deploy. Powered by a turbofan engine, the Tomahawk has a maximum range of some 1,000 miles at subsonic speed. It normally flies a low-level flight profile, climbing just before reaching its target, then climbing to dive on the target. Over land the early TLAMs employed terrain contour matching (TERCOM).

The accuracy of the early TLAM missiles was on the order of 33 feet (10 meters); later missiles employing the global positioning satellite (GPS) system have more accuracy. Press reports have cited approximately 3 feet (1 meter).

The nuclear variant was designated BGM-109A and carried the W80-0 warhead.

An air-launched Tomahawk competed unsuccessfully with the Boeing air-launched cruise missile (ALCM) for use on B-52 strategic bombers. Tomahawk launch trials were carried out successfully from A-6 Intruder aircraft.

The BGM-109G ground-launched cruise missile (GLCM) variant was selected as a theater nuclear weapon for deployment in western Europe under Air Force control, but those weapons were withdrawn and destroyed under the Intermediate-range Nuclear Forces (INF) treaty. See Gryphon missile entry (pages 167–68).

There were several conventional TLAM/TASM variants; their warheads were a 1,000-pound Bullpup unitary warhead, bomblets, and carbon-fiber spools for disrupting electric power generators and transmission wires.[17]

Operational: IOC TASM in surface ships was fitted in 1982, TASM in submarines in 1983, TLAM in surface ships in 1984, and TLAM(N) in 1987.

A Tomahawk cruise missile was successfully launched from the submarine *Barb* (SSN 696) on 1 February 1978; the *Guitarro* (SSN 665) was the first submarine armed with Tomahawk. The *Merrill* (DD 976) was fitted with the first Tomahawk surface installation in October 1982 for at-sea evaluation; the battleship *New Jersey* (BB 62) was the second ship, receiving the Tomahawk in March 1983.

The first operational use of the Tomahawk was in the 1991 Gulf War (Operation Desert Storm). The U.S. Navy fired 288 Tomahawks—276 from surface combatants and 12 from submarines. Of the 288 missiles, approximately 80 percent were daylight attacks and 20 percent night. Ninety-eight percent of the missiles transitioned to a cruise profile for a successful strike. According to the official Department of Defense report on the Gulf War:

> The TLAM played an important role in the air campaign as the only weapon system used to attack central Baghdad in daylight. The cruise missile concept—incorporating an unmanned, low-observable platform able to strike accurately at long distances—was validated as a significant new instrument for future conflicts.[18]

Over the next decade, additional TLAMs were launched against targets in Iraq, Afghanistan, Bosnia, and Sudan.

Only the conventional TLAM is now operational.

Operational:	1987–1992 nuclear variant
Design:	General Dynamics/Convair
Weight:	2,650 lb
Length:	18 ft 3 in
Diameter:	1 ft 8½ in
Wingspan:	8 ft 8 in

17 The TLAM-D missile dispenses 166 BLU-97 bomblets weighing 3.4 pounds in packets of 24; these submunitions can be armor-piercing, fragmentation, or incendiary.

18 U.S. Department of Defense, "Conduct of the Persian Gulf War: Final Report to Congress" (Washington, D.C.: April 1992), p. 244.

Propulsion:	1 Williams F107-WR-400 turbofan; 600 lbst
	1 solid-propellant rocket booster
Speed:	approx 550 mph
Range:	approx 1,000 miles
Guidance:	inertial and TERCOM
Warhead:	1 W80-0 nuclear (5–150 KT)
Launch mode:	21-inch torpedo tubes and vertical-launch cells in submarines
	vertical launch cells in surface ships

Trident I C-4 UGM-96

A final version of the Polaris/Poseidon series was the so-called EXPO missile (extended-range Poseidon), which entered advanced development in the early 1970s. This missile was soon renamed the Trident I/C-4, with 12 of the original 41 Polaris submarines being further converted to carry this weapon. In addition, the first eight submarines of the *Ohio* class were armed with the Trident C-4 missile.

The Trident I SLBM program evolved from the Department of Defense STRAT-X study of the late 1960s that proposed an advanced SLBM with a range of 6,000 nautical miles to be carried in a new class of submarines. Subsequently, the Navy proposed a two-phase program: the Trident C-4 based on the EXPO missile with a range of some 4,000 nautical miles, and the later Trident D-5, to be developed at a later date with the longer range.

Design: The C-4 missile had a MIRV warhead with eight Mk 4 independently targeted reentry vehicles. It had double the yield and twice the accuracy of the previous Poseidon C-3 missile. The missile carried the Mk 4 reentry body; it was designed to alternatively carry the Mk 500 Evader maneuvering reentry vehicle (MaRV) warhead; this was designed to overcome ballistic missile defenses and was not developed. Missile throw weight was more than 3,000 pounds.

The C-4 was a three-stage missile. After it reached a certain altitude an aerospike extended from the nose. This spike cut the friction of the air flowing past the missile, extending its range by about 350 miles.

The missile's CEP had been estimated at approximately 500 yards.

Operational: The first test flight of a Trident C-4 was on 6 March 1974. The first test flight from a submarine, the *Francis Scott Key* (SSBN 657), was on 10 April 1979.

Lockheed Trident I C-4 SLBM. (U.S. Air Force)

The first test firing from the USS *Ohio* was on 17 January 1982. Trident C-4 IOC was in October 1979.

Twelve *Lafayette*-class submarines (16 tubes) and the first eight *Ohio*-class submarines (24 tubes) were fitted with this missile. Of the latter submarines, four were subsequently converted to transport/cruise missile submarines (SSGN) and four were modified to carry the Trident D-5 missile.

Operational:	20 Oct 1979–15 Dec 2003
Design:	Lockheed
Weight:	73,000 lb
Length:	34 ft
Diameter:	6 ft 2 in
Wingspan:	(ballistic)
Propulsion:	3-stage solid-propellant rocket motors

Range:	4,600 miles
Guidance:	inertial and stellar
Warhead:	8 MIRV W76 nuclear (100 KT)
Launch mode:	12 submarines of the *Lafayette* class (12 tubes)
	8 submarines of the *Ohio* class (24 tubes)

Trident II D-5 UGM-133A

The Trident D-5 is the ultimate SLBM developed by the United States and now arms all 14 Trident ballistic missile submarines of the *Ohio* class. The D-5 missile evolved from the Department of Defense STRAT-X study of 1966.

President Reagan announced on 2 October 1981 that the Navy would fund development of the Trident D-5 with a December 1989 IOC as part of his Strategic Modernization Program. On 29 April 1982 Secretary of the Navy John Lehman decided to place the D-5 missile on the 9th *Ohio*-class SSBN instead of the 12th as planned.

Design: The D-5 is similar in several respects to the Trident C-4 but is significantly larger with a larger payload and increased accuracy. The range is similar to that of the C-4.

President George H. W. Bush cancelled production of the missile's W88 warhead on 28 January 1992. Production had already been suspended some two years earlier because of concerns about safety at the plant producing the plutonium triggers for the weapon. Unofficial estimates indicate that about 400 W88 warheads were produced by 1990. Accordingly, the Navy employs two warheads on Trident D-5 missiles: the new Mk 5 reentry vehicle with the 455-KT W88 warhead and the Mk 4 reentry vehicle with the 100-KT W76 warhead that was used on the C-4 missiles. Each D-5 missile carries only one type of warhead. The warhead complement on individual submarines have varied with a submarine loaded with some Mk 4 MIRV warheads and some carrying Mk 5 MIRV warheads.

Lockheed Trident II D-5 SLBM. (U.S. Navy)

The D-5 CEP is estimated to be in the 400- to 500-foot range.

The START I counting rule for Trident D-5 missiles was eight RVs whether they carry Mk 4s or Mk 5s.

The first eight *Ohio*-class submarines carried the C-4 missile. Four were converted to non-SSBN configurations and the second four (SSBN 730–733) have been modified to carry the D-5 missile.

Operational: The first development test flight from Kennedy Space Center at Cape Canaveral, Florida, occurred on 15 January 1987. The first submarine launch (a failure) took place on 21 March 1989, from the USS *Tennessee* (SSBN 734). The D-5 achieved IOC aboard the *Tennessee* in March 1990.

On 11 March 1982, Great Britain acquired the Trident D-5 missile for its second-generation ballistic missile submarines of the *Vanguard* class. The first D-5 patrol was undertaken by HMS *Vanguard* in December 1994.

Operational:	29 Mar 1990–current
Design:	Lockheed
Weight:	approx 130,000 lb
Length:	44 ft
Diameter:	6 ft 11 in
Wingspan:	(ballistic)
Propulsion:	3-stage solid-propellant rocket motors
Range:	4,600 miles
Guidance:	inertial and stellar
Warhead:	8 MIRV W76 nuclear (100 KT)
or	8 MIRV W88 nuclear (455 KT)
Launch mode:	14 submarines of the *Ohio* class (24 tubes)
	4 submarines of the *Vanguard* class (16 tubes)

Tritino

This U.S. Navy program was intended to provide a precise nuclear strike cruise missile for use from submarines; secondary launch platforms considered were aircraft and surface ships of destroyer size and larger. The missile was cancelled early in the design stage.

Design: This was to have been a Mach 4 missile. The warhead was to be nuclear with a weight of 150 pounds.

Operational: The planned IOC was in 1965.

Operational:	cancelled
Design:	
Weight:	7,050 pounds
Length:	24 ft 1 in
Diameter:	
Wingspan:	4 ft 6 in
Propulsion:	1 ramjet engine
Range:	2,070 miles
Guidance:	
Warhead:	1 nuclear (0.1 KT)
Launch mode:	aircraft surface ships submarines

Triton SSM-N-2

The Triton was envisioned as a high-speed successor to the Rigel missile (which, in turn, was the planned successor to the Regulus). Whereas Regulus and Rigel had aircraft-type configurations and were the products of the Navy's Bureau of Aeronautics, the Triton was part of the "Bumblebee" family of guided missiles developed under the aegis of the Navy's Bureau of Ordnance (see entries for Talos, Terrier).

Early plans called for a Mach 1.6 to Mach 2.5 speed with a range of 230 miles and a weight of 4,000-pounds. A CEP of 600 yards was specified.

Full-scale development was approved in 1955. Within a year the requirements were modified and the planned IOC pushed ahead from about 1960 out to 1965. The new specifications called for a speed of Mach 2.4 or more and delivery of a 1,500-pound nuclear warhead to a target 1,380 miles away.

At the time of cancellation the Triton was listed at Mach 3.5 with a range of 1,725 miles. Delays occurred in the program, caused in part by funding constraints imposed by the Navy's participation in the Army-initiated Jupiter IRBM program. The Triton was formally cancelled in September 1957.

Design: The missile had a cylindrical configuration with a tandem booster and small canards forward, X-fitted wings on the rear of the missile, and two larger tail fins with end plates.

The radar map-matching scheme proposed for this missile was an early version of the terrain contour matching (TERCOM) concept used in the later Tomahawk land-attack missile (TLAM).

The missile was planned for use on submarines, cruisers, and aircraft carriers, that is, the same platforms as the Regulus and Rigel.

Operational: The Triton project was cancelled in December 1955.

Operational:	cancelled
Design:	McDonnell
Weight:	
Length:	
Wingspan:	
Height:	
Propulsion:	
Range:	
Guidance:	
Warhead:	1 Mk 5 or W27 nuclear
Launch mode:	aircraft carriers, cruisers, and submarines

AIR-LAUNCHED MISSILES

Missile	Type[a]	Service[b]	IOC	Warhead
ACM	ALCM	AF	1991	W80-1
ALCM	ALCM	AF	1982	W80-1
Crossbow	ALCM	AF	cancelled	Mk 7/W31
Hound Dog	ALCM	AF	1960	W28
Mastiff	ALCM	AAF	cancelled	—
Rascal	ALCM	AF	1957	Mk 5/W27
SCAD	ALCM	AF	cancelled	—
Skybolt	ALBM	AF	cancelled	W59
SRAM-A	ALCM	AF	1972	W69
SRAM-II	ALCM	AF	cancelled	W89
TASSM	ALCM	AF/MC/N	cancelled	—

a ALCM = air-launched cruise missile; ALBM = air-launched ballistic missile
b AAF = Army Air Forces; AF = Air Force; MC = Marine Corps; N = Navy

ACM AGM-129

The advanced cruise missile (ACM) was a low-observable (stealth), air-to-surface missile intended for launching by strategic bombers. The ACM used both composite materials and design features to reduce its radar cross section. The missile had significant improvements over the earlier ALCM-B with respect to range, accuracy, and stealth.

The missile was carried externally by B-52H Stratofortress bombers.

General Dynamics Advanced Cruise Missile (ACM). (General Dynamics)

Design: The forward-swept, pop-out wings were fitted just aft of midbody, high on the fuselage, and there were folding control surfaces. The missile's flat bottom provided additional lift.

AGM-129A	Production missile.
AGM-129B	Modified AGM-129A with structural and software changes and an alternate nuclear warhead "for accomplishing a classified cruise missile mission," according to official statements. Development began in 1988 with first flight planned for December 1992; the program was cancelled on 22 November 1991.

The Northrop-developed guidance system is classified but is probably similar to that in the AGM-86 ALCM with a larger memory, a radar altimeter with a low probability of intercept, a Kearfott inertial navigation system (INS), and a Hughes Aircraft laser Doppler velocimeter ranging system for navigation and targeting. A GPS is provided in the conventional missile. An article in the Soviet military journal *Krasnaya Zvezda* (Red Star) on 9 October 1991 estimated ACM's CEP as 16 feet.

Operational: The first flight was in July 1985. The first operational ACM was delivered to the Air Force in January 1991.

Production: The initial ACMs were manufactured by General Dynamics/Convair. In November 1987 the Air Force selected McDonnell Douglas as a second source in an effort to overcome some of the development problems. Low-rate production began in 1986 but was curtailed in 1989 due to quality control problems discovered during flight tests.

Procurement of approximately 2,000 AGM-129s was originally planned; the number was reduced to 1,461 after further study and in April 1991 was reduced to 1,000 missiles. On 25 July 1991, the Defense Acquisition Board reinstated the goal of some 1,500 missiles. In his State of the Union address on 28 January 1992, President George H. W. Bush announced that production of the ACM would end at 640 missiles. One month later, the Air Force terminated the program at 520 missiles.

The ACM was withdrawn from service in 2007. The W80-1 warheads are being dismantled, and the Air Force is studying whether to destroy the missiles or convert them to conventional weapons.

Production totaled 461 missiles delivered from June 1990 to August 1993.

Operational:	Jan 1991–2007
Design:	General Dynamics/Convair
Weight:	3,700 lb
Length:	20 ft 10 in
Diameter:	2 ft 3¾ in
Wingspan:	10 ft 2 in
Propulsion:	1 Williams F112-WR-100 turbofan
Speed:	550 mph
Range:	1,865 miles
Guidance:	inertial/TERCOM
Warhead:	1 W80-1 nuclear (5–150 KT)
or	conventional
Launch mode:	B-52H

ALCM AGM-86

The air-launched cruise missile (ALCM) is a long-range cruise missile for attacking strategic targets. It has many of the attributes of the earlier Hound Dog cruise missile but with greater range and accuracy, although the later missile is slower.

Development began in January 1977 with the ALCM being selected in March 1980 over the air-launched AGM-109 Tomahawk to arm B-52G/H aircraft.

Following cancellation of the U.S. tri-service standoff missile (TSSAM) in January 1995, the Air Force announced plans to convert some AGM-86 weapons to conventional warheads, possibly as many as 500. The conventional ALCM is generally referred to as the CALCM.

Design: The missile's slender, pop-out wings are swept at 25°; the graphite/epoxy elevons operate together for pitch, separately for roll. Neither the vertical tail nor the

Boeing Air-Launched Cruise Missile (ALCM). (U.S. Air Force)

mainplanes have movable surfaces. Additional lift is generated by the missile's "beluga whale" nose.

The sensor and guidance system includes TERCOM software, a gimballed inertial navigation platform, radar altimeter, and three rate gyros and pitch accelerometers. The carrying aircraft's inertial navigation equipment aligns the ALCM's guidance at the beginning of a mission; position updates are provided automatically every 60 seconds until launch. The ALCM flies at a very low altitude, relying on its small radar cross section and ground clutter to defeat enemy air defenses.

CEP for AGM-86B is estimated at 100 feet.

The variants are

ALCM-A	Prototype missile.
ALCM-B	Stretched missile with 30 percent of increase for fuel to achieve a 100 percent increase in range. The missile is too long to fit the standard weapon bays of B-1B or B-52H bombers and is carried on wing pylons on the B-52G/H and on the B-1B; subsequently the B-52H and B-1B weapons bay designs have been modified to accommodate eight ALCMs internally on a rotary launcher.
ALCM-C	Converted ALCM-B with conventional blast/fragmentation warheads; warhead weight is 2,000 pounds.
ALCM-D	Fitted with hard-target-penetration warhead.

Operational: IOC with nuclear warhead was in December 1982 and with conventional warhead in January 1988. The Air Force planned to have 528 nuclear ALCMs in inventory by the end of 2012.

Conventional variants of ALCM were launched by B-52s against Iraqi targets in Operation Desert Storm in 1991.

Production: A total of 1,715 missiles were produced through October 1986. Ninety-eight B-52G aircraft were modified to carry 12 ALCM-B missiles on external pylons; 90 B-52Hs were modified to carry a total of 20 ALCMs and SRAMs; the 97 B-1B bombers could each carry up to 22 ALCMs.

Operational:	Dec 1982–1992 nuclear variant
Design:	Boeing
Weight:	3,144 lb
Length:	20 ft 9 in
Diameter:	2 ft 3 in
Wingspan:	12 ft
Height:	2 ft 2 in
Propulsion:	1 Williams/Teledyne CAE F107-WR-10 turbofan; 600 lbst
Speed:	550 mph
Range:	1,550 miles
Guidance:	inertial/TERCOM
Warhead:	1 W80-1 nuclear (5—150 KT) in AGM-86B
	conventional 2,000-lb in AGM-86C/D
Launch mode:	B-1B
	B-52G/H

Crossbow GAM-67

The Crossbow was a bomber-launched missile intended to strike enemy air-defense radars. Strategic bombers were to carry up to four of the weapons. Initiated with an Air Force contract to the Radioplane Corp. in March 1953, the Crossbow was originally to carry a 1,000-pound conventional warhead.

A study of possible nuclear warheads for the Crossbow missile was started in September 1953. The Mk 7 was approved for the missile on 23 September 1954; subsequently, the more-capable W31 warhead was selected in place of the Mk 7.

The Crossbow was found to fly 50 mph slower than planned and had less range than existing Soviet air defense radars. It was cancelled on 24 January 1957.

The Crossbow was to have been replaced by WS-121B Longbow, but that project was also cancelled.

Design: The missile had an aircraft-type fuselage with nose air intake for its turbojet engine. Short, straight wings were fitted with a twin-fin tail assembly although only the left vertical fin had a moveable rudder. The missile was to be launched outside the range of hostile radars at an altitude of about 34,000 feet. It would then climb to 40,000 feet and fly at some 550 mph to within seven miles of the target radar installation, then make a 30° power-on dive into the target, reaching Mach 0.9 speed.

Designation: The weapon was initially designated B-67 in the Air Force bomber series.

Operational: The first powered flight was in March 1956, the first successful flight (lasting 18 minutes) on 2 July 1956, and first guided flight in May 1957 (i.e., after program cancellation).

Production: Fourteen XGAM-67 missiles were built.

Operational:	cancelled
Design:	Radioplane
Weight:	2,700 lb
Length:	19 ft 1 in
Diameter:	2 ft 6 in
Wingspan:	12 ft 7 in
Height:	5 ft
Propulsion:	1 YJ69 turbojet
Speed:	550 mph (planned)
Range:	285 miles
Guidance:	radar homing
Warhead:	1 W31 nuclear
or	1,000 lb conventional
Launch mode:	aircraft

Hound Dog GAM-77/AGM-28

The Hound Dog was a long-range, air-to-surface standoff missile carried externally by later-model B-52 bombers (one under each wing). It was capable of Mach 2 speeds.

On 15 March 1956, the Air Force issued a requirement for a standoff air-to-surface missile to help B-52 bombers penetrate Soviet air defenses. On 16 October 1958 the Air Force awarded a contract to North American Aviation for Hound Dog development and production.

The Hound Dog's engines could be started by the B-52 crew during takeoff and flight to provide the carrier aircraft additional thrust. The missile could then be refueled from the aircraft's own fuel supply while in flight and attached to the B-52. However, carrying two five-ton missiles on the B-52 degraded flight performance, and missile reliability was a constant concern.

North American Hound Dog ALCM mounted on B-52. (U.S. Air Force)

Design: The Hound Dog had a canard (forward), a delta wing, and a J52 turbojet engine mounted under a slender, streamlined fuselage.

The missile could be launched at altitudes up to 45,000 feet, or as low as 5,000 feet. The missile changed altitude and course while in flight, and could take evasive action if necessary. At minimum range the missile had a CEP of 3,600 feet, and at maximum range the CEP was some 6,500 feet.

Designation: Originally designated GAM-77, the Hound Dog was redesignated AGM-28 in 1963.

The missile was named for a popular song by Elvis Presley—"Hound Dog."

Operational: The first test flight was on 23 April 1959. The first missile delivery was in December 1959. The first B-52 wing with Hound Dog missiles became operational in July 1960; the Strategic Air Command had a peak strength of 29 B-52 wings armed with 593 Hound Dog missiles in August 1963.

The last Hound Dog was taken off alert on 15 June 1975.

Production: A total of 722 missiles were purchased —47 for research and development, and 675 for operational use.

Operational:	July 1960–15 June 1975
Design:	North American
Weight:	10,147 lb
Length:	42 ft 6 in
Diameter:	2 ft 3½ in
Wingspan:	12 ft 2 in
Propulsion:	1 Pratt and Whitney J52-P-3 turbojet; 7,500 lbst

Speed:	1,200 mph (Mach 2)
Altitude:	200 ft to 56,200 ft
Range:	750 miles
Guidance:	inertial
Warhead:	1 W28Y1/Y2 nuclear (1.1 MT, 350 KT)
Launch mode:	B-52G/H

Mastiff

Even before World War II was over—within hours of the atomic bombing of Hiroshima—the Army Air Forces initiated the Mastiff missile program to rapidly mate an atomic warhead to a guided missile. But the proposal was premature because AAF access to nuclear data was severely restricted.

The Mastiff concept became a major issue when General of the Army H. H. Arnold, head of the Army Air Forces, in an "off-the-cuff" remark to a senator on 18 October 1945, alluded to the possibility of immediately producing a guided missile carrying an atomic bomb. By adding wings, a propulsion unit, and television guidance, an atomic bomb could be launched from a bomber against targets up to 300 miles away, declared Arnold.[19]

The classification of nuclear weapons data by the Manhattan Project delayed the Mastiff project for almost two years. To further complicate the project, the newly established Atomic Energy Commission in 1947 also insisted that the existing atomic bombs could not be modified for placing components in optimum positions within the missile. For such use as proposed by the AAF, the bomb, less its tail assembly, had to be fitted into the missile intact.

This initial attempt to mate the atomic bomb and an air-launched guided missile, however, was continued into 1948 before the project was finally laid to rest. It never progressed beyond the preliminary stage.

Operational:	cancelled
Design:	Army Air Forces/Air Force
Weight:	
Length:	
Wingspan:	
Diameter:	
Propulsion:	turbojet
Range:	300 miles
Guidance:	TV guidance from launching aircraft
Warhead:	nuclear
Launch mode:	B-29

Rascal B-63/GAM-63

The Rascal was a U.S. bomber-launched, air-to-surface missile with rocket propulsion. It was the world's first supersonic air-to-surface missile and was to deliver both conventional and nuclear warheads.

The Rascal was cancelled before becoming fully operational.

Design: This was a rocket-propelled cruise missile designed to carry a 3,000-pound nuclear warhead. It could attain Mach 2.95 speeds. The missile's terminal dive began about 20 miles from the target. During this final phase of flight the Rascal's course could be altered by signals from the launching or a "director" aircraft.

Several aircraft were proposed as carriers of the Rascal —the B-36, B-50, B-52, and B-47. Only the B-47 was an active candidate when the program was terminated in 1958.

The first warhead considered for the Rascal was the Mark 5, proposed by the Military Liaison Committee of the Atomic Energy Commission on 16 March 1950. This warhead was cancelled in March 1956 in favor of the W27.

The missile was to have a CEP of 1,500 feet.

Tests indicated that the missile was unreliable.

Designation: Development project MX-776.

The name Rascal was derived from the guidance system used during the terminal phase of the attack—Radar Scanning Link.

Operational: The first launch of a guided Rascal took place in October 1953 from a DB-47 aircraft.

The Rascal was declared operational on 30 October 1957. But a year later, on 29 November 1958, the Rascal program was abruptly cancelled in favor of the Hound Dog attack missile, then under development.

Operational:	Oct 1957–Nov 1958
Design:	Bell
Weight:	18,169 lb
Length:	32 ft
Diameter:	4 ft

19 Discussed in Max Rosenberg, *The Air Force and the National Guided Missile Program 1944–1950* (Washington, D.C.: USAF Historical Division Liaison Office, June 1964), p. 87.

Bell Rascal air-to-surface missile. (Bell Aircraft)

Wingspan:	16 ft 8 in
Propulsion:	1 Bell LR67-BA-9 rocket motor; 10,440 lbst
Speed:	1,950 mph
Range:	100 miles
Guidance:	inertial
Warhead:	1 W27 nuclear
Launch mode:	B-47

SCAD AGM-86A

The subsonic cruise armed decoy (SCAD) was developed as a means of increasing B-52 bomber survivability in the face of increasing Soviet air defenses. In January 1968 the Air Force drew up a requirement for the SCAD decoy to be carried by B-52 and B-1A bombers. It would simulate the bombers on enemy radar to disrupt and confuse air-defense systems. Unlike the previous GAM-72/ADM-20 Quail decoy carried by B-52s, the SCAD would carry a nuclear warhead.

Full-scale development of SCAD was approved in August 1970 as an unarmed decoy. In 1971 Congress directed the Air Force to also develop SCAD to carry a nuclear warhead. Full-scale engineering development began in mid-1972. Despite all attempts to control costs, the projected price of even the unarmed version of SCAD rose sharply.

Some members of Congress believed that the Air Force was reluctant to weaponize the SCAD because it would undermine justification for a new penetrating bomber (i.e., the B-1).[20]

In June 1973 development was halted after it had become clear that it was more cost-effective to develop a pure attack cruise missile without any decoy capability. Some of SCADs airframe and propulsion technologies were used for subsequent cruise missiles. Thus, SCAD was the progenitor of the AGM-86 ALCM.

Design: As the SCAD was being developed, the rotary launcher being fitted in the B-52 for the AGM-69A SRAM determined the dimensions of the SCAD and fostered its triangular or trapezoidal cross section to provide greater volume within the launcher constraints.

Operational:	cancelled
Design:	Boeing
Weight:	
Length:	14 ft
Diameter:	
Wingspan:	
Propulsion:	1 William WR19 turbofan
Speed:	
Range:	
Guidance:	inertial/TERCOM
Warhead:	nuclear
Launch mode:	B-52 B-1A

Skybolt GAM-87/AGM-48

The Skybolt was the most ambitious bomber-launched missile developed by the United States. It was a ballistic missile, planned as a successor to the Hound Dog, an air-breathing cruise missile.

By the late 1950s several American firms had demonstrated the feasibility of a bomber-launched ballistic missile that could strike targets with a nuclear weapon at significantly higher speeds than could a cruise missile. On 2 January 1958, the Air Force directed a feasibility study of an air-launched ballistic missile, and on 17 July 1958, the Strategic Air Command established a requirement for an advanced air-to-surface missile.

The British government had expressed interest in the Skybolt from its inception, seeking a standoff nuclear weapon that would permit the Avro Vulcan B.2 bomber to remain an effective deterrent/strike weapon from about 1965—the expected IOC of Skybolt—until about 1970. The interest was exacerbated by British cancellation of the Blue Streak Mk-2 standoff missile intended for V-bombers and the subsequent cancellation of the ground-launched Blue Streak medium-range ballistic missile (MRBM).

In March 1960 President Eisenhower and British prime minister Harold Macmillan agreed that the Skybolt would arm the Vulcan B.2 bombers. At the time, the

20 Dr. John L. McLucas, *Reflections of a Technocrat: Managing Defense, Air, and Space Programs during the Cold War* (Maxwell AFB, Ala.: Air University Press, August 2006), p. 146.

Mock-ups of four Douglas Skybolt missiles are mounted on a B-52. The Skybolt was the most ambitious air-to-surface missile ever attempted in the West. It was to be provided to Britain to extend the service life of RAF strategic bombers; with cancellation of the Skybolt in December 1962 the British were promised the Polaris missile, marking the end of the RAF strategic capabilities. (U.S. Air Force)

British government planned a purchase of 100 Skybolts to be fitted with a British warhead known as the Super Octopus, intended as the thermonuclear primary for RE.179 nuclear warhead. Planned British procurement was later changed to 144 missiles then, in July 1962, changed to 100 missiles and 90 nuclear warheads.

Each Vulcan B.2 was to carry two Skybolts.[21]

By early 1961 both the B-52F and the Vulcan B.2 had begun carrying tests with the Skybolt, with the former initiating aerial launch tests in April 1962. These flight tests revealed continuing technical problems although they certainly appeared to be solvable within a reasonable time. Technical problems—exemplified by five failures in five flight tests—coupled with Secretary of Defense McNamara's negative attitude toward manned strategic bombers led the Kennedy administration to announce cancellation of the Skybolt program on 7 December 1962.

This action proved embarrassing for British–American relations. In meetings on 20–21 December 1962, President Kennedy offered Macmillan the submarine-launched Polaris missile (less warhead) as an alternative for arming the British nuclear deterrence force. The offer was accepted and planning began immediately to build a nuclear-submarine force modeled on the U.S. Polaris program. (At the 20–21 December 1962 talks in Nassau, the U.S. government also offered the Hound Dog missile to the British. The modifications needed to enable the Vulcan and Victor strategic bombers to carry the Hound Dog would require four to five years, hence that offer was rejected.)

The cancellation of the Skybolt missile presaged the end of the manned bomber as the British strategic/nuclear deterrent.

Design: The Skybolt had a streamlined shape with rear tail cone fitted to reduce aerodynamic drag. When the missile dropped from the carrying aircraft, the tail cone was ejected and the first-stage rocket engine ignited. After first-stage burnout the missile coasted briefly before the second-stage ignited. First-stage control was accomplished by movable tail fins and second-stage control by a gimballed rocket nozzle.

In the attack mode, after launching from an altitude of 40,000 feet, the missile's ballistic flight trajectory would take it to an altitude of 240 miles before plunging down to its target. Skybolt could carry a 600-pound nuclear

21 Some documents indicated that the Handley Page Victor B.2 bomber could also be fitted with the Skybolt missile; however, those aircraft would not have been armed with Skybolt.

warhead to a range of approximately 1,100 miles when launched from an altitude of 40,000 feet.

Each B-52H bomber was to carry four Skybolt missiles on wing pylons (compared to two Hound Dogs carried by B-52s).

Designation: The Skybolt was developed as weapon system WS138A. It was initially designated XGAM-87A and changed to XAGM-48A in June 1963.

Operational: Air launches of ballistic test vehicles from B-47 bombers began in March 1958. The first fully successful Skybolt flight occurred on 19 December 1962—12 days after the program was cancelled. On 22 December 1962, an XGAM-87A missile launched from a B-52 made a successful 900-mile flight; this was the last Skybolt test flight.

An RAF Vulcan B.2 flew to the United States in January 1961 for compatibility tests with the Skybolt.

Production: Douglas had produced fewer than 100 Skybolt missiles when the program was cancelled.

Operational:	cancelled
Design:	Douglas
Weight:	11,353 lb
Length:	38 ft 6 in
Diameter:	2 ft 11 in
Wingspan:	7 ft (first-stage tail fins)
Propulsion:	2-stage Aerojet General solid-propellant rocket motors 1st stage 24,950 lbst
Speed:	5,000 mph
Range:	1,100 miles
Guidance:	inertial
Warhead:	W59 nuclear (1.2 MT)
Launch mode:	B-52H Vulcan B.2

SRAM-A AGM-69A

The short-range attack missile (SRAM) was a guided nuclear weapon launched from strategic bombers against air defense systems during a bomber's final run-in to its target. The SRAM was originally conceived as a standoff weapon for fighter-attack aircraft, but its operational use was restricted to strategic bombers and the FB-111A/F-111G deep-strike aircraft.

Boeing began SRAM studies in December 1963 and, following intense competition between Boeing and Martin, the firm was selected in November 1965 to produce the SRAM.

Design: The SRAM was a solid-fueled, supersonic weapon designed to attack enemy surface-to-air missile sites and radars. The SRAM had either an air burst or contact fuze. The missile could be launched at subsonic or supersonic speed, from high or low altitudes. Four trajectories were available: semiballistic, terrain following, terrain following with tight terminal pullup from behind radar screening terrain, and a combination of the first three.

Tail fairings are fitted for external carriage.

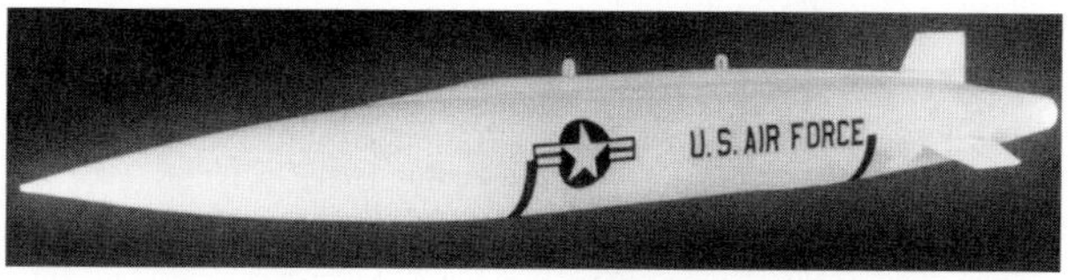

Boeing Short-Range Attack Missile (SRAM). (U.S. Air Force)

Typical SRAM loadings were:

B-1B	12 to 16 SRAMs on weapons bay rotary launcher
B-52G/H	8 SRAMs on weapons bay rotary launcher 12 SRAMs in three-round clusters under wings
FB-111A	4 SRAMs under wings 2 SRAMs in weapons bay

Designation: The missile was referred to as SRAM-A to distinguish it from the later AGM-131 SRAM B.

Operational: Live flight tests of the SRAM began in 1969 from B-52s with the first powered flight in July 1969. The first successful launch from an FB-111 occurred on 27 March 1970.

The first production line SRAM was delivered to SAC on 1 March 1972, and the first B-52 bomb wing became operational with SRAMs on 15 September 1972.

Concerns over the safety of the missile's rocket engine and its warhead forced SAC to remove the missile from alert status in mid-1990 and by the end of the year from alert duty all together. The warhead was retired and all were dismantled by September 1994.

Production: The Air Force accepted 1,500 production SRAM missiles plus 41 for research and development, the last delivered in August 1975.

Operational:	Sep 1972–1990
Design:	Boeing

SRAMs mounted on the rotary bomb bay of a B-52 Stratofortress. The B-52 has become a classic in aviation because of its longevity in service and the variety of weapons—conventional and nuclear—that could be carried by the aircraft. Unlike its predecessors—the B-29, B-50, and B-47—the B-52 was never adapted for specialized roles. (U.S. Air Force)

Weight:	2,212, lb
Length:	14 ft
Diameter:	1 ft 5¾ in
Wingspan:	2 ft 6 in
Propulsion:	1 Lockheed LPC-415 solid-propellant rocket motor
Speed:	Mach 3.2
Range:	35–50 miles with low-altitude launch 100–135 miles with high-altitude launch
Guidance:	inertial with terrain clearance sensor
Warhead:	1 W69 nuclear (170–200 KT)
Launch mode:	B-1B B-52G/H FB-111A

SRAM-II AGM-131A

The short-range attack missile SRAM-II was designed as the successor to the AGM-69 SRAM and intended for launch from strategic bombers against air defense systems during a bomber's final run-in to its target. The SRAM-II was a lighter, longer-range missile with a new nuclear warhead, making it capable of penetrating improved enemy air defenses more effectively than its predecessor.

Boeing was selected as prime contractor in December 1986.

Initial operational capability for the SRAM-II was planned for 1994. However, on 28 September 1991, the Department of Defense cancelled SRAM-II development as part of President George H. W. Bush's reduction of air-launched and tactical nuclear weapons inventories.

A tactical derivative designated AGM-131B SRAM-T was planned to arm various tactical aircraft. This program was also terminated on 28 September 1991.

Design: The SRAM-II used composite materials for lighter weight and to provide reduced susceptibility to enemy radars. The rocket casing was made of graphite epoxy composites.

The B-1B and B-2 bombers were expected to each carry 12 SRAM-IIs on six-round rotary launchers fitted in the weapons bay. Initially FB-111As had been slated to carry up to six SRAM-IIs; their conversion to F-111G and a February 1991 decision to withdraw F-111s from British bases meant that F-111s would not carry SRAM-IIs.

Operational: The first SRAM-II test launch was scheduled for September 1990 but was postponed until April 1991 and then until January–February 1992. Twenty-five test launches were planned.

Production: Air Force plans initially envisioned a total of 1,633 missiles, a figure later cut to 700.

Operational:	cancelled
Design:	Boeing
Weight:	1,945 lb
Length:	14 ft
Diameter:	1 ft 3½ in
Wingspan:	

Boeing Short-Range Attack Missile (SRAM) II. (U.S. Air Force)

Propulsion:	1 Hercules solid-propellant rocket motor
Speed:	supersonic
Range:	185–215 miles
Guidance:	inertial
Warhead:	1 W89 nuclear (200 KT)
Launch mode:	B-1B B-2A B-52G/H

TASSM

In the late 1970s the Air Force initiated development of the tri-service standoff attack missile (TSSAM) to provide a long-range standoff missile for penetrating heavily defended areas with strategic bombers and tactical strike aircraft. It was to have a nuclear warhead and be carried by a variety of Air Force, Marine Corps, and Navy aircraft.

It was cancelled early in development because of technical difficulties and cost increases on the order of threefold early in its development. In 1996 the AGM-158A joint air-to-surface standoff missile (JASSM) was initiated in its place.

AIR/BALLISTIC MISSILE DEFENSE MISSILES

The two nuclear-armed air-to-air missiles developed by the United States—the MB-1/AIR-2A Genie and GAR-11/AIM-26A Falcon—are listed with other nuclear air-to-air weapons in Chapter 6 of this volume.

Missile	Service[a]	IOC	Warhead
BOMARC	AF	1959	W40
Nike-Hercules	USA	1958	W31
Nike-Zeus	USA	cancelled	W50
Spartan	USA	1975	W71
Sprint	USA	1975	W66

a AF = Air Force; USA = Army

BOMARC F-99/IM-99/CIM-10

The BOMARC was an all-weather, supersonic, long-range, two-stage, interceptor missile. It was deployed at eight sites in the northern and northeastern United States and at two sites in Canada.

Originally conceived by Boeing in 1949 as the XF-99, the BOMARC was an Air Force project while other land-based surface-to-air missiles were pursued under the aegis of the Army.

Boeing BOMARC surface-to-air missile. (U.S. Air Force)

BOMARC being prepared for launch. (U.S. Air Force)

On 27 January 1953, the Joint Chiefs of Staff notified the Atomic Energy Commission of a requirement for a nuclear warhead for an air defense missile.

Design: Two variants of the BOMARC were produced, the liquid-fuel booster CIM-10A and the solid-propellant booster CIM-10B. The BOMARC resembled a needle-nose aircraft with triangular wings. The sustainer engines were mounted under the wings next to the fuselage, the booster in the missile's tail.

The missiles were housed in concrete shelters. To launch, the shelter roof opened and the missile raised on its launcher-erector to stand vertically.

The missile was ground controlled by the SAGE (semiautomatic ground environment) air-defense system, shifting to a target-homing system for terminal guidance.

On 23 March 1961, a CIM-10B intercepted a Regulus II target drone at 100,000 feet.

Designation: The missile was originally designated F-99 and then IM-99 (for interceptor missile) in the fighter aircraft series, that is, between the F-97 and F-100 Super Sabre. The F-98 was an air-to-air missile, redesignated GAR-1.

BOMARC was an acronym from Boeing and Michigan Aeronautical Research Center.

Operational: The first launch of a primitive IM-99 test missile occurred on 10 September 1952. (During that first test from Cape Canaveral the missile rose 500 feet, pinwheeled, crashed, and exploded.)

In late 1952 the Air Force's Air Defense Command thought it would require 3,000 BOMARC missiles in 30 squadrons by 1960; by 1956 the Air Force proposed 40 squadrons with 120 missiles per squadron—a total of 4,800 missiles to be deployed between 1959 and 1964. In June 1958 the Air Force plan was for 31 BOMARC bases: two with 56 launchers, and 28 launchers each at the 29 other bases—a total of 924 launchers. A year later the plan was reduced to 18 sites (including two in Canada). Each site would have 60 missiles for a total of 1,080 missiles. The plan was further cut, and the eventual number of BOMARCs procured was 700 including 130 missiles for research and development.

At the BOMARC's peak force level in early 1964 there were 56 CIM-10As and 242 CIM-10Bs deployed at eight sites in the United States and two in Canada.

An official history said of the missile's 1959 IOC status:

> By severely straining the concept of operational readiness it was possible to declare the BOMARC unit at McGuire as "ready" on 1 September 1959. This was the shakiest sort of readiness, however, because as late as mid-December 1959 only one missile was available at McGuire.[22]

On 17 August 1963, the United States and Canada agreed to the storage of nuclear warheads in Canada. By the end of the year there were 28 BOMARC missiles at North Bay, Ontario, and another 28 at La Macaza, Quebec, with nuclear warheads deployed in early 1964.

The last BOMARC launch sites that stood down were at Langley AFB, Virginia, on 1 October 1972.

Following withdrawal from operational service, some CIM-10 missiles were reconfigured as high-altitude, high-speed, nonrecoverable target drones for Air Force use with

22 Aerospace Defense Command, *An Overview of ADC Weapons, 1946–1972*, p. 118.

the designation CQM-10; they continued to serve in that capacity until 1985.

What was probably the most serious accident in the United States involving a nuclear weapon occurred on 7 June 1960 when a BOMARC erupted in a fuel-fed fire on its launch pad at McGuire AFB. The warhead did not detonate but there was radioactive leakage. Contamination was restricted to an area immediately beneath the weapon and an adjacent, elongated area approximately 100 feet long, caused by drain-off of fire-fighting water.

Operational:	Sep 1959–Oct 1972	
Design:	Boeing	
Weight:	CIM-10B	15,637 lb
Length:	CIM-10B	46 ft 10 in
Diameter:	2 ft 11 in	
Wingspan:	18 ft 2 in	
Propulsion:	CIM-10B	2 Marquardt RJ43 ramjet sustainer engines; 12,000 lbst each
	CIM-10A	1 Aerojet General liquid-propellant rocket booster
	CIM-10-B	1 Thiokol solid-propellant rocket booster; 50,000 lbst
Speed:	CIM-10A	1,800 mph
	CIM-10B	2,100 mph
Altitude:	CIM-10A	65,000 ft
	CIM-10B	82,000 ft
Range:	250+ miles	
Guidance:	radio control and active radar terminal homing	
Warhead:	1 W40Y1 nuclear (10 KT)	
Launch mode:	ground erector-launcher	

Nike-Hercules SAM-A-25/MIM-14

The U.S. Army's Nike air-defense program began in May 1945 when Bell Labs proposed an anti-aircraft missile system in response to a War Department request in 1944 for an advanced system to defend against jet-propelled aircraft. The subsequent Nike program delivered the first operational missile—the Nike-Ajax—in December 1953. That site was established at Fort George G. Meade, Maryland, between Washington, D.C., and Baltimore.[23]

Western Electric/Douglas Nike-Hercules surface-to-air missile. (U.S. Army)

Almost 250 Nike sites were constructed in the United States during the 1950s and 1960s to defend major cities and military installations against the threat of Soviet bomber attacks. Western Electric was the prime contractor for the Nike system while Douglas produced the intercept missiles.

The original Douglas-built MIM-3 missile of the Nike-Ajax system carried a conventional warhead. The subsequent Nike-Hercules was a dual-capable, medium-range, surface-to-air missile that carried an optional nuclear warhead. Most Nike-Hercules missiles had conventional fragmentation warheads. The missile was credited with having a secondary surface-to-surface capability.

23 The Nike-Ajax was designated SAM-A-7 and, subsequently, MIM-3.

In addition to the United States, the Nike-Hercules was deployed in Greenland, Guam, Okinawa, West Germany, Belgium, Greece, Italy, and South Korea. Belgian and Dutch troops operated Nike-Hercules sites in West Germany. The weapon was in service for more than 30 years.

Design: The Nike-Hercules was a two-stage missile. It could carry the T-45 high-explosive warhead weighing 1,106 pounds containing 600 pounds of the explosive HBX-6 or a nuclear warhead.

Designation: The Nike missiles were named for Greek mythological characters. "Nike" was the Greek goddess of victory.

The Nike-Hercules was initiated as the Nike-B.

Operational: The Nike-Hercules achieved IOC in June 1958 with the first installation near Chicago. By 1960 the U.S. Air Defense Command operated 88 Nike-Hercules missile batteries as well as 174 Nike-Ajax batteries; peak Nike-Hercules deployment was in 1963 with 134 batteries.

Nike sites were disestablished beginning in 1965 as the Soviet missile threat to the United States overtook the bomber threat. The last Nike-Hercules nuclear warheads were removed from overseas (Greece) at the end of 1988. (The Nike-Ajax missile was in use through 1964.)

Production: A total of 9,302 missiles—including 302 research and development missiles—were procured.

Operational:	June 1958–1988 nuclear variants
Design:	Western Electric and Douglas
Weight:	10,396 lb
Length:	41 ft 6 in
Diameter:	2 ft 7½ in
Wingspan:	6 ft 2 in
Propulsion:	1 Thiokol M30 solid-propellant sustainer rocket motor; 10,000 lbst 4 Hercules M42 solid-propellant booster motors; 59,000 lbst each
Speed:	Mach 2.25
Altitude:	approx 100,000 ft
Range:	
Guidance:	radar tracking/command
Warhead:	1 W31Y1/Y2/Y4 nuclear (2, 20, or 40 KT)
Launch mode:	fixed ground erector-launcher

Nike-Zeus LIM-49

Formal U.S. efforts to develop an anti-ballistic missile (ABM) system began with the Army's Nike-Zeus missile. Initiated in 1956 as the Improved Nike-Hercules, the Nike-Zeus had a more powerful booster capability.

A test-firing of a Western Electric/Douglas Nike-Zeus, the first U.S. attempt at an anti-ballistic missile system. Although its predecessors—the Nike-Ajax and Nike-Hercules—were highly successful, the Nike-Zeus was cancelled in favor of an entirely new ABM system named Sentinel with two interceptor missiles. (U.S. Army)

When the Kennedy administration entered the White House in January 1961 the Army proposed a 70-battery Nike-Zeus system with nearly 7,000 missiles to protect 27 major American cities and military installations at a five-year cost of $7.8 billion. After lengthy deliberations the project was deferred in 1961 and then cancelled in 1963 in favor of a new ABM system initially designated Nike-Zeus-B and then renamed Sentinel.

The Nike-Zeus did demonstrate a hit-to-kill capability during tests.

Design: The Nike-Zeus was a development of the Nike-Hercules missile.

Designation: The missile was also referred to as the Nike-Zeus-A.

Operational: It was planned for the Nike-Zeus to become operational by 1963–1964. By that time it would have been obsolete in terms of intercepting contemporary Soviet ICBM and SLBM warheads.

The altitude listed below is based on anti-satellite (ASAT) test launches.

Operational:	cancelled
Design:	Western Electric and Douglas
Weight:	approx 11,000 lb
Length:	44 ft 3 in
Diameter:	3 ft
Wingspan:	9 ft 9 in
Propulsion:	1 solid-propellant sustainer rocket motor; approx 50,000 lbst 1 Thiokol TX-135 solid-propellant rocket booster motor; 400,000 lbst
Speed:	approx Mach 4
Altitude:	174 miles
Guidance:	radar tracking/command
Warhead:	1 W50 nuclear
Launch mode:	ground launcher

Western Electric/Douglas Spartan ABM missile. (U.S. Army)

Spartan LIM-49A

The Spartan missile was part of the Safeguard ABM system. It was intended to intercept ballistic missile reentry vehicles above the atmosphere at long range with a high-yield nuclear warhead.

In 1967 Secretary of Defense McNamara announced the decision to deploy a ballistic missile defense system designated Sentinel. The plan called for missile deployment at 17 sites for area defense of the continental United States, Alaska, and Hawaii. The plan proved unpopular with the public, and when President Nixon took office in January 1969 he halted the program pending a review. In March 1969 an alternative plan was announced—the Safeguard. Safeguard reduced the number of deployment sites from 17 to 12 and shifted them away from cities. Safeguard would proceed in phases, with the first phase consisting of interceptor missile sites near Malmstrom AFB, Montana, and Grand Forks AFB, North Dakota.

The SALT Interim Agreement of 1972 and the ABM Treaty of 1972 limited deployments to two widely separated areas, the national capital and one ICBM site with up to 100 missiles total. A protocol two years later reduced deployment to one site, either the capital city or an ICBM field. The Soviet Union selected Moscow, and the United States selected Grand Forks AFB.

A full-scale mock-up of the Spartan ABM missile shows the three stages of the weapon, the long-range interceptor component of the Sentinel system. (U.S. Army)

The Spartan—derived from the Nike-Zeus-B system—was the only U.S. anti-ballistic missile system to be deployed (albeit briefly) with a nuclear warhead. Spartan was also proposed by the Navy for use in a sea-based ballistic missile intercept system (SABMIS).

Design: Apparently the Nike-Zeus-B was an "interim" development between the (operational) Nike-Zeus and the Spartan. The "B" missile shared the guidance and first-stage booster of the "A" missile but otherwise was a new weapon with a third stage provided in both the "B" missile and the Spartan. The relatively large nuclear warhead was intended to kill incoming RVs by thermal X-ray flux, not by blast.

A subtype of the "B" missile was the Nike-X, which had an endoatmospheric intercept capability for use against submarine-launched ballistic missiles.

An improved Spartan was planned, which would have a faster launch time than the reported 30 seconds of the basic Spartan with higher acceleration and velocity.

Designation: The designation LIM-49 was apparently used for the Nike-Zeus and LIM-49A used for Spartan. However, available documents are confusing.

Spartan was originally known as Nike-EX, a component of the so-called Nike-X ABM program.

Operational: Launch tests of the Nike-Zeus-B began on 28 April 1960. In May 1963 a modified Nike-Zeus-B missile intercepted an orbiting satellite. From June 1963 until May 1966 a "B" missile with a nuclear warhead was kept on alert status at the Kwajalein atoll's launch complex for the possible intercept of Soviet satellites.

Launch tests of the Spartan began on 30 March 1968, at Kwajalein. Fifteen flight tests were conducted at Kwajalein followed by 20 system tests from April 1970 to June 1973. In the first Spartan operational launch on 28 August 1970, there was an intercept of a Minuteman reentry vehicle at a range of 100 miles.

Operational:	1 Oct 1975–1976

Design:	Western Electric and Douglas	
Weight:	33,500 lb	
Length:	55 ft 2½ in	
Wingspan:	9 ft 10 in	
Diameter:	3 ft 7 in	
Propulsion:	3-stage Thiokol rocket motors	
	1st stage:	450,000 lbst
Speed:		
Altitude:	350 miles	
Range:	460 miles	
Guidance:	command	
Warhead:	1 W71 nuclear (5 MT)	
Launch mode:	ground silo	

Sprint

The Sprint was the most ambitious missile ever developed by the United States, intended as part of the Safeguard ABM system. It was intended to accelerate to a speed of Mach 10 in five seconds to make a terminal intercept of incoming reentry vehicles.

The decision to develop the Sprint was announced in January 1963, and Martin Marietta was selected as the development contractor on 18 March 1963.

In contrast to the Spartan missile, the Sprint had a small, enhanced-radiation warhead, probably on the order of 1 to 2 KT.

Design: Sprint was a cone-shaped missile with the second stage fitted with four small fins. Through its flight the missile was subjected to extreme acceleration loads and high aerodynamic temperatures and pressure.

The Sprint was deployed in a dormant state in an underground vertical launch station made of reinforced steel 8 feet in diameter and 31 feet long. When the order was given the missile was readied for launch, the launch cover was explosively opened, a gas generator ejected the missile, and, when a few feet above ground, the first stage ignited. After a 1.2-second first stage burn it separated, and the second stage ignited.

The missile was guided by command signals from the ground radar that tracked the target and made fine adjustments in trajectory. The missile warhead was detonated on ground command. The attacking reentry vehicles were to be destroyed by both the nuclear blast and thermal X-ray flux. The interval from launch to intercept was to be less than 15 seconds.

Operational: The first Sprint flight test was on 17 November 1965. Forty-two flight tests were conducted at White Sands Missile Range, New Mexico, between November 1965 and August 1970. The program was then shifted to Kwajalein Missile Range, Marshall Islands, where another 34 flights took place and Sprint was integrated with other components of the Safeguard system. The Kwajalein tests took place from December 1970 to December 1973.

On 23 December 1970 the reentry vehicle of a surplus Minuteman IB ICBM was launched from Vandenberg AFB, California, and was successfully intercepted by a Sprint missile launched from Kwajalein.

Operational:	1 Oct 1975–1976	
Design:	Martin Marietta	
Weight:	7,600 lb	
Length:	27 ft	
Diameter:	4 ft 6 in	
Wingspan:		
Propulsion:	2-stage Hercules solid-propellant rocket motors	
	1st stage	X-265 650,000 lbst
	2nd stage	X-271
Speed:	7,500 mph	
Altitude:	100,000 ft	
Range:	25 miles	
Guidance:	command	
Warhead:	1 W66 nuclear (2 KT)	
Launch mode:	vertical underground launch cell	

Martin Marietta Sprint ABM missile. (U.S. Army)

CHAPTER SIX

Tactical Missiles and Rockets

AIR-TO-AIR MISSILES AND ROCKETS

Missile	Service[a]	IOC	Warhead
Eagle	N	cancelled	W42
Falcon	AF	1960	W54
Genie	AF	1957	W25
Phoenix	N	cancelled[b]	—
Quetzalcoatl	AF	cancelled	W30
Sparrow	AF-N	cancelled[b]	W42

a AF = Air Force; N = Navy
b Nuclear variant cancelled

Eagle AAM-N-10

The Eagle was to be a high-speed, long-range air-to-air missile to intercept Soviet bombers with standoff missiles attacking U.S. carrier groups. Both nuclear and conventional warheads were envisioned. It was the predecessor of the later AAM-N-11/AIM-54 Phoenix, which, although produced only with a conventional warhead, at one point was to be fitted with a nuclear warhead.

The Eagle evolved from a Bell Labs fleet air defense study in 1955. In December 1958 the Bendix Corp. was selected as prime contractor for the missile. Subsequently, on 21 July 1961, Douglas Aircraft received a contract to design an aircraft to launch the missile. Known as the Douglas F6D-1 Missileer, the fighter was to have had the capability of remaining aloft for several hours, loitering some 150 miles from a task force with eight Eagle missiles on wing pylons. Endurance was obtained at the expense of speed with the twin-jet aircraft being subsonic. Gross aircraft weight would be about 50,000 pounds.

The missile's jettisonable rocket booster would have accelerated it to Mach 3.5; the sustainer engine would have pushed the missile to Mach 4.5.

In January 1960 the Department of Defense requested a nuclear warhead study for the Eagle missile. The obvious candidate was the W42. However, that effort was cancelled on 22 June 1961 after the Eagle missile itself had been cancelled in March 1961 because of massive cost increases. The F6D aircraft was cancelled on 25 April 1961.

Design: The Eagle was to have been a two-stage missile that would have used boost-glide propulsion.

Operational:	cancelled	
Design:	Bendix	
Weight:	missile	650 lb
	booster	634 lb
Length:	11 ft 7½ in + 4 ft 6 in booster	
Diameter:	1 ft 2 in + 4 ft 4½ in booster	
Wingspan:	2 ft 10 in	
Propulsion:	1 solid-propellant rocket motor	
Speed:	Mach 4.5	
Altitude:	100,000 ft	
Range:	100+ miles	
Guidance:	preset (autopilot), command, active terminal radar	
Warhead:	W42 nuclear	
Launch mode:	F6D	

Falcon GAR-11/AIM-26A

The AIM-4 Falcon was a conventional missile that entered production in 1954 and was widely used by the U.S. Air Force and foreign services. The enlarged GAR-11—later AIM-26A—model was fitted with the W54, an improved version of the Genie rocket's warhead (W25), and had an active-radar proximity fuse to detonate the warhead. This missile was carried only by later variants of the F-102A Delta Dagger fighter in the air-intercept role. Subsequently a conventional version of the nuclear Falcon with a 40-pound warhead was produced—the GAR-11A/AIM-26B. That missile was also produced in Sweden as the RB.27.

Hughes Falcon air-to-air missile. (U.S. Air Force)

Design: The Falcon had four distinctive, elongated, triangular fins.

Designation: The Falcon was developed as Air Force project MX-904.

The conventional Falcon missiles were originally designated GAR-1/2/3/4; in 1963 they were redesignated AIM-4 with suffix letters; the advanced GAR-9 became the AIM-47. Development of the XGAR-5/6 variants intended for use against high-speed missiles and aircraft were cancelled.

Operational: Falcon IOC was in 1960.

On 27 October 1962, at the height of the Cuban missile crisis, two U.S. Air Force F-102 fighters with AIM-26A Falcon missiles were scrambled from Eielson AFB, Alaska, to escort a U-2 spyplane that had accidently entered Soviet air space.[1] Soviet MiG fighters were scrambled to intercept the U-2, and the American fighters had orders to prevent such an intercept. The U-2 was able to regain its correct course before there was a confrontation between the U.S. and Soviet fighters.

Production: Hughes produced almost 2,000 AIM-26A missiles. Another 800 of the conventional model—the AIM-26B—were also produced.

Operational:	1960–1972
Design:	Hughes
Weight:	203 lb
Length:	7 ft
Diameter:	11 in
Wingspan:	2 ft ½ in
Propulsion:	1 Thiokol solid-propellant rocket motor; approx 6,000 lbst
Speed:	Mach 2
Altitude:	50,000+ ft
Range:	5 miles
Guidance:	semiactive radar homing
Warhead:	1 W54Y1 nuclear (0.25 KT)
Launch mode:	F-102

Genie MB-1/AIR-2A

The Genie was a nuclear-only, air-to-air, unguided rocket developed for the air defense of the United States and Canada. It was intended specifically for use against Soviet bomber formations.

In January 1953 the Air Force issued a tentative requirement for the use of atomic weapons in air defense. As a result of preliminary studies, in October 1953 efforts began to determine the feasibility of such weapons in defense against Soviet bombers with a contract awarded to Douglas Aircraft in mid 1954 for development of Genie.

The Genie, which became operational in 1957, and the AIM-26A Falcon, which entered service in 1960, are believed to have been the world's only air-to-air weapons fitted with nuclear warheads.

Design: Development of the Genie was begun when the Los Alamos National Laboratory could predict the development of a subkiloton nuclear warhead suitable for a rocket.

The relatively small Genie had four retractable tail fins.

The flight time varied between 4 and 12 seconds at ranges of 1½ to approximately 6 miles at a speed of 2,100 mph (Mach 3).

Operational: The first successful Genie flight tests from an F-89D and YF-102 were conducted in March

1 The U-2 had flown to the North Pole on an atmospheric sampling mission and was returning to Alaska when it strayed off course, toward Soviet Siberia.

Douglas Genie air-to-air rocket. (U.S. Air Force)

and May 1956, respectively. An F-89D/Genie IOC was declared on 1 January 1957.

A nuclear-armed Genie was tested on 19 July 1957 in Operation Plumbob. An F-89J launched the rocket at an altitude of some 15,000 feet over Yucca Flat, Nevada. After travelling some three miles in less than five seconds, the nuclear warhead was detonated by a signal from the ground; the test yield was under 2 KT.

The Genie was only fitted to fighters of the U.S./North American Air Defense Command, including three squadrons of Canadian CF-101B Voodoo aircraft. Nuclear Genies were first deployed to Canada in 1965 with the last ones were removed in 1984; they were stored under U.S. custody at Comox, Chatham, and Bagotville, Canada.

Production: Approximately 10,000 Genie rockets were produced between 1955 and 1962.

Operational:	Jan 1957–1984
Design:	Douglas
Weight:	822 lb
Length:	9 ft 7 in
Diameter:	1 ft 5½ in
Wingspan:	3 ft 4 in
Propulsion:	1 Aerojet General solid-propellant rocket motor; 36,000 lbst
Speed:	2,100 mph (Mach 3)
Altitude:	
Range:	6 miles
Guidance:	none
Warhead:	1 W25 nuclear (1.5 KT)
Launch mode:	F-89J F-101B/F F-106A/B CF-101B/F

Phoenix AIM-54

The U.S. Navy's Phoenix was developed for long-range fleet air defense against attacking Soviet bomber aircraft. It was the most sophisticated and longest-range air-to-air missile in service with any nation. It evolved from the aborted Eagle air-to-air missile. The Phoenix was developed specifically to be carried by the multipurpose Grumman F-14 Tomcat fighter.

While fitted with a conventional, 133-pound warhead, there were several efforts to provide a nuclear warhead to the Phoenix. In the end, no nuclear warhead was fitted to the weapon.

The Phoenix, a Mach 5 missile, had a range of more than 100 miles and could reach an altitude of 100,000 feet.

Operational: Phoenix IOC was in 1974; the missile was retired from U.S. Navy service in October 2004. (The F-14 was retired in September 2006.) No U.S. F-14 ever fired a Phoenix missile against an enemy.

The AIM-54A was compromised by having been provided to the Iranian Air Force prior to the fall of the Shah in 1979.

Production: Hughes produced more than 2,500 AIM-54A missiles and more than 1,000 of the advanced AIM-54C/C+ models.

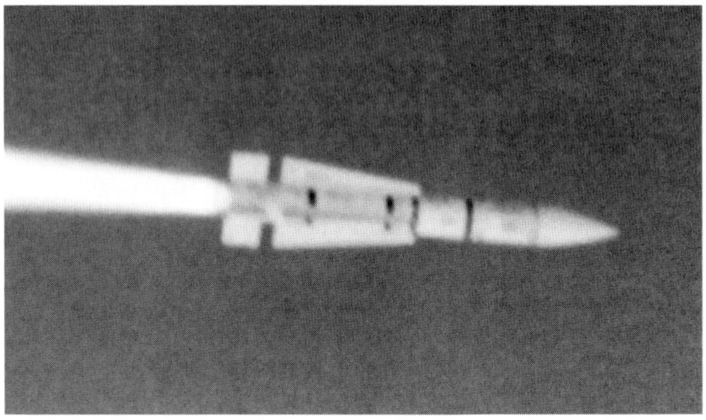

Hughes Phoenix air-to-air missile. (U.S. Navy)

Quetzalcoatl AIM-68

Usually referred to as "Big Q," the Quetzalcoatl was to be a small, air-to-air guided missile, intended to replace the unguided AIR-2A Genie nuclear-armed rocket. Although smaller than Genie, the Quetzalcoatl would have greater range and speed in addition to onboard guidance.

Development of the missile began in 1963. It was cancelled because of high costs and the priority of other programs for available funds.

The Raytheon Sparrow III air-to-air missile as well as the Hughes Phoenix and the Navy's cancelled Eagle were planned for nuclear warheads at various times. Here an F-15 Eagle is armed with four Sparrow III missiles mounted on the fuselage (outboard of the centerline drop tank); the aircraft also carries a Sidewinder missile under the left wing and an advanced medium-range air-to-air missile (AMRAAM) on the right wing (inboard of the drop tank). (U.S. Air Force)

Design: The missile had canards forward for flight control; the four main fins/wings extended in flight. It had a dual-thrust rocket motor that served as both booster and sustaining engine.

The missile's infrared guidance was adopted from the GAR-2/AIM-4 Falcon.

Designation: Quetzalcoatl was an Aztec serpent god.

Operational: No flight tests of the missile were made. The aircraft listed below were probable carriers.

Operational:	cancelled
Design:	U.S. Air Force Weapons Laboratory
Weight:	500 lb
Length:	9 ft 7 in
Diameter:	1 ft 2 in
Wingspan:	2 ft 10 in (extended)
Propulsion:	1 dual-thrust solid-propellant rocket motor
Speed:	Mach 4
Altitude:	
Range:	40 miles
Guidance:	semiactive radar, infrared
Warhead:	W30 nuclear (0.5 KT)
Launch mode:	F-4C F-101B F-106A

Sparrow III AIM-7

The U.S. Navy–developed Sparrow is an all-weather, medium-range air-to-air missile. After the AIM-9 Sidewinder, the Sparrow was the most widely used air-to-air missile in the West. It has been replaced as the principal U.S. air-to-air radar-guided missile by the AIM-120 AMRAAM.[2]

Fitted with a conventional 90-pound warhead, the Sparrow was a candidate for the W42 nuclear warhead. However, the Sparrow/W42 program was cancelled on 13 January 1958.

The Sparrow has also been adopted for surface-ship launch in the Sea Sparrow variants for use in the anti-ship missile defense role from ships. Unlike the Phoenix missile (see above), the Sparrow can be carried by a large number of U.S. and allied fighter-attack aircraft.

The AIM-7R model is among the latest variants, combining infrared terminal guidance with semiactive radar guidance (the latter fitted in earlier versions). The basic Sparrow, a Mach 4 missile, has a range of approximately 35 miles.

2 AMRAAM = advanced medium-range air-to-air missile.

Operational: The basic Sparrow became operational in 1958.

Until 1991 the weapon had minimal use in air-to-air engagements, including the Vietnam War. In the 1991 Persian Gulf War, the Sparrow III was credited with 68 percent of the air-to-air kills, that is, 28 of the 41 air-to-air kills against Iraqi aircraft.[3]

Production: Raytheon and General Dynamics/Pomona have produced thousands of Sparrows for U.S. and foreign service. The missile was also manufactured in Japan by Mitsubishi.

SURFACE-TO-AIR MISSILES

Missile	Service[a]	IOC	Warhead
HAWK	USA/MC	cancelled	W42
Standard SM-2	N	cancelled	W81
Talos	N	1959	W40
Terrier	N	1962	W45
Typhon	N	cancelled	W60

a MC = Marine Corps; N = Navy; USA = Army.

Raytheon HAWK surface-to-air missiles. (U.S. Army)

HAWK SAM-A-18/MIM-23

The Raytheon HAWK surface-to-air missile was developed for the U.S. Army and Marine Corps to provide a mobile, medium-range air-defense system. During almost 50 years of service the HAWK system—the missile, radars, fire-control equipment—has been updated numerous times, with capabilities added to permit the intercept of high-speed aircraft and tactical missiles. The HAWK was used by more than two dozen other nations.

In September 1956 the Department of Defense initiated a request for a nuclear warhead for the HAWK missile. The W42 warhead was proposed, but the HAWK/W42 effort was cancelled on 20 December 1960.

Designation: HAWK was originally an acronym for Homing All-the-Way Killer.

Operational: The HAWK's IOC in U.S. service was in 1960. It served—in updated versions—in the U.S. Army until 1994 (when it was replaced by the Patriot system) and in the U.S. Marine Corps until 2002 (when it was replaced by the Stinger missile).

It remains in service with several nations.

Production: Approximately 40,000 missiles were produced by Raytheon.

Standard SM-2 RIM-66C

The Standard SM-2 medium-range (MR) missile was the successor to the Terrier surface-to-air missile in U.S. cruisers and destroyers. The SM-2(N) version was to have had the W81 nuclear warhead.

The W81 warhead intended for the SM-2(N) entered Phase 3 development in October 1977. However, the warhead program did not advance further; schedules slipped and doubts arose as to the need for a nuclear anti-aircraft weapon when the Navy was unable to successfully articulate the need for the weapon and a skeptical Congress cut its funding. The SM-2(N) program was formally terminated in May 1984. But four months later the Navy's leadership requested a reinstatement of funding for a nuclear Standard SM-2. Congress provided a small amount in fiscal year 1986 but subsequently refused to fund the program.

The Standard missile series—SM-1, SM-2, and SM-3—succeed both the medium-range Terrier and the short-range Tartar in U.S. and foreign naval service.

3 U.S. Air Force F-15C Eagles used Sparrow III missiles to shoot down 22 high-performance Iraqi aircraft and 3 helicopters; in addition, a U.S. Navy F/A-18 Hornet shot down an Iraqi fighter with a Sparrow and a Saudi F-15C Eagle used Sparrows to kill two Iraqi fighters in the same engagement. (Ten Iraqi aircraft were brought down by Sidewinder missiles.)

General Dynamics Standard SM-2 surface-to-air missile on the USS *Norton Sound* (AVM 1). (General Dynamics)

Talos SAM-N-6/IM-90/RIM-8

Talos was a long-range, surface-to-air missile developed by the U.S. Navy. It was installed in seven U.S. cruisers with both conventional and nuclear warheads. The missile had a secondary surface-to-surface capability, and in the Vietnam War, according to historian Malcolm Muir Jr.,

> Talos also saw some action in another guise. Even though details are still sketchy, development started in 1965 on an anti-radar version of the big missile, and by 1967 one hundred conversion kits had entered the Navy inventory. Seventh Fleet began planning for its use that fall. A potent weapon, the Talos ARM [anti-radiation missile] flew a high-altitude trajectory to a maximum range of 120 miles and, once in action, reportedly shut down North Vietnamese radar installations for an entire week.[4]

The U.S. Air Force and Army considered adapting the Talos for use as a land-based missile system, with the Air Force assigning the interceptor missile designation IM-90 (in the fighter series). The Air Force interest in Talos ended with adoption of the BOMARC missile.

Design: The Talos used ramjet sustainer propulsion with a solid-propellant rocket booster. It was a beam-riding missile with terminal guidance provided by a semi-active radar seeker.

The large size of the Talos missile and its magazine, handling, and fire control equipment only permitted its installation in seven large, cruiser-type ships; the Long Beach carried both the Talos and Terrier systems.

Ships	Type/Class	Comm.[a]	Missiles	Launchers[b]
1	CGN 9 *Long Beach*	1961	52	1 Mk 12-0
3	CG 10 *Albany*	1962–1964	104	2 Mk 12-1
3	CLG 3 *Galveston*	1958–1960	46	1 Mk 7-0

a Date of commission as missile ship.
b All were twin-rail launchers.

There were several subtypes and improvements of Talos, which reached the designation RIM-8H. These variants included improved seekers and anti-radar homing guidance. The nuclear version was originally designated SAM-N-6c1 and had a W30 warhead.

The USS *Long Beach* was the world's first nuclear-propelled surface warship. She was preceded as the world's first nuclear surface ship by the Soviet icebreaker *Lenin*, which became operational in 1959.

Historical: The Talos was one of three shipboard surface-to-air missiles that evolved from the U.S. Navy's World War II–era Project Bumblebee. In response to the German guided bombs used against ships in the European theater and the Japanese kamikaze (suicide) attacks in the Pacific, the Navy directed John Hopkins University's Applied Physics Laboratory, under the aegis of the Bureau of Ordnance, to develop the technology essential for advanced anti-aircraft weapons.

The effort was dubbed Project Bumblebee because of the technological challenges—in theory, the Bumblebee's wings are too small to support its weight in flight.

The project led to development of a family of missiles designated as the Three-T series: the long-range Talos, the medium-range Terrier, and the short-range Tartar. The first two were produced with nuclear as well as conventional warheads. Although the Project Bumblebee was

4 Quoted in Malcolm Muir Jr., *Black Shoes and Blue Water: Surface Warfare in the United States Navy, 1945–1975* (Washington, D.C.: Naval Historical Center, 1996), p. 158. Dr. Muir provides official sources for the statement.

The Bendix Talos surface-to-air missile, shown here on the USS *Little Rock* (CLG 4), was the U.S. Navy's longest-range surface-to-air missile of the Cold War. With conventional warheads it was employed with considerable success in the Vietnam War. The large "searchlights" above the Talos twin-arm launcher are AN/SPG-49 target illuminating and tracking radars. (U.S. Navy)

established to counter World War II–type aerial threats, the first of these missiles—the Terrier—became operational in the mid-1950s, almost simultaneous with the Soviet Union introducing long-range strike aircraft and air-launched anti-ship missiles.

The Three-T missiles have been succeeded by a series of missiles known by the uninspiring name "Standard," initially the Standard-ER and -MR variants, indicating "extended range" and "medium range."

Operational: The first "full" Talos missile flew in October 1952. Talos IOC was in 1958.

During the Vietnam War the cruiser *Long Beach* was credited with shooting down two North Vietnamese MiG-type aircraft with Talos missiles (one at a range of some 75 miles); the cruiser *Chicago* (CG 11) shot down one MiG with a Talos missile.

The last ship to carry Talos was the *Long Beach*; her Talos system was removed in 1979 (the ship was decommissioned and stricken in 1995).

The surviving Talos missiles were expended as high-speed targets designated MQM-8G Vandal; the last was used in about 2005.

Production: About 300 nuclear warheads for the Talos were produced.

Operational:	1958–1979 with nuclear warhead	
Design:	Bendix	
Weight:	missile	3,400 lb
	booster	4,400 lb
Length:	38 ft	
Diameter:	2 ft 4 in	
Wingspan:	9 ft 2 in	
Propulsion:	1 Bendix ramjet sustainer engine	
	1 Mk 11 solid-propellant booster motor	
Speed:	Mach 2.5	
Altitude:	80,000 ft	
Range:	70 miles	
Guidance:	beam riding with semiactive terminal radar homing	
Warhead:	1 W30 nuclear (sub-KT, 5 KT)	
Launch mode:	cruisers	

Terrier SAM-N-7/RIM-2

This was a medium-range, surface-to-air missile developed under Project Bumblebee (see above). Terrier was installed in U.S. aircraft carriers, cruisers, and destroyer-type ships as well as in Italian cruisers.

The nuclear-capable (W45) missiles were designated Terrier-BTN for beam-riding Terrier, nuclear.

Design: Like the larger Talos, the Terrier was a two-stage missile but with a solid-propellant rocket booster and sustaining engine.

The first shipboard Terrier installation was in the missile test ship *Mississippi* (AG 128, ex-BB 41); subsequently, operational Terrier systems were installed in 40 U.S. warships:

Ships	Type/Class	Comm.[a]	Missiles	Launchers[b]
3	CVA 63 *Kitty Hawk*	1961–1965	80	2 Mk 10-3/4
2	CAG 1 *Boston*	1955–1956	144	2 Mk 4-0/1
3	CLG 6 *Providence*	1959–1960	60	1 Mk 9-1
1	CGN 9 *Long Beach*	1961	80	2 Mk 10-1/2
10	DLG 6 *Farragut*	1960–1961	40	1 Mk 10-0
9	DLG 16 *Leahy*	1962–1964	80	2 Mk 10-5/6[c]
1	DLGN 25 *Bainbridge*	1963	80	1 Mk 10-5/6[c]
9	DLG 26 *Belknap*	1964–1967	60	1 Mk 10-7[c]
1	DLGN 35 *Truxtun*	1967	60	1 Mk 10-8

a Date of commission as missile ship.
b All were twin-rail launchers.
c Forward launchers also fired the ASROC anti-submarine rocket.

General Dynamics/Convair Terrier surface-to-air missiles. (U.S. Navy)

The DLGs (guided missile frigates) were large, destroyer-type ships; all were redesignated as cruisers (CG/CGN) or destroyers (DDG) in 1975. In addition, the destroyer *Gyatt* (DDG 1, ex-DD 712) was fitted with a Terrier twin launcher but was not capable of handling nuclear warheads for the weapon.

The Terrier had a nominal surface-to-surface capability. But a commanding officer of the cruiser *Topeka* (CLG 8) recalled, "Our beam-rider Terrier had a supposed limited anti-surface capability, but we did not give it much credence and, as I recall, never trained in its use."[5] Still, the Terrier system initially proved very reliable, although the improved BT-3 variant encountered technical difficulties when it was introduced (doubling the range of the original missile to 20-plus miles). Reportedly, the *Boston* and her sister ship *Canberra* (CAG 2) could launch four Terrier missiles at half-minute intervals.

The Italian Navy also fitted the Terrier missile (without nuclear capability) in several cruisers.

The Terrier and smaller Standard missiles evolved into the so-called Standard series of surface-to-air missiles.

Operational: Terrier IOC was in 1955 and the nuclear variant IOC in 1962.

During the Vietnam War at least three North Vietnamese MiG-type aircraft were shot down by Terrier missiles, two launched from the frigate *Biddle* (DLG 34) and one from the *Sterett* (DLG 31); the later ship was also credited with a "possible" kill.

Production: Approximately 8,000 missiles of all versions were produced when production ended in 1966. About 750 nuclear warheads were produced for these missiles. The RIM-2D had enhanced surface-to-surface performance without prejudice to its surface-to-air effectiveness.

Operational:	1962–1989 with nuclear warhead	
Design:	General Dynamics/Convair	
Weight:	missile	1,180 lb
	booster	1,820 lb
Length:	missile	13 ft 6 in
	booster	12 ft 4 in
Diameter:	1 ft 1 in	
Wingspan:	1 ft 8 in	
Propulsion:	1 solid-propellant sustainer rocket motor	
	1 solid-propellant booster rocket motor	

5 Muir, *Black Shoes and Blue Water*, p. 149.

Speed:	Mach 3
Altitude:	80,000 ft
Range:	20+ miles
Guidance:	radar beam-riding with semiactive terminal homing
Warhead:	1 W45Y1 nuclear (1 KT, 5 KT)
Launch mode:	aircraft carriers cruisers destroyers

Typhon SAM-N-8/RIM-50A

Typhon was a planned advanced surface-to-air missile system intended as successor to the Three-T missiles for a new generation of surface warships. In particular, Typhon was designed to counter mass attacks by Soviet bombers and air-launched anti-ship cruise missiles.

The system was developed by the Johns Hopkins University's Applied Physics Laboratory and Westinghouse, under the direction of the Navy's Bureau of Ordnance.

There were two missiles in development for the Typhon system—the long-range SAM-N-8/RIM-50A and the medium-range SAM-N-9/RIM-55A. The former missile, as large as the Talos, was to carry both nuclear (W60) and conventional warheads.

The Typhon program was begun in 1958 to provide for fleet air defense against advanced Soviet bomber aircraft. The entire program was cancelled in December 1963 because of the complexity of the system and the high cost of the planned nuclear-propelled frigates (DLGN) that were to carry the system.

It was succeeded by the Navy's advanced surface missile system (ASMS), which made use of some Typhon technology and sought to use existing missiles—the Terrier, Tartar, and their descendent SM-series. Although development was slower than expected, ASMS became the basis for the Aegis combat system, whose development formally began in December 1969. Aegis, in turn, became the basic area air-defense system of the U.S. Navy. Aegis was fitted in 27 U.S. cruisers of the *Ticonderoga* (CG 47) class and 62 destroyers of the *Arleigh Burke* (DDG 51) class with the distinctive AN/SPY-1 phased-array radar, an advanced weapon control system, and the Standard-series missiles.

The missile test ship *Norton Sound* (AVM 1, ex-AV 11) was fitted with components of the Typhon system during a 1962–1964 modernization. Subsequently portions of the later ASMS/Aegis combat system were fitted in the ship.

Typhon surface-to-air missile. (U.S. Navy)

Operational:	cancelled	
Design:		
Weight:	missile	1,700 lb
	booster	1,920 lb
Length:	missile	15 ft 5 in
	booster	12 ft 2 in
Diameter:	missile	1 ft 4 in
	booster	1 ft 6½ in
Wingspan:	missile	3 ft 4 in
	booster	5 ft 2 in
Propulsion:	1 Bendix ramjet sustainer engine 1 solid-propellant rocket booster motor	
Speed:	Mach 4	
Altitude:	90,000 ft	
Range:	125+ miles	
Guidance:	command and terminal radar homing	
Warhead:	1 W60 nuclear	
Launch mode:	frigates (DLG/DLGN)	

AIR-TO-SURFACE TACTICAL MISSILES

Missile	Service[a]	IOC	Warhead
BOAR	N	1956	Mk 7
Bullpup-B	AF	1959	W45
Condor	N	cancelled	W73
Corvus	N	cancelled	W40
Hopi	N	cancelled	W50
Walleye II	AF	1967	W72

a AF = Air Force; N = Navy.

BOAR Mk 1 Mod 1

The Bureau of Ordnance atomic rocket (BOAR) was an unguided, rocket-boosted, nuclear bomb. It carried the Mk 7 nuclear weapon, which had become available in January 1962.

BOAR was developed in response to Navy efforts to arm all fighter/attack-type aircraft with nuclear strike weapons. The weapon was delivered with the so-called loft bombing technique: the aircraft would go into a steep climb, release the bomb at an appropriate point to send it on a high arching trajectory, and the aircraft would complete a half loop for a quick escape. In addition to avoiding flying directly over the target's air defenses, the technique provided time for the aircraft to escape the blast of a nuclear explosion.

The Naval Ordnance Test Station at China Lake, California (now Naval Weapons Center), began development of the BOAR in 1952 when the Mk 7 nuclear warhead became available.

The BOAR was known as the 30.5-inch Mk 1 Mod 1 rocket.

The Naval Ordnance Test Station also developed a conventional version of the BOAR, a 30½-inch-diameter, air-to-surface weapon sans nuclear warhead, confusingly referred to as the "bombardment aircraft rocket"—also called BOAR.

Operational: The BOAR was flight tested from June 1953 and was approved for production in 1955.

The principal delivery aircraft was the piston-engine AD (A-1) Skyraider.

Production: A total of about 225 BOARs were produced.

Operational:	1956–1963
Design:	Naval Ordnance Test Station
Weight:	2,000 lb
Length:	15 ft 3 in
Diameter:	2 ft 6½ in
Wingspan:	4 ft 6 in (fins)
Propulsion:	1 solid-propellant rocket motor; 15,000 lbst
Speed:	480 mph
Range:	7½ miles
Guidance:	unguided
Warhead:	1 Mk 7 nuclear (22 KT)
Launch mode:	AD/A-1

Naval Ordnance Test Station BOAR air-to-surface rocket launched from an F2H Banshee. (U.S. Navy)

Martin Marietta and Maxson Electronics Bullpup air-to-surface missile. (U.S. Navy)

Bullpup-B ASM-N-7/GAM-83/AGM-12

This missile was developed from the earlier Navy Bullpup-A air-to-surface weapon, which carried only a conventional warhead. The Bullpup was one of the first tactical standoff weapons used by U.S. aircraft, enabling the attacking aircraft to remain outside of the range of anti-aircraft defenses. It was designed to attack small targets such as pillboxes, bridges, and rail marshalling yards.

Several versions of the Bullpup were developed with different guidance and warheads. Only the Air Force's ASM-7B/GAM-83B/AGM-12C variant was fitted with a nuclear warhead.

Design: The bullet-shape missile had four small canards and four small tail fins. The missile was treated as a "wooden round," requiring no support while in storage.

The AGM-12D version of the Bullpup-B had interchangeable high-explosive, fragmentation, and nuclear (W45) warheads.

Designation: Originally given the Navy designation ASM-N-7, it was changed briefly to GAM-83 and then to AGM-12 in 1963.

Operational: The Bullpup-A variant became operational with Navy FJ-4B Fury fighter aircraft on board the aircraft carrier *Lexington* (CVA 16) when she deployed to the Western Pacific in April 1959. The conventional Bullpup was used extensively by Navy and Air Force aircraft in the Vietnam War.

The AGM-12D Bullpup-B nuclear missile was in the Air Force inventory from 1964 to 1976.

Conventional Bullpup flight tests included launches from a Marine HUS-1 (H-34) Seahorse helicopter and from a Navy S2F (S-2) Tracker ASW aircraft.

Production: The Bullpup was originally designed and produced by Martin; Maxson Electronics was responsible for development and production of the Bullpup-B. The conventional missile was also produced and used by several NATO countries (conventional only). Deliveries of the basic Bullpup-B model began in April 1960.

Almost 40,000 Bullpup missiles of all types were produced.

Operational:	1964–1976 with nuclear warhead
Design:	Martin Marietta and Maxson Electronics
Weight:	1,785 lb
Length:	13 ft 7 in
Diameter:	4 ft
Wingspan:	4 ft
Propulsion:	1 Thiokol LR62-RM-2 storable liquid-propellant rocket motor; 33,000 lbst
Speed:	Mach 2
Range:	10+ miles
Guidance:	command from launching aircraft
Warhead:	1 W45Y1 nuclear (5–10 KT)
or	conventional 990 lb
Launch mode:	F-4C
	F-100
	F-105

Condor ASM-N-11/AGM-53

The Condor was to be an air-to-surface missile. Development began in the early 1960s with the goal of designing a missile with a longer range than the Bullpup missile.

The Condor was approved for production in July 1975, and the Navy planned to procure the missiles beginning in 1976. However, the program was cancelled in March 1976 because of the high cost and technical difficulties.

Design: The Condor had cruciform, steerable delta wings fitted on the midbody, and fixed cruciform tail fins.

North American/Rockwell Condor air-to-surface missile. (U.S. Navy)

The dual-thrust rocket motor served as both booster and sustaining engine.

The missile had a television seeker that communicated with the launch aircraft over a two-way datalink. The operator would select the target and lock the seeker onto a high-contrast element.

Either a nuclear or conventional warhead was to be fitted.

Operational:	cancelled
Design:	North American/Rockwell
Weight:	2,100 lb
Length:	13 ft 10 in
Diameter:	1 ft 5 in
Wingspan:	4 ft 5 in
Propulsion:	1 dual-thrust solid-propellant rocket motor
Speed:	Mach 2.9
Range:	70 miles
Guidance:	inertial, television
Warhead:	1 W73 nuclear
or	conventional 630 lb
Launch mode:	aircraft

Corvus ASM-N-8

Corvus was a Navy effort initiated in 1955 to develop a supersonic air-to-surface missile to be launched by carrier-based aircraft against heavily defended targets, including surface ships and radar installations.

The ASM-N-8 designation was originally assigned to a missile concept known as Raven; that program was incorporated into the Corvus program. A development contract for Corvus was awarded to Temco in January 1957.

The Navy cancelled the Corvus program in July 1960.

Design: The missile had midset delta wings and cruciform tail surfaces, indexed at 45° to the wings. Both conventional and nuclear warheads were planned.

The Corvus had radar-homing guidance and could also home on the reflected radar transmission from the launching aircraft.

Operational: Twenty missiles were built and successfully tested. The first test launch was from an A4D Skyhawk at Point Mugu, California, on 18 July 1959.

Operational:	cancelled
Design:	Temco
Weight:	1,750 lb
Length:	16 ft
Diameter:	1 ft 7 in
Wingspan:	5 ft
Propulsion:	1 Reaction Motors packaged liquid-propellant rocket motor; 1,000 lbst
Speed:	Mach 0.8
Ceiling:	50,000 ft
Range:	195 miles
Guidance:	radar homing
Warhead:	1 W40 nuclear (10 KT)
Launch mode:	aircraft

Hopi

The Navy's Hopi missile was initiated as an air-to-surface weapon under development at the Naval Ordnance Test Station, China Lake, California. It was rocket propelled and to be launched by carrier-based aircraft and was to carry the W50 warhead.

Hopi was flight tested in 1958 from A4D Skyhawks. It was cancelled on 15 December 1958.

Official records are not clear with respect to the Hopi's designation; it may have been assigned designation ASM-N-9.

Walleye II AGM-62

The Walleye was an unpowered glide bomb. However, it is listed in the U.S. missile designation series and was operationally considered as such for use against surface ships and hardened ground targets.

Naval Ordnance Test Station/Martin Walleye air-to-surface missile. (U.S. Air Force)

The U.S. Navy and Marine Corps used the Walleye only with conventional warheads, as did the Israeli Air Force; the U.S. Air Force had both nuclear and conventional versions. The Navy and Marine Corps ceased using the missile in 1991 after the Persian Gulf War, when Navy planes launched 131 of the missiles and Marine planes launched 2. The nuclear version was phased out of Air Force service in 1979, followed a short time later by the conventional missile.

The missile was developed by Naval Ordnance Test Station at China Lake, California. Martin Marietta developed the final design and produced the weapon; Hughes Aircraft became a second producer.

Design: The Walleye was one of the first "smart bombs" to enter U.S. service. It was developed to take advantage of then-new television guidance technology that would yield the accuracy of a guided weapon without requiring the launch aircraft to fly toward the target until the bomb's impact.

The Walleye had no propulsion, but glided to its target on cruciform cropped delta wings. A hydraulic pump for the control surfaces and an alternator for the onboard electronics were powered by a windmilling turbogenerator housed in the bomb's tail.

The nose contained a gyrostabilized TV camera that could be locked onto a high-contrast spot on a target before launch by the launch aircraft's pilot or weapons officer. The operator focused the camera on the target, locked that image in the camera, and launched the bomb. Once the weapon was released, it was self-homing because the camera remained locked on the target until impact. If necessary, the pilot could provide update commands through a radio link in midflight. In combat the updates were often provided by a second aircraft several miles away, which reduced the vulnerability of the launching aircraft. The launching or controlling aircraft carried the AN/AWW-9B datalink pod.

The CEP was 15 feet on the target range; it was greater in combat.

Operational: Walleye I IOC was in 1967. The U.S. Air Force used the Walleye I in the Vietnam War.

Production: Walleye missiles were manufactured by Martin Marietta from 1966 to 1970. A second production source was Hughes Aircraft beginning in 1967 to meet Vietnam War requirements. About 300 W72 warheads were produced for the Walleye II.

Operational:	1967–1979 nuclear variant	
Design:	Naval Ordnance Test Station/ Martin Marietta	
	Walleye I	Walleye II
Weight:	1,100 lb	2,340 lb
Length:	11 ft 3 in	13 ft 3 in
Wingspan:	3 ft 9 in	4 ft 3 in
Diameter:	1 ft 3 in	1 ft 6 in
Propulsion:		(glide bomb)
Speed:	subsonic	subsonic
Range:	16 miles	35 miles
Guidance:	optical homing (television) with operator updates through datalink	
Warhead:	1 W72 nuclear (100 KT)	
or	conventional 850 lb in Walleye I and 2,000 lb in Walleye II	
Platforms:	aircraft	

BATTLEFIELD/TACTICAL MISSILES

The Davy Crockett recoilless rifle, sometimes referred to as a missile, is listed in Chapter 7 of this volume.

Missile	Service[a]	IOC	Warhead
Corporal	USA	1953	Mk 7
Hermes	USA	cancelled	—
Honest John	USA	1954	Mk 7/W31
Lacrosse	MC/USA	1959	W40
Lance	USA	1972	W70
LittleJohn	USA	1961	W45
Mace	AF	1959	W28
Matador	AF	1954	Mk 5
Pershing I	USA	1964	W50
Redstone	USA	1958	W39
Sergeant	USA	1962	W52

a AF = Air Force; MC = Marine Corps; USA = Army.

Corporal SSM-A-17/MGM-5

The Corporal was the first ballistic missile to enter U.S. service and was the U.S. Army's first nuclear weapon. It was a corps-level support weapon.

In May 1944 the U.S. Army requested the California Institute of Technology's Jet Propulsion Laboratory (CIT/JPL) to undertake a study of the German V-2 ballistic missile. A contract was placed with CIT/JPL on 22 June 1944 to conduct research and engineering in connection with developing long-range missiles. This contract led to development and testing of the Private and then WAC-Corporal series of liquid-propellant rocket test

Army Redstone Arsenal/Firestone Corporal battlefield ballistic missile. (U.S. Army)

vehicles. In February 1949 the Army "married" a WAC-Corporal rocket to the top of a German V-2 and launched the so-called Bumper-WAC, with the missile attaining an altitude of 250 miles. These projects led to design of the operational Corporal missile.

Design: The needle-like missile, only 30 inches in diameter with a length of 46 feet, carried a Mk 7 nuclear warhead to a range of some 75 miles.

The missile was mobile although it was not a quick-reaction weapon, requiring seven hours to be prepared for launch.

In late 1954 the U.S. government agreed to furnish Britain with 113 Corporal missiles and associated ground equipment. These were the first U.S. guided missiles destined for service in a foreign country and used by a foreign military service. British Army personnel underwent training at Ordnance Guided Missile School, Redstone Army Arsenal, at Fort Bliss, Texas, and at White Sands Proving Ground, New Mexico.

Operational: The Corporal IOC was on 1 January 1953. It was deployed in Europe from February 1955 with U.S. and British forces; U.S. troops controlled the warheads for the British missiles. By 30 June 1956, 12 Corporal battalions had been activated, of which 8 had been deployed to Europe. A one-missile firing battalion required 15 vehicles and 250 men.

On 31 March 1963, the first Corporal battalion in Europe was inactivated, and by 25 June 1964, the last Corporal artillery unit was inactivated to be replaced by the Sergeant ballistic missile.

Production: The Corporal was produced by the Firestone Tire & Rubber Co.

Operational:	1 Jan 1953–25 June 1964
Design:	Redstone Army Arsenal/Firestone
Weight:	11,250 lb
Length:	45 ft 5 in
Diameter:	2 ft 6 in
Wingspan:	(ballistic)
Propulsion:	1 liquid-propellant rocket motor
Range:	100 miles
Guidance:	preset and command
Warhead:	Mk 7 nuclear (2–40 KT)
Launch mode:	mobile erector/launcher

Hermes SSM-A-13

The Hermes was an early U.S. Army ballistic missile project. It was to have a solid-propellant rocket motor.

Because of technical problems, the Hermes effort was cancelled at an early stage and the Army instead developed the Corporal test missile into an operational system. The Hermes effort also contributed to the Army's Redstone missile program.

Honest John M31/M50/MGR-1/MIM-1

The Honest John was a short-range, unguided, nuclear-capable missile. The basic M31 Honest John was deployed in 1954 as a U.S. Army battlefield nuclear weapon system. Seven years later the improved M50 Honest John with the MGR-1B/MIM-1B reached the field. This version was almost 2½ feet shorter and had improved performance.

Design: The Honest John was "warmed" with special electric blankets for 24 to 48 hours prior to firing to attain a temperature of 77°F for proper and even propellant burn.

McDonnell Douglas Honest John battlefield ballistic missile. (U.S. Army)

Operational: Honest John IOC was in November 1954. Honest John batteries—each with four launchers—were assigned to Army infantry, armored, and airborne divisions as well as higher-level "missile commands."

The last active U.S. Army Honest John battalion, in South Korea, retired in 1979, and missiles and support equipment were turned over to South Korean forces.

Nonnuclear-capable Honest John systems were operated by several allied armies with U.S. troops in control of their nuclear warheads.

Operational:	Nov 1954–1979 with nuclear warhead	
Design:	McDonnell Douglas	
Weight:	4,332 lb	
Length:	MGR-1	27 ft
	MGR-1B	24 ft 9½ in
Diameter:	2 ft 6 in	
Wingspan:	8 ft 8 in	
Propulsion:	1 solid-propellant rocket motor	
Range:	3 to 24 miles	
Guidance:	ballistic	
Warhead:	1 Mk 7 nuclear	
or	1 W31Y1 nuclear (1 KT)	
or	1 W31Y2 nuclear (20 KT)	
or	1 W31Y3 nuclear (40 KT)	
Launch mode:	truck-mounted launch rail	

Lacrosse SSM-A-12/M4/MGM-18

The Lacrosse missile was initiated by the U.S. Marine Corps, probably the only nuclear weapon to be started for that service. It was developed in response to a Marine Corps requirement in 1951—during the Korean War—as a short-range, highly accurate missile that, controlled by a forward observer, could be used against caves and protected positions. The concept of such a weapon dated back to studies begun in 1947 under the aegis of the Navy's Bureau of Ordnance. In 1950 the effort was transferred to the Army's Ordnance Corps.

The missile never achieved the accuracy wanted by the Marines, but the U.S. Army termed the Lacrosse "the most accurate of all Army missiles."

Fitted with a nuclear (W40) or conventional warhead, the missile began deployment to western Europe in 1960. It was also used—with conventional warheads—by Canadian troops in western Europe.

Cornell Aeronautical Laboratory/Martin Lacrosse battlefield cruise missile. (Martin)

Design: The small, streamlined missile had four swept wings and four large tail fins, the latter positioned 45° from the plane of the wings.

Designation: When under development for the Marine Corps, the missile was designated SSM-N-9. The Army designated the weapon M4, with MGM-18 being assigned in 1963 shortly before the weapon was discarded.

Operational: The first flight tests were in 1954. IOC in U.S. Army was in July 1959. The Marine Corps withdrew from the project in 1959.

Eight Army battalions had the Lacrosse missile: six deployed to West Germany and one to South Korea, and one was retained in the United States.

Production: Martin produced the missile from 1956 to 1960.

Operational:	July 1959–1964
Design:	Cornell Aeronautical Laboratory/ Martin
Weight:	2,360 lb
Length:	19 ft 2 in
Diameter:	1 ft 8½ in
Wingspan:	
Propulsion:	1 solid-propellant rocket motor
Speed:	Mach 1.3
Range:	19 miles
Guidance:	command
Warhead:	1 W40Y1 nuclear (2 KT)
or	1 W40Y2 nuclear (10 KT)
or	conventional 400 lb
Launch mode:	M398 truck-mounted launcher

Lance MGM-52

The Lance was a dual-capable, short-range, all-weather ballistic missile developed for the U.S. Army. Lance replaced the Honest John and Sergeant missiles in the U.S. and several other NATO armies.

LTV Lance battlefield ballistic missile. (U.S. Army)

The basic Lance missile program began in November 1962 and was originally known as Missile B.

Beginning in 1981 some Lance/W70 warheads were converted to an enhanced radiation configuration; however, because of political sensitivity they were not deployed to western Europe but were retained in the United States.

Operational: The first test firing took place in March 1969.

In 1976 Lance began deployment. Approximately 950 missiles with 88 launchers were deployed in western Europe in the 1970s and 1980s, of which six battalions with 6 launchers each were deployed with U.S. Army units in West Germany. Lance missiles with conventional warheads were also deployed with British (12 launchers), Belgian (4 launchers), Dutch (6 launchers), Italian (6 launchers), and West German (24 launchers) troops.

Production: A total of 2,314 missiles were produced, including 181 development missiles, from September 1971 through October 1980.

Operational:	July 1972–1992
Design:	LTV

Weight:	2,834 lb
Length:	20 ft 3 in
Diameter:	1 ft 10¼ in
Wingspan:	4 ft 6 in
Propulsion:	1 storable liquid-propellant rocket motor; 49,500 lbst
Range:	nuclear 71 miles
	conventional 57 miles
Guidance:	unguided/ballistic
Warhead:	1 W70-1/2/3 nuclear (1–100 KT)
or	conventional
Launch mode:	M752 self-propelled tracked launcher

LittleJohn MGR-3

The LittleJohn was developed as a smaller version of the Honest John battlefield nuclear weapon for use by U.S. airborne forces. In May 1953 the Army outlined planes for a family of nuclear rocket weapons based on the Honest John. Douglas Aircraft began engineering studies of the Honest John Junior, which became the progenitor of the LittleJohn. Preliminary design of LittleJohn was begun at the Redstone Arsenal in February 1955. The project, reoriented several times before the final design was approved on 14 July 1955, was formally established as a separate Army program.

A truck-launched "free rocket," the weapon could be easily lifted by existing helicopters. However, it was severely limited in range.

It could be used with a nuclear (W45) or a conventional warhead.

Designation: The rocket was designated M47.

Operational: The first LittleJohn rocket was successfully fired at White Sands Proving Ground on 27 July 1956.

Six LittleJohn tactical battalions were deployed beginning in November 1961: the first to Okinawa; then to Hawaii; Fort Campbell, Kentucky (101st Airborne Division); Fort Bragg, North Carolina (82d Airborne Division); Fort Lewis, Washington; and Fort Sill, Oklahoma (training site). The airborne divisions discarded their LittleJohn units in 1965; the other battalions were gone by August 1969.

Production: The Army procured 2,404 LittleJohn missiles, including 61 for research and development. Emerson Electric Manufacturing Co. produced the missiles.

Operational:	Nov 1961–Aug 1969
Design:	Redstone Army Arsenal/Emerson
Weight:	730 lb
Length:	14 ft 5 in
Diameter:	1 ft ½ in
Wingspan:	1 ft 9 in (fins)
Propulsion:	1 solid-propellant rocket motor; 32,000 lbst
Range:	12½ miles
Guidance:	unguided/ballistic
Warhead:	1 W45Y2 nuclear (10 KT)
or	1 W45Y3 nuclear (15 KT)
Launch mode:	M34 towed launcher

Redstone Army Arsenal/Emerson LittleJohn battlefield ballistic missile being lifted by an Army H-34 Choctaw helicopter. (U.S. Army)

Mace TM-61B/TM-76/MGM-13A/CGM-13B

The Mace surface-to-surface missile was an improved version of the TM-61 Matador developed for the U.S. Air Force, which it replaced. Fitted with an inertial navigation system, the Mace had a range exceeding 1,300 miles.

Although the Mace was superior to its antecedent, the TM-61A/C Matador, it also suffered some reliability problems. The Mace was the only success in the U.S. Air Force's extensive attempt to develop surface-to-surface missiles. (The Army's efforts, both in short-range battlefield missiles and longer-range missiles, would enjoy far more success.)

Design: The Mace was an enlarged version of the Matador. Fitted with radar-mapping guidance and other improvements, the TM-61B variant of the Matador missile was renamed Mace. The missile could operate to maximum range if adequate radar maps were available for the target area. This was a predecessor of TERCOM (terrain contour matching) guidance.

Martin Marietta Mace tactical cruise missile. (U.S. Air Force)

The MGM-13A was mounted on a launch vehicle; the CGM-13B was launched from a fixed, hardened launcher.

Range data given below are for high-altitude (more than 30,000 feet) flight profile.

Designation: Originally designated B-61 in the Air Force bomber series.

Operational: This missile first flew in 1956.

The TM 76 Mace began entering the Air Force missiles units in Europe in August 1959, replacing the TM-61 Matador. By the end of 1962 the Mace had replaced the Matador, and nearly 100 Mace missiles were in the inventory. By the end of 1964 there were 112 Mace missiles in Europe.

Eventually there were six squadrons of Matadors and Maces with a maximum of almost 200 missiles in western Europe comprising the 38th Tactical Missile Wing. (The Matador was phased out in 1962). Matadors and then Mace missiles also were based in South Korea from January 1959, followed by Mace missiles at sites on Okinawa in December 1961.

The Mace squadrons would remain in Europe and the Far East until deactivated in 1969. The last Mace unit was inactivated at Bitburg Air Base, West Germany, on 30 April 1969.

Operational:	1959–1969
Design:	Martin Marietta
Weight:	18,418 lb
Length:	44 ft 6 in
Diameter:	4 ft 6 in
Wingspan:	22 ft 9 in
Height:	9 ft 7 in
Propulsion:	1 Allison J33-A-41 turbojet; 5,200 lbst 1 Thiokol M-16E3 solid-propellant booster rocket motor; 100,000 lbst
Speed:	650 mph
Range:	MGM-13A 700+ miles CGM-13B 1,300 miles
Guidance:	inertial with midcourse ATRAN (automatic terrain recognition and navigation) map-matching from lock-on to target
Warhead:	1 W28Y1 nuclear (1.1 MT)
or	1 W28Y2 nuclear (350 KT)
or	1 W28Y3 nuclear (70 KT)
Launch mode:	MGM-13A mobile launcher CGM-13B ground launched from zero-length launcher at hardened sites

Matador SSM-A-1/TM-61A/C

The Matador was the first successful cruise missile developed by the U.S. Air Force. The third cruise missile developed by that service in the postwar period, it was the least ambitious with respect to performance but the most successful.

In August 1945, along with intercontinental cruise missiles, the Army Air Forces put forth a requirement for a surface-to-surface cruise missile with a 175- to 500-mile range and a speed of 600 mph. Martin received a contract to study both subsonic and supersonic variants, with the latter requirement being deleted in December 1946. The missile was given a high priority in September 1950 because of the Korean War, but a lengthy flight test process was still required as structural problems were encountered. Another problem was the ground guidance requirement (see below).

The missiles were deployed in West Germany, South Korea, and Okinawa.

Design: The Matador resembled a swept-wing, jet propelled aircraft. Warhead weight was 3,000 pounds.

The initial guidance system required a ground-based operator to track and guide the missile, limiting its range

Martin Matador tactical cruise missile. (U.S. Air Force)

to 250 miles. In late 1954, the Shanicle guidance system was added. There were still range and guidance limitations, that is, line-of sight, and a third type of guidance known as ATRAN (automatic terrain recognition and navigation), a radar map-matching system, was installed in the TM-61B variant, which was redesignated as the Mace.

The overall missile reliability was estimated at 71 percent with a CEP of 2,700 feet for the TM-61A; the TM-61C had a CEP of 1,750 feet.

Designation: The missile was developed as weapon system MX-771A. It was originally designated B-61A in the Air Force bomber series.

Operational: The first flight of a Matador test vehicle occurred on 19 January 1949—and ended in a crash; the first successful flight was in March 1949.

The Tactical Air Command's 1st Pilotless Bomber Squadron deployed to West Germany in March 1954 with its first Matador missiles arriving in May 1954. The squadron became operational in 1955 at Hahn Air Base, West Germany.

Six squadrons—comprising the 38th Tactical Missile Wing— served in Europe with just under 200 missiles (both Matador and, subsequently, Mace).

The 58th Tactical Missile Group became operational with 60 TM-61Cs in South Korea in January 1959. A second Pacific site was established on Okinawa in December 1961 when the 498th Tactical Missile Group deployed two squadrons of Matador missiles.

Test firings were carried out in Florida and in Libya.

Production: Martin produced about 1,200 missiles through 1957.

Operational:	9 Mar 1954–Apr 1969
Design:	Martin
Weight:	11,5503 lb
Length:	39 ft 8 in
Diameter:	
Wingspan:	28 ft 7 in
Height:	9 ft 8 in
Propulsion:	1 Allison J33-A-37 turbojet; 4,600 lbst 1 T-50 solid-propellant rocket booster motor; 57,000 lbst
Speed:	512 mph
Range:	620 miles
Guidance:	radar tracking/command
Warhead:	1 Mk 5 nuclear (10–47 KT)
Launch mode:	mobile zero-length launcher or hardened shelter

Pershing I MGM-31A/B

The Pershing I ballistic missile was conceived as a replacement for the U.S. Army's Redstone missile. The Army initiated a feasibility study in November 1956 of a ballistic missile with a required range of 575 miles. Subsequently, on 28 March 1958, the Martin Co. was awarded a contract for research, development, and initial production of the Pershing I missile.

The Pershing Ia (MGM-31B) was a modification for making the missile a quick-reaction alert (QRA) weapon with the secondary role of an Army field support weapon. The major changes were the incorporation of the ability to fire from unsurveyed firing positions and a change in transport/launch vehicle (see below). By September 1969 conversion from Pershing 1 to Pershing 1a for first U.S. European battalion was complete and by August 1971 conversion from Pershing I to Ia for West German units was completed.

The Pershing I missiles were banned by the Intermediate-range Nuclear Forces (INF) treaty. The last Pershing Ia was destroyed on 6 July 1990.

The Pershing I/Ia was the basis for the improved, long-range Pershing II missile (described in Chapter 5).

Design: The Pershing I was carried and launched from the M474 tracked vehicle; the Ia from the M656.

Operational: The first of 56 Pershing I research and development test launches was on 25 February 1960.

Martin Marietta Pershing Ia battlefield ballistic missile. (U.S. Army)

The first Pershing I unit to be deployed overseas became operational in West Germany in June 1964. The U.S. Army deployed 108 missiles in West Germany.

The Pershing I was also used by the West German Army, with U.S. control of their nuclear warheads; the German IOC was in July 1964. The Germans had 72 launchers with some 100 missiles. The German Pershing missiles were not covered by the INF treaty; however, German government agreed to retire them, the last on 4 October 1990.

Production: Production of the Pershing Ia was originally completed in 1975, but Martin Marietta reopened the production line from 1977 to 1979 to replace missiles expended in training and exercises.

Operational:	June 1964–1990	
Design:	Martin Marietta	
Weight:	10,275 lb	
Length:	34 ft 6 in	
Diameter:	3 ft 4 in	
Wingspan:	(ballistic)	
Propulsion:	2-stage solid-propellant rocket motors	
	1st stage	27,450 lbst
	2nd stage	19,200 lbst
Range:	115–460 miles	
Guidance:	inertial	
Warhead:	1 W50Y1 nuclear (60 KT)	
or	1 W50Y2 nuclear (200 KT)	
or	1 W50Y3 nuclear (400 KT)	
Launch mode:	mobile erector-launcher	

Redstone SSM-A-14/PGM-11

The Redstone was a battlefield ballistic missile developed for the U.S. Army. It was a large missile with a launch weight of 31 tons, five times as much as the Corporal missile, and stood 70 feet high compared to the Corporal's 45½ feet. But the larger weapon had twice the range and could deliver a significantly larger nuclear (W39) warhead.

This was the first American land-based missile able to strike Soviet territory, albeit only the western extremity of the country.

The Redstone in turn served as the basis for the Army's next ballistic missile project, the longer-range Jupiter, classified as an intermediate-range ballistic missile (IRBM).

Design: The Redstone was derived in part from the Hermes missile project. Early discussions had addressed a Redstone range of 500 miles, but this was soon reduced to 200 to 250 miles carrying a W39 warhead.

It had an estimated CEP of 1,300 feet. When launched the Redstone reached an altitude of 57 miles to maximum range and 34 miles to minimum range of 58 miles.

Designation: At one point the missile was named Major, to fit in with the Private, Corporal, and Sergeant name series.

Operational: The Redstone was a particularly successful missile with the first definitive test missile launched

Redstone Army Arsenal/Chrysler Redstone battlefield ballistic missile. (U.S. Army)

from Cape Canaveral on 20 August 1953. Flight tests covered the period from August 1953 to October 1963, during which 89 launches occurred.

The first Redstone missile units became operational in West Germany in June 1958. For the next five years the missiles stood on alert at U.S. bases in West Germany.

The Redstone was deployed from June 1958 to the end of June 1964 and was officially retired on 30 October 1964.

A Redstone missile was launched from Johnston Island to an altitude in excess of 200,000 ft, and a W39 warhead was detonated on 1 August 1958, in Operation Hardtack/Teak. A similar test was conducted on 12 August 1958, again from Johnston Island, to carry out a high-altitude W39 detonation in Hardtack/Orange. These tests were intended to determine the effects of high-altitude nuclear detonations related to development of the Nike-Zeus ABM system.

A Redstone was used on 31 January 1958 as the first stage to launch into orbit the *Explorer 1*, the West's first scientific Earth satellite. Subsequently, in 1961, Redstone missiles were used for the two manned suborbital Mercury flights flown by astronauts Alan Shepard and Virgil "Gus" Grissom.

Production: Total Redstone procurement was 120 missiles—35 for research and development and 85 operational missiles.

Operational:	18 June 1958–June 1964
Design:	Redstone Army Arsenal/Chrysler
Weight:	61,346 lb
Length:	69 ft 4 in
Diameter:	5 ft 10 in
Wingspan:	(ballistic)
Propulsion:	1 liquid-propellant rocket motor; 78,000 lbst
Range:	200 miles
Guidance:	inertial
Warhead:	1 W39Y2 nuclear (3 MT)
Launch mode:	ground

Sergeant SSM-A-27/M15/MGM-29

The Sergeant was a second-generation ballistic missile, replacing the U.S. Army's Corporal weapon as a battlefield nuclear weapon. The later weapon was the first solid-fuel surface-to-surface missile deployed by the U.S. Army.

In November 1947 the California Institute of Technology's Jet Propulsion Laboratory (CIT/JPL) and Thiokol Corp. studied solid-propellant rocket designs for the Army under the name Sergeant, but the effort did not produce a suitable engine for tactical missile applications. The Sergeant missile program was officially started in January 1955 with CIT/JPL as prime contractor. The missile could only deliver a nuclear warhead (W52).

Design: The Sergeant was half as large as the Corporal and required less than one-third of the ground support equipment. The weapon system was air transportable. Also, the Sergeant's solid-fuel rocket motor was safer and more reliable than the Corporal's liquid-fueled motor. The guidance method was also different; in place of the Corporal's command guidance, the Sergeant had the AN/DJW-8 inertial guidance system developed by Sperry Gyroscope. This was more resistant to countermeasures and required less ground support equipment.

Because the solid-rocket motor could not be shut off once ignited, the missile's range was controlled by drag brakes, which were opened by the guidance system about halfway through the planned flight path. These drag brakes neutralized engine thrust, causing the missile to follow a ballistic flight path to its target.

CIT/JPL and Sperry Rand Sergeant battlefield ballistic missile. (U.S. Army)

A complete Sergeant system could be moved with significantly less effort than a Corporal system, and the first missile launch could be achieved less than 90 minutes after the launch site had been reached (compared to nine hours for Corporal).

A high explosive warhead for Sergeant was cancelled, and the biological and chemical warheads were developed but never approved for production. Thus, it was a nuclear-only weapon.

Operational: The first experimental Sergeant missile was launched on 19 January 1956. IOC was in July 1962. From June 1962 to September 1964 the U.S. Army established seven artillery battalions with the Sergeant missile; four were deployed to West Germany, one to Italy, and one to South Korea; one was retained in the United States.

In 1972 the Lance missile began replacing the Sergeant with the last Sergeant being retired from U.S. Army service in May 1977.

Four German battalions were also equipped with the Sergeant missile from 1963 to 1966; their nuclear warheads were under U.S. control. The missile remained in service with the West German Army until 1979.

Production: The Sperry Rand Corp. was the primary CIT/JPL subcontractor for missile production and, from 1960, the prime contractor of the weapon system. About 500 missiles were produced for the U.S. Army.

Operational:	July 1962–May 1977
Design:	CIT/JPL and Sperry Rand
Weight:	10,100 lb
Length:	34 ft 6 in
Diameter:	2 ft 7 in
Wingspan:	(ballistic)
Propulsion:	1 Thiokol XM100 solid-propellant rocket motor; 45,000 lbst
Range:	100 miles
Guidance:	inertial
Warhead:	1 W52Y1/Y2 nuclear (200 KT)
Launch mode:	mobile transporter/erector

CHAPTER SEVEN

Artillery

Guns are classified by their inside barrel diameter and barrel length. Diameters traditionally have been listed in inches for weapons larger than one-inch diameter and in millimeters for smaller weapons. Thus, an 8-inch/40-cal gun has a barrel length of 320 inches (8.128 m). In the following list the weapons are listed according to size.

Few details of nuclear ammunition are available; accordingly, unless otherwise indicated, performance data for conventional munitions are provided, approximating nuclear data.

Special artillery abbreviations:

HE	high explosive
RAP	rocket-assisted projectile

16-in/50-cal (406-mm) Naval Gun

These guns were the main battery of the four U.S. Navy battleships of the *Iowa* (BB 61) class. They were the world's largest guns fitted to fire nuclear projectiles and the only U.S. shipboard guns with that capability. Each *Iowa* had nine guns in three triple turrets. Nuclear projectiles were available for these guns from December 1956 to October 1962.

The practical value of battleships firing their 16-inch nuclear rounds to their maximum range of 23 nuclear miles was questionable at best.

These 16-inch guns were the largest guns to be mounted in warships except for the 18.1-inch (460-mm) guns of the two Japanese *Yamato*-class battleships of World War II and the single 18-inch (457-mm) gun of the British battle cruiser-carrier *Furious* of World War I.

Two additional *Iowa*-class battleships were begun but not completed; the 16-inch/50-cal guns were also intended for the five never-built battleships of the *Montana* (BB 67) class. These guns were an improvement of the 16-inch/45-cal naval guns fitted in nine previous U.S. battleships, none of which was in commission after 1947.

Each 16-inch round was fired with up to six 110-pound powder bags.

Designation: The 16-inch/50-cal gun barrel was designated Mk 7 Mod 0.

Operational: When the Navy took delivery of its first 16-inch atomic projectiles, there were three *Iowa*-class battleships armed with those guns in active service, but all three were laid up in reserve by early 1958. Thus, the actual nuclear capability was available for these battleships for little more than one year.

Although one ship was reactivated in 1968–1969 (*New Jersey*/BB 62) and when all four ships were again placed in active service briefly in the 1980s, nuclear projectiles were no longer available for them. All four *Iowa*s have been stricken.

Operational:	1943 with conventional ammunition
Type:	naval gun
Crew:	74 (27 in turret and 47 in below-deck spaces) for three-barrel turret
Weight:	1,700 tons for three-barrel turret
Length:	66 ft 9 in barrel
Range:	40,185 yds for armor-piercing round at 45° elevation 41,622 yds for high-capacity round at 45° elevation
Muzzle velocity:	2,425 ft/sec for armor-piercing round 2,690 ft/sec for high-capacity round

The forward 16-inch/50-cal turrets of the battleship *New Jersey* (BB 62) of the *Iowa* (BB 61) class indicate the tremendous firepower of these ships. However, there were battleships in commission for only a short period when nuclear rounds were available for their 16-inch guns. All four ships of the class are now memorials/museums. (U.S. Navy)

Rate of fire:	2 rounds per minute per barrel
Projectile weight:	2,700 lbs armor-piercing round 1,900 lbs high-capacity round
Warhead:	W23 nuclear (15–20 KT) conventional

280-mm M65 Gun

This was the first U.S. nuclear-capable artillery piece, generally referred to as the "atomic cannon" or "atomic Annie." By 1950 the U.S. Army sought the development of an atomic artillery shell as a stopgap measure until the missiles would become available. The Navy gave support to the Army proposal, with both services eventually developing such weapons. Because of the delays in missile development, the U.S. Air Force chief of staff, General Hoyt S. Vandenberg, reluctantly agreed to support the Army's nuclear artillery development, although he considered the delivery of nuclear weapons by tactical aircraft a far more efficient means of supporting ground operations than by nuclear artillery.[1]

In 1952 the Army received its first atomic artillery projectiles for the specially developed 280-mm atomic cannon. The shells carried the Mk 9 warhead; three years later the improved Mk 19 warhead became available for the cannon.

Army publicists extolled the virtues of the atomic cannon, especially praising such features as the ability to set up and fire the gun in only 15 minutes. Later an Army historian would write:

> The achievement was astonishing. Yet the 280-mm atomic cannon was absurdly obsolete as soon as it arrived in the field. It possessed *none* [emphasis in original] of the qualities that the Army deemed necessary for the new battlefield of the 1950s. Its limited 17-mile range gave it precious little capability to reach worthwhile targets. Commanders would have to place the cannon precariously near the line of contact for it to have any use at all. The further forward the cannon was deployed, the more

1 Memorandum from the Chief of Staff USAF to the Joint Chiefs of Staff, "Artillery Delivered Atomic Weapons," 13 April 1950.

A 280-mm M65 "atomic cannon" in firing position. These weapons were highly touted in the 1950s, although the practicality of such weapons was highly questionable. One of these cannon fired the only nuclear round to be fired by U.S. nuclear-capable artillery, either by the Army or Navy; this ocurred on 25 May 1953. (U.S. Army)

A 280-mm M65 "atomic cannon" in traveling position. (U.S. Army)

> vulnerable it would become to an enemy who would surely spare no effort to eliminate U.S. nuclear artillery. The mere possession of such a weapon would impose heavy security requirements on the local ground commander, hampering his ability to deal freely with the enemy.[2]

Design: The U.S. 280-mm cannon was developed from a 240-mm weapon that had been under development by the Army since the fall of 1944. The 280-mm gun concept was adopted from a German 170-mm gun scheme that employed an ingenious double-recoil system with the gun recoiling across the trail. However, whereas the German concept provided for hauling the gun separate from the mounting/recoil system, the U.S. Army employed two special truck-tractor/pusher vehicles that were attached to each end of the gun and carriage, hoisted it up, and moved the entire assembly at 35 mph on paved roads; its turning diameter was 80 feet.

Each truck-tractor/pusher had a 375-horsepower engine and was fitted with interdriver radio communications for maneuvering the weapon. The tractor/pusher vehicles were developed with the designation T10; in service they were designated transporters M249 and M250.

The gun could be emplaced from the travel configuration ready to fire in less than 15 minutes or readied for road transport in the same time after firing.

When in firing position the weapon had a 360° traverse; seven men pushing could traverse the gun around its firing baseplate. The gun had a hydraulic power rammer for loading the separate projectile and powder charge into the breach.

Designation: The gun weapon was developed with the designation T131; it was placed in service as the M65.

Operational: The atomic cannon figured prominently in President Eisenhower's first inaugural parade in January 1953 in Washington, D.C.

On 25 May 1953 it became the first known artillery piece to fire a nuclear round in shot Grable of Operation Upshot/Knothole test at the Frenchman Flat test site in Nevada. The Mk 9 round detonated at a range of seven miles with an air burst at an altitude of 524 feet with a yield of 15 KT. This was the only nuclear round fired by a U.S. atomic cannon.

Within months of the firing the Army deployed a half dozen of the cannon to West Germany to provide nuclear fire support for NATO ground forces. The 280-mm gun remained in Army service until 1963. They were also assigned to U.S. Army units in Okinawa and South Korea.

The Army's 280-mm gun field artillery battalion had three gun batteries with each battery assigned two atomic cannon.

Production: Twenty 280-mm cannon were manufactured.

Operational:	1952–1963
Type:	towed gun
Crew:	28
Weight:	94,000 lb in firing configuration 166,638 lb in transport configuration

2 Lt. Col. A. J. Bacevich, USA, *The Pentomic Era* (Washington, D.C.: National Defense University Press, 1986), pp. 82–83.

Length:	42 ft 8½ in barrel
	38 ft 5 in firing configuration
	84 ft 1¾ in transport configuration
Range:	31,400 yds HE round
	26,300 yds Mk 9 round
	32,700 yds Mk 19 round
Muzzle velocity:	2,500 ft/sec
Rate of fire:	1 round per minute
Projectile weight:	550 lb nuclear
	600 lb HE round T122 + powder charge
Warhead:	Mk 9 nuclear (15 KT)
	Mk 19 (approx 15 KT)
	conventional

240-mm (9.5-in) T92 Gun

The 240-mm self-propelled was considered for use as a nuclear-capable cannon, but there are no available data to indicate that it was in fact nuclear capable. The 240-mm T92 was the largest self-propelled gun to be built during World War II. It was developed as a potential "bunker buster" for use in the planned 1945–1947 campaign by U.S. forces in the Japanese home islands of Kyushu and Honshu.

Development began in November 1944. The Army planned to produce 115 guns, but only a few were completed before the program was cancelled with the end of World War II.

In the spring of 1950 the Army directed that the bores of the weapons be drilled out to provide for firing an 280-mm (11-inch) nuclear projectile. However, the outbreak of the Korean War in June 1950 led to acceleration of the 280-mm M65 gun. Accordingly, interest in providing a nuclear capability to the 240-mm weapon soon ended.

8-in (203-mm) M110 Howitzer

The 8-inch self-propelled howitzer was the heaviest field artillery piece deployed by the U.S. Army and Marine Corps in the Cold War era except for the 280-mm atomic cannon.

The M110 fired a variety of conventional and nuclear munitions, including the M422 with the W33 warhead and the rocket-assisted M753 with the W79 enhanced radiation warhead.

Design: The M201 barrel was 40 calibers in length.

The tracked M110 gun system was propelled by a liquid-cooled, two-stroke, turbocharged V-8 diesel engine; maximum power was 405 hp. Speed on paved roads was 35 mph; cross-country speed was 9 mph.

Two rounds are carried in the chassis.

Operational: IOC for the M110 was in 1963.

In September 1991, when the decision was made to withdraw tactical nuclear warheads, the U.S. Army had 49 artillery battalions with M110A2 howitzers: 12 in western Europe (288 tubes), 36 in the United States (active and reserve with 468 tubes), and 1 in South Korea (24 tubes).[3] The Marine Corps had M110A2 howitzers in two artillery regiments (24 tubes).

British Army M110s and Dutch Army M110s were configured to fire nuclear projectiles until 1987 and 1991, respectively.

Production: Approximately 2,000 M110s were produced for the U.S Army and Marine Corps and about 15 other countries; Japan produced 201 M110A2 models.

Operational:	1956–1991 with nuclear capability
Type:	self-propelled howitzer (tracked)
Crew:	5 in vehicle + 8 in accompanying vehicles
Weight:	62,500 lb
Length:	27 ft ½ in barrel
	18 ft 9 in vehicle
	35 ft 2 in vehicle with gun forward
Range:	23,293 yds HE round
	32,808 yds RAP round
Muzzle velocity:	2,333 ft/sec
Rate of fire:	1 round per 2 minutes
	2 rounds per minute max
Projectile weight:	approx. 200 lb HE
Warhead:	W33 nuclear (0.5–12 KT)
	W79 nuclear (variable approx 0.5, 1, 2.5 KT)
	conventional

155-mm (6.1-in) M44 Howitzer

The M44 was an immediate post–World War II development that used several components of the M41 light tank and the M52 105-mm self-propelled howitzer, which were also developed shortly after the war.

The M44 fired a variety of ammunition including nuclear rounds.

Design: Fitted with M45 barrel.

3 The nuclear-capable weapons were, at various times, in West Germany, Greece, Italy, the Netherlands, and Turkey.

An 8-inch M110 self-propelled howitzer firing a conventional projectile. Except for the 280-mm "atomic cannon" this was the heaviest artillery piece deployed by the U.S. Army during the Cold War. It was also used—with a nuclear capability—by the U.S. Marine Corps and the British and Dutch Armies. (U.S. Army)

It was propelled by a Continental AOS-895 air-cooled, 6-cylinder gasoline engine; maximum power was 500 hp; speed on paved roads was 35 mph.

The vehicle could carry 24 rounds.

Operational: IOC in the early 1950s (initially with the designations T99E1 and modified to T194); 608 were produced. The M44 was replaced in U.S. service by the M109 beginning in 1962. It was used by several other nations (some of which still employ the M44).

Operational:	1963
Type:	self-propelled howitzer (tracked)
Crew:	5 in vehicle
Weight:	62,500 lb
Length:	20 ft 2 in vehicle
Range:	15,967 yds
Muzzle velocity:	1,850 ft/sec
Rate of fire:	1 round per minute
Projectile weight:	approx 100 lb
Warhead:	W48 nuclear (0.1 KT) conventional

155-mm (6.1-in) M109 Howitzer

This howitzer is used in large numbers by the U.S. Army and Marine Corps. During the later decades of the Cold War it was the most widely used self-propelled howitzer in the world.

The M109 fires a variety of ammunition; it was nuclear capable prior to the U.S. government decision in 1991 to dispense with nuclear artillery. It fired the M454 projectile with the W48 0.1-KT yield warhead.

The much improved M109A6 Paladin entered U.S. service in 1992, that is, after the decision to dispense with nuclear munitions. The following data are for the earlier models.

The M109 was developed to replace the 155-mm M44 howitzer.

Design: The early M109s had the M185 barrel with a length of 23 calibers; the M109A2 and later models have M284 barrels that are 39 calibers in length.

They are propelled by a liquid-cooled, turbocharged V-8 diesel engine; maximum power is 405 horsepower. Speed on paved roads is 35 mph.

155-mm M109A6 self-propelled howitzer. (U.S. Army)

There are 36 to 39 rounds carried in the vehicle, depending upon model.

Operational: IOC for the M109 was in 1963, M109A1 in 1973, M109A2 in 1979, and M109A6 Paladin in 1992. It has been used by the U.S. Army and Marine Corps and some 30 other nations.

In September 1991, when the decision was made to withdraw tactical nuclear warheads, the U.S. Army had 82 artillery battalions with M109 howitzers: 15 in Europe (360 tubes), 65 in the United States (active and reserve with 1,188 tubes), and 2 in South Korea (48 tubes). The Marine Corps had M109A3 howitzers in two artillery regiments (24 tubes).

The M109 was used extensively in combat in the Vietnam conflict, the Iran–Iraq war, and various Arab–Israeli campaigns.

Production: About 4,000 M109s were built.

Operational:	1963–1992 with nuclear capability
Type:	self-propelled howitzer (tracked)
Crew:	4 in vehicle + 3 in accompanying vehicles
Weight:	56,000 lb M109A model
Length:	29 ft 11 in vehicle
Range:	19,790 yards HE round 25,693 yards RAP round
Muzzle velocity:	2,244 ft/sec
Rate of fire:	1 round per minute sustained 4 rounds per minute in first 3 minutes
Projectile weight:	approx 100 lb
Warhead:	W48 nuclear (0.1 KT) conventional

155-mm/39-cal (6.1-in) M198 Howitzer

The M198 155-mm howitzer replaced the M114 as the principal towed 155-mm howitzer in the U.S. Army and Marine Corps. The M198 was 25 percent heavier than its predecessor, creating major mobility problems for the Marine Corps, which lacks the heavy trucks and gun movers available to Army units. Indeed, the only U.S. helicopter that could lift the M198 is the Sikorsky CH-53E Super Stallion.

The M198 fires a variety of ammunition; it was nuclear capable prior to the U.S. decision in September 1991 to dispense with nuclear artillery. It fired the M454 projectile with the W48 warhead.

Beginning in 2006 the M198 is being replaced in Army and Marine artillery units by the lightweight 155-mm M777 howitzer.

Operational: IOC for M198 was in 1979.

In September 1991 when the decision was made to withdraw tactical nuclear warheads from the U.S. armed forces, the U.S. Army had 35 artillery battalions with M198 howitzers: 34 in the United States (both active and reserve with 654 tubes) and 1 in South Korea (24 tubes). The Marine Corps had the M198 in three artillery regiments (270 tubes).

Production: Production began in 1978. More than 1,800 M198s have been built for the U.S. Army and Marine Corps and about ten other countries.

155-mm M198 towed howitzer. (U.S. Marine Corps)

Operational:	1963–1992 with nuclear capability
Type:	towed howitzer (wheeled)
Crew:	11
Weight:	15,791 lb in travelling and firing configurations
Length:	19 ft 10 in barrel
Range:	19,029 yds HE round 32,808 yds RAP round
Muzzle velocity:	2,245 ft/sec HE round 2,710 ft/sec RAP round
Rate of fire:	4 rounds per minute in first 3 minutes 2 rounds per minute sustained
Projectile weight:	approx 95.6 lb HE
Warhead:	W48 nuclear (0.1 KT) conventional

Davy Crockett M28/M29

In 1961 the U.S. Army fielded the Davy Crockett, a nuclear weapon at the battalion level, to provide infantry units with an effective weapon against Soviet armored forces.

There were two Davy Crockett systems: The M28 was a lightweight, three-man, portable 120-mm recoilless rifle with a range of 1,150 to 6,560 feet. The M29 was a vehicle-transported weapon using a 155-mm recoilless rifle with a range of 1,150 to 13,125 feet. Both systems fired the M388 projectile fitted with the W54 nuclear warhead. The projectile had a diameter of 11 inches, a length of 30 inches, and a weight of 76 pounds.

The M28 launcher weighed 185 pounds and the Mk 29 launcher was 440 pounds. The M388 projectile was fitted to the end of the gun barrel, similar to a rifle grenade, and was not fired through the barrel.

Operational: The first deployed Davy Crockett weapons became operational with the U.S. Seventh Army in West

Davy Crockett M29 recoilless rifle mounted on a light vehicle. (U.S. Army)

Germany on 9 June 1962. Davy Crockett was also deployed with U.S. Army units in Guam, Hawaii, Okinawa, and South Korea. Withdrawal from service began in July 1967 and the last weapons were retired in early 1971.

The Davy Crockett systems twice fired live nuclear rounds—Little Feller II on 7 July 1962, and Little Feller I on 17 July 1962.

Production: The Army procured 400 nuclear rounds through February 1965.

Operational:	May 1961–1971
Weight:	51 lb
Length:	2 ft 6 in
Diameter:	11 in
Wingspan:	(ballistic)
Propulsion:	solid propellant
Range:	see text
Guidance:	unguided
Warhead:	W54Y1 nuclear (0.25 KT)
Launch mode:	120-mm and 155-mm recoilless rifles

Davy Crockett M28 recoilless rifle fitted to a tripod mount. (U.S. Army)

CHAPTER EIGHT

Anti-Submarine Weapons

The U.S. Navy's first nuclear anti-submarine weapons were nuclear depth bombs. These were:

Weapon	Warhead	Operational
Mk 90 Betty	Mk 7	1955–1963
Mk 101 Lulu	W34	1958–1971
B57	B57	1963–1992

Nuclear depth bombs are described in Chapter 2.

Four nuclear-capable weapons were initiated specifically for the anti-submarine warfare (ASW) role to (1) provide a high probability of kill, (2) compensate for imprecise targeting data, and (3) overcome acoustic decoys and jamming devices used by submarines. The "down side" of nuclear ASW weapons included the political-military constraints and, when employed, the "blue out" effect that could create environmental conditions that prevented the use of sonar for many minutes or possibly hours.

Of the four weapons described in this chapter, two, the anti-submarine torpedo (ASTOR) and submarine rocket (SUBROC) carried only nuclear warheads; the anti-submarine rocket (ASROC) and the never-produced Sea Lance could deliver either a nuclear or conventional warhead.

Weapon		IOC	Launch Platform	Warhead
ASROC	RUR-5A	1961	surface ships	W44
ASTOR	Mk 45	1958	submarines	W34
Sea Lance	UUM-125B	cancelled	submarines, surface ships	W89
SUBROC	UUM-44A	1965	submarines	W55

ASROC RUR-5A

The ASROC (anti-submarine rocket) was a widely used ship-launched ballistic ASW weapon fitted with a conventional Mk 46 homing torpedo or nuclear depth bomb. From 1961 through 1980 ASROC was installed in 262 existing and new construction U.S. Navy warships—35 cruisers, 162 destroyers, and 65 frigates.[1] In addition, ASROCs fitted with only conventional ASW homing torpedoes were installed in warships of 11 allied navies.

There had been periodic proposals to upgrade ASROC and, especially, to increase its range. However, no improvement programs were undertaken. An improved, increased-range vertical-launch ASROC (VLA) was developed for use in warships with vertical launch systems (VLS), but that variant did not have a nuclear capability although it was originally envisioned.

1 The U.S. Navy's ship designations were changed in 1975: the large "frigates" (DLG/DLGN) became cruisers (CG/CGN) or destroyers (DDG), and the escort ships (DE/DEG/AGDE) became frigates (FF/FFG/AGFF). The ASROC-armed ships were:

Number	Original Type	Post-1975
1	CLK/DL	(stricken 1974)
1	CGN	CGN
8	CG	CG
8	DLGN	CGN
18	DLG	CG
10	DLG	DDG
33	DDG	DDG
118	DD	DD
6	DEG	FFG
58	DE	FF
1	AGDE	AGFF

Honeywell ASROC anti-submarine rocket fired by the frigate *Brooke* (DEG/FFG 1). (U.S. Navy)

The last nuclear warheads were removed from the fleet in 1989; the conventional ASROC remained in some U.S. and foreign warships for several more years.

Design: The ASROC was a ballistic missile that could be employed with two warhead options, a W44 nuclear depth bomb or a lightweight ASW homing torpedo with a conventional warhead.

After launch the rocket booster would fall away and the nuclear depth bomb would enter the water and be detonated by a hydrostatic fuse. (The Mk 44 or Mk 48 ASW torpedo would be lowered to the water by parachute.)

The ASROC payloads were:

Mod 3	Mk 44 homing torpedo
Mod 4	Mk 46 Mod 1 homing torpedo
Mod 5	W44 nuclear depth bomb

The ASROC was fired from the Mk 16 eight-tube "pepper box" launcher and from Mk 10 and Mk 26 twin-arm, Terrier/Standard surface-to-air missile launchers.

Initial targeting information for ASROC was from the ship's AN/SQS-23 sonar.

Designation: RUR was a rocket vice missile designation.

Operational: ASROC was declared operational on 20 January 1961 when it was fired from the guided missile frigate *Mahan* (DLG 11).

The ASROC was tested with a nuclear warhead fired from the destroyer *Agerholm* (DD 826) on 11 May 1962 in the Pacific in a test known as Swordfish as part of the Dominic test series. The *Algerholm*, steaming in an area about 370 n.miles southwest of San Diego, fired the ASROC at a target raft some 4,000 yards away. The W44 warhead produced a low-yield, underwater detonation, about 350 yards beyond the nominal aim point.

The test was intended to determine the effects of a nuclear explosion on the sonar performance of surface ships and submarines as well as being an ASROC/W44 proof test.

Shot Swordfish was the last of only five underwater nuclear tests conducted by the United States.

Production: Honeywell manufactured 12,000 ASROCs from 1960 through 1970.

Operational:	20 Jan 1961–1989 with nuclear capability
Design:	Honeywell
Weight:	1,000 lb
Length:	15 ft
Wingspan:	1 ft 6 in
Diameter:	1 ft ½ in
Propulsion:	1 solid-propellant rocket
Range:	1 to 7 miles
Guidance:	unguided ballistic trajectory (with terminal homing conventional torpedo)
Warhead:	1 W44 nuclear (5 KT)
or	conventional Mk 44 or Mk 46 homing torpedo
Launch mode:	cruisers (Mk 16 8-cell launcher
or	Mk 10 or Mk 26 multipurpose launcher)
	destroyers (Mk 16 launcher)
	frigates (Mk 16 launcher)

ASTOR Mk 45

The Mk 45 ASTOR (anti-submarine torpedo) was the only torpedo with a nuclear warhead developed in the

Westinghouse Electric ASTOR anti-submarine torpedo being loaded aboard a submarine. (U.S. Navy)

West.[2] The ASTOR was in service in U.S. submarines—carried in both attack and strategic missile submarines—from 1963 to 1976 when it was replaced by the Mk 48 torpedo with a conventional warhead. The later torpedo's range and accuracy made it as effective with a moderate conventional warhead in most tactical situations.

In a form of gallows humor, U.S. submariners often cited the ASTOR as having a probability of kill (Pk) of 2.0—the target submarine and the launching submarine!

An attempt was made by the Navy to rehabilitate Mk 45s into conventional torpedoes—designated the Freedom torpedo—for foreign sales. The effort was not successful.

Design: The ASTOR was wire-guided by the launching submarine and had no homing guidance. The warhead was detonated by (wire) command from the submarine, having no contact or influence exploder.

The weapon was launched from standard 21-inch-diameter torpedo tubes.

Operational: Development began in 1957 and was completed in 1960–1961 and approved for service use in 1961. Production deliveries began in 1963.

In 1966 a submarine launched what may still be the longest U.S. torpedo run in history when a Mk 45 exercise (unarmed) torpedo travelled more than 15,000 yards.

Two Mk 45 torpedoes were on board the U.S. nuclear-propelled attack submarine *Scorpion* (SSN 589) when she sank in the eastern Atlantic on 27 May 1968. The *Scorpion*'s loss was at one point believed to have been caused by a malfunctioning Mk 37 conventional torpedo that exploded within the submarine; it was later determined that was not the cause.

2 The Soviet Navy deployed several nuclear torpedoes beginning with the Type 53–58 in 1958; see N. Polmar and K. J. Moore, *Cold War Submarines: The Design and Construction of U.S. and Soviet Submarines* (Dulles, Va.: Brassey's/Potomac Books, 2004), pp. 28–29.

Operational:	1963–1976	
Design:	Westinghouse Electric	
Weight:	Mod 0	2,330 lbs
	Mods 1, 2	2,213 lbs
Length:	Mod 0	18 ft 9 in
	Mods 1, 2	18 ft 11 in
Diameter:	19 in	
Propulsion:	electric motor powered by seawater activated battery; 160 hp; contra-rotating propellers	
Speed:	40 knots	
Range:	15,000 yards	
Guidance:	command (wire)	
Warhead:	1 W34 nuclear (11 KT)	
Launch mode:	submarines (SSN/SSBN)	

Artist's concept of the Boeing submarine-launched Sea Lance ASW missile. (U.S. Navy)

Sea Lance UUM-125B

Originally known as the ASW standoff weapon (SOW), the Sea Lance was to be a submarine-launched weapon that provided a rocket booster for a Mk 50 anti-submarine torpedo. Although often labeled a successor to SUBROC, the Sea Lance was initially planned to have only a conventional (torpedo) warhead, whereas the SUBROC carried only a nuclear depth bomb. The Sea Lance warhead would thus inhibit its use at longer ranges because of the limited target localization capability of the Mk 50. A nuclear option was planned, but that effort was delayed almost from the outset.

When the Navy asked Congress to begin funding the program in 1979, Vice Admiral David Emerson told the solons that

> the urgency of the requirement [for the new weapon] is severe. . . . It is vital that our SSNs be weapon equipped so that they can take advantage of their sensor superiority and not be placed in the situation wherein they will have to close the enemy [submarine] within his localization an weapon delivery envelope in order to launch a torpedo.[3]

Following several changes in direction of the program, including whether or not to provide a nuclear warhead, the Navy cancelled the Sea Lance on 3 January 1990. It had been in development for almost ten years.

Design: The Sea Lance was designed for attacks out to the third sonar convergence zone (CZ), that is, approximately 90–100 nautical miles. However, when fitted with the conventional Mk 50 torpedo, the effective range would probably be only the first CZ, or some 30–35 nautical miles.

The weapon was to be stowed and launched from a standard 21-inch torpedo tube in a canister, much like the Harpoon anti-ship missile and the CAPTOR encapsulated mine. When the capsule reached the surface, the missile booster would ignite, launching the missile on a ballistic trajectory toward the target area. At a designated point the torpedo or nuclear warhead would separate from the booster and be slowed by a parachute to reenter the water.

During the concept stage the Navy envisioned a common ASW standoff weapon for surface ships and submarines. The technical and program difficulties proved too great, however, and the surface-launched weapon became the vertical-launch ASROC (VLA).

Name: Boeing had used the name Seahawk for the weapon before the Navy designated it Sea Lance.

Operational: Sea Lance flight testing was to begin in mid-1989, with a limited production start planned for December 1990 and full production to be initiated in late 1991.

3 Vice Adm. David Emerson, USN, Director, Office of Research, Development, Test and Evaluation, Office of the Chief of Naval Operations, testimony before the Subcommittee of the Committee on Appropriations, House of Representatives, 25 April 1979.

Procurement: The Navy's procurement goal for Sea Lance was some 2,400 missiles.

Operational:	cancelled
Design:	Boeing
Weight:	3,100 lb
Length:	20 ft 6 in
Diameter:	19 in (21 in encapsulated)
Propulsion:	solid-propellant rocket motor
Range:	100+ nautical miles
Guidance:	inertial (acoustic homing with Mk 50 torpedo)
Warhead:	Mk 89 nuclear
or	conventional Mk 50 homing torpedo
Launch mode:	submarines (SSN)

SUBROC UUM-44A

The SUBROC (submarine rocket) was a rocket-propelled nuclear depth bomb that could be launched from standard 21-inch submarine torpedo tubes. It could not be fitted with a conventional warhead.

The weapon was analog and hence not compatible with U.S. attack submarines having the Mk 117 digital fire-control system. Thus, only about 25 submarines of the *Permit* (SSN 594) and later classes fitted with the Mk 113 fire control system carried the weapon. The missile, which became operational in 1965, was taken out of Navy service in 1989. The Sea Lance ASW standoff weapon was to have replaced the SUBROC, with alternative nuclear or conventional warheads; in the end the Sea Lance was cancelled.

In the late 1950s, with the cancellation of the Regulus II land-attack missile, the SUBROC was considered for use against shore targets with an air burst warhead.

Design: After being launched from the submarine, the missile streaked to the surface, left the water, and traveled on a ballistic trajectory for a predetermined time. At that point the booster fell away and the bomb reentered the water at supersonic speed, detonating at a preset depth.

Launching the SUBROC required the submarine to produce an active acoustic "ping" to determine the range to the target submarine, which was normally detected by passive sonar. The use of active sonar would have alerted the enemy submarine to the launching submarine's presence in the area.

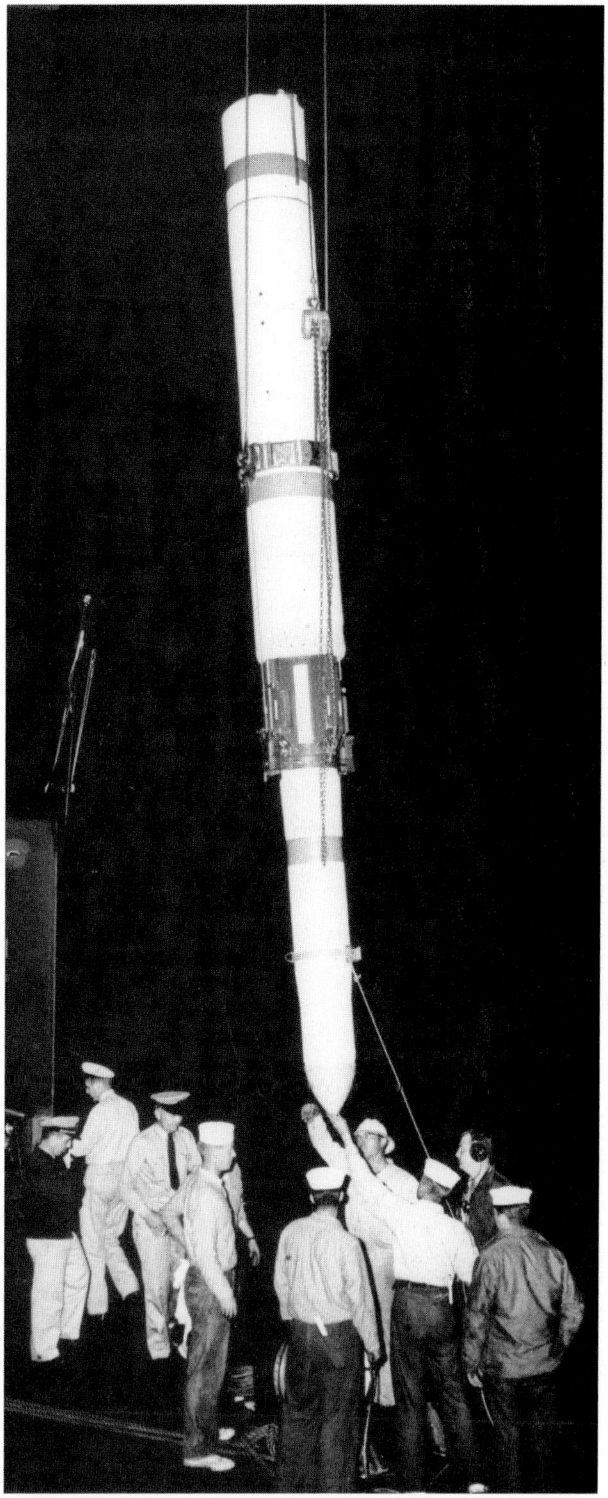

Goodyear SUBROC anti-submarine missile being loaded aboard the submarine *Permit* (SSN 594). (Goodyear)

Underwater launch of a SUBROC anti-submarine missile. (Goodyear)

Operational:	1965–1989	Speed:	Mach 1+
Design:	Goodyear Aerospace	Range:	25–30 miles
Weight:	4,000 lb	Guidance:	inertial
Length:	21 ft	Warhead:	1 W55 nuclear (1–5 KT)
Diameter:	1 ft 9 in	Launch mode:	submarines (SSN)
Propulsion:	2-stage Thiokol solid-propellant rocket motors		

APPENDIX A

U.S. Nuclear Stockpile

The first several years after World War II saw very few bombs being built, but after the Soviet test in August 1949 and the expansion of the military–industrial complex to research, develop, test, and produce weapons that took place in the early 1950s, the U.S. nuclear stockpile grew rapidly. An intense interservice rivalry among the Air Force, Navy, and, to some extent, Army drove the American side, which in time became inextricably linked with the capabilities of their Soviet counterparts and produced a mighty action–reaction engine that fueled the arms race. Adding to the intensity was a fierce competition between the national laboratories at Los Alamos and (after 1952) Lawrence Livermore to design new and improved warheads of every shape and size and for every conceivable mission. The emergence of China as a new enemy added to the massive accumulation of megatons that threatened the globe.

The U.S. nuclear stockpile grew from less than 500 warheads in 1951 to more than 22,000 a decade later. Remarkably, in 1959 and 1960 the number of warheads built was in excess of 7,000 per year. The U.S. stockpile reached a peak of approximately 32,000 warheads/weapons in 1967. For the entire period 1945 to 1990, the United States produced approximately 70,000 nuclear warheads of 105 types, according to official documents. Depending upon how the "types" are counted, they were deployed on approximately 120 kinds of delivery systems.[1] The Air Force adopted 58 of these types and the Navy 39, the Marine Corps 18, and the Army 28. Los Alamos was responsible for designing 83 warhead types and Lawrence Livermore 26 types.

With respect to megatonnage, the peak was reached in 1960 with 20,491 megatons, according to the Department of Energy (i.e., the equivalent of more than 20 billion tons or 40 trillion pounds of TNT)—the equivalent of more than a million Hiroshima explosions. The authors estimate the current stockpile megatonnage to be approximately 2,390 megatons, a little more than one-tenth of the 1960 level—still the equivalent to 159,000 Hiroshimas.

There was never a precise answer to the question of "How Much Is Enough?" with respect to nuclear weapons. No efficient analytical process was developed to determine how many ICBMs, missile submarines, nuclear artillery shells, or whatever delivery system should be built to serve the goal of deterrence, or if that failed, to fight a nuclear conflict. In fact, the process was exceedingly arbitrary and capricious. The stockpile could easily have been twice as large or half as large as what it turned out to be if

1 Stephen I. Schwartz, ed., *Atomic Audit: The Costs and Consequences of U.S. Nuclear Weapons since 1940* (Washington, D.C.: Brookings Institution Press, 1998), p. 42; and Robert S. Norris, Thomas B. Cochran, and William M. Arkin, "History of the Nuclear Stockpile," *The Bulletin of the Atomic Scientists* (August 1985), pp. 106–9.

U.S. military and civilian leaders made different, arbitrary decisions. The size and scale of each weapon system was determined in part by the turbulent congressional authorization process in which the Executive Branch, especially the Pentagon and Atomic Energy Commission (after 1977 the Department of Energy), requested new programs that were then filtered through key committees and powerful members. Those individuals were, in turn, subject to pressures by a military–industrial complex anxious to win contracts, with the Soviet threat always looming over the process.[2]

After reaching a peak in the late 1960s, the military establishment's enthusiasm for nuclear weapons began to wane, and the 1970s saw the overall nuclear stockpile decline by almost 20 percent even as the number of warheads on strategic missiles increased dramatically—through MIRV technology—and tactical, or nonstrategic weapons decreased. In this period the impetus to reduce nuclear weapons, especially nonstrategic ones, came in large part from the military leadership. And, on 27 September 1991, with a new relationship building between the United States and the soon-to-disappear Soviet regime, President George H. W. Bush announced that he had ordered the withdrawal of all nuclear artillery shells and short-range missiles from overseas bases along with several types of naval tactical nuclear weapons and bombs. After the unilateral cuts and treaty reductions, there were approximately 11,000 warheads left in the active stockpile by the mid-1990s.

The strategic arms agreement known as START I, the planning that went into START II and START III, and the Comprehensive Test Ban Treaty helped to determine the size and composition of the U.S. nuclear stockpile in the post–Cold War era. Some warheads removed from missiles and bombers to comply with START I were placed in storage as a "hedge," able to be redeployed if the political–military competition with Russia were somehow rekindled. This "lead and hedge" policy, adopted by the Clinton administration in their 1994 Nuclear Posture Review, provided a large reserve of warheads that had to be maintained, serviced, and secured.

In 2002, the United States signed an agreement with Russia to reduce "operationally deployed" strategic warheads to between 1,700 and 2,200 by the end of 2012. The United States reached the halfway mark toward implementing the Moscow Treaty (also known as the Strategic Offensive Reductions Treaty, or SORT) in 2007.

The administration of George W. Bush took a further step in June 2004 when it declared that the U.S. nuclear stockpile would be cut "nearly in half" by 2012. Under this plan the stockpile was scheduled to decrease from about 10,500 to about 5,400. A related development took place on 18 December 2007, when the White House announced that President Bush had "approved a significant reduction in the U.S. nuclear weapons stockpile to take effect by the end of 2007." In effect this reaffirmed the June 2004 cut of "nearly 50 percent" but accelerated the timetable by five years—from the end of 2012 to the end of 2007. A related announcement stated that an additional 15 percent of the stockpile would be cut by 2012, leaving an estimated 4,600 nuclear warheads in the U.S. stockpile as of that date.

2 J. Ronald Fox, *Arming America: How the U.S. Buys Weapons* (Cambridge, Mass.: Harvard University Press, 1974); and Jacques S. Gansler, *The Defense Industry* (Cambridge, Mass.: MIT Press, 1980).

APPENDIX B

Effects of Nuclear Weapons

Nuclear weapons are fundamentally different from conventional weapons because of the vast amounts of explosive energy they can release and the effects they produce, such as high temperatures and radiation. The immediate effects of a nuclear explosion and fallout are well known through data gathered from the bombings of Hiroshima and Nagasaki, Japan; from the more than 500 atmospheric and more than 1,500 underground nuclear tests conducted worldwide; and from extensive calculations and computer modeling.

Longer term effects on human health and the environment are less certain but have been extensively studied. The impacts of a nuclear explosion depend on many factors, including the design of the weapon (fission or fusion); its yield; whether the detonation takes place in the air (and at what altitude), on the surface, underground, or underwater; the meteorological and environmental conditions at the site of the explosion; and whether the target is urban, rural, or military.

When a nuclear weapon detonates, a fireball occurs with temperatures similar to those estimated to occur at the center of the Sun. The energy emitted takes several forms. Approximately 85 percent of the explosive energy produces air blast (shock) and thermal radiation (heat); the remaining 15 percent is released as initial radiation produced within the first minute or so and residual (or delayed) radiation emitted over a period of time, some of which can be in the form of local area fallout.

BLAST

The expansion of intensely hot gases at extremely high pressures in a nuclear fireball generates a shock wave that expands outward at high velocity. The "overpressure," or crushing pressure, at the front of the shock wave can be measured in pascals (or kilopascals, kPa) or in pounds per square inch (psi). The greater the overpressure, the more likely that a given structure will be damaged by the sudden impact of the wave front. A related destructive effect comes from the "dynamic pressure," or high-velocity wind, that accompanies the shock wave. An ordinary two-story, wood-frame house will collapse at an overpressure of 34.5 kPa (5 psi). A one-megaton weapon exploded at an altitude of 10,000 feet will generate overpressure of this magnitude out to about four miles from the point of detonation. The winds that follow will hurl a standing person against a wall with several times the force of gravity. Within five miles, few people in the open or in ordinary buildings will likely be able to survive such a blast. Enormous amounts of masonry, glass, wood, metal, and other debris created by the initial shock wave will fly at velocities greater than 100 miles per hour, causing further destruction.

THERMAL RADIATION

Approximately 35 percent of the total energy yield of an airburst is emitted as thermal radiation—light and heat capable of causing skin burns and eye injuries and starting fires of combustible material at considerable distances.

The shock wave, arriving later, may spread fires farther. If the individual fires are extensive enough, they can coalesce into a mass fire known as a firestorm, which generates a single convective column of rising hot gases that sucks in fresh air from the periphery. The inward-rushing winds and the extremely high temperatures generated in a firestorm consume virtually everything combustible. At Hiroshima the incendiary effects were quite different from those at Nagasaki, in part because of differences in terrain. The firestorm that raged over the level terrain of Hiroshima left 4.4 square miles severely damaged—roughly four times the area burned in the hilly terrain of Nagasaki.

INITIAL RADIATION

A special feature of a nuclear explosion is the emission of nuclear radiation, which may be separated into initial radiation and residual radiation. Initial radiation, also known as prompt radiation, consists of gamma rays and neutrons produced within a minute of the detonation. Beta particles (free electrons) and a small proportion of alpha particles (helium nuclei, i.e., two protons and two neutrons bound together) are also produced, but these particles have short ranges and typically will not reach Earth's surface if the weapon is detonated high enough above ground. Gamma rays and neutrons can produce harmful effects in living organisms, a hazard that persists over considerable distances because of their ability to penetrate most structures. Although their energy is only about 3 percent of the total released in a nuclear explosion, they can cause a considerable proportion of the casualties.

RESIDUAL RADIATION AND FALLOUT

Residual radiation is defined as radiation emitted more than one minute after the detonation. If the fission explosion is an airburst, the residual radiation will come mainly from the weapon debris. If the explosion is on or near the surface, the soil, water, and other materials in the vicinity will be sucked upward by the rising cloud, causing early (local) and delayed (worldwide) fallout. Early fallout settles to the ground during the first 24 hours; it may contaminate large areas and be an immediate and extreme biological hazard. Delayed fallout, which arrives after the first day, consists of microscopic particles that are dispersed by prevailing winds and settle in low concentrations, possibly over extensive portions of Earth's surface.

A nuclear explosion produces a complex mix of more than 300 different isotopes of dozens of elements, with half-lifes from fractions of a second to millions of years. The total radioactivity of the fission products is extremely large at first, but it falls off at a fairly rapid rate as a result of radioactive decay. Seven hours after a nuclear explosion, residual radioactivity will have decreased to about 10 percent of its amount at 1 hour, and after another 48 hours it will have decreased to 1 percent. (The rule of thumb is that for every sevenfold increase in time after the explosion, the radiation dose rate decreases by a factor of ten.)

ELECTROMAGNETIC PULSE

A nuclear electromagnetic pulse (EMP) is the time-varying electromagnetic radiation resulting from a nuclear explosion. The development of the EMP is shaped by the initial nuclear radiation from the explosion—specifically, the gamma radiation. High-energy electrons are produced in the environment of the explosion when gamma rays collide with air molecules (a process called the Compton effect). Positive and negative charges in the atmosphere are separated as the lighter, negatively charged electrons are swept away from the explosion point and the heavier, positively charged ionized air molecules are left behind. This charge separation produces a large electric field. Asymmetries in the electric field are caused by factors such as the variation in air density with altitude and the proximity of the explosion to Earth's surface. These asymmetries result in time-varying electrical currents that produce the EMP. The characteristics of the EMP depend strongly on the height of the explosion above the surface.

EMP effects were first noticed in the United States in the 1950s, when electronic equipment failed because of induced currents and voltages during some nuclear tests. In 1960 the potential vulnerability of U.S. military equipment and weapons systems to EMP was officially recognized. EMP can damage unprotected electronic equipment, such as radios, radars, televisions, telephones, computers, and other communication equipment and systems. EMP damage can occur at distances of tens, hundreds, or thousands of kilometers from a nuclear explosion, depending on the weapon yield and the altitude of the detonation. For example, in 1962 a failure of electronic components in street lights in Hawaii and activation of numerous automobile burglar alarms in Honolulu were attributed to a high-altitude U.S. nuclear test at Johnston Atoll, some 800 miles to the southwest. For a high-yield explosion of approximately ten megatons detonated at an altitude 200 miles above the center of the continental United States, almost the entire country, as well as parts of Mexico and Canada, would be affected by EMP. Procedures to improve the ability of networks, especially military command and control systems, to withstand EMP are known as "hardening."

APPENDIX C

U.S. Strategic Offensive Forces, 1945–2008

	ICBMs		SLBMs		Bombers*		Total	
Year	Launchers	Warheads	Launchers	Warheads	Launchers	Warheads	Launchers	Warheads
1945	—	—	—	—	15	6	15	6
1946	—	—	—	—	148	11	148	11
1947	—	—	—	—	319	32	319	32
1948	—	—	—	—	556	100	556	100
1949	—	—	—	—	525	200	525	200
1950	—	—	—	—	520	330	520	330
1951	—	—	—	—	669	500	669	500
1952	—	—	—	—	857	720	857	720
1953	—	—	—	—	762	878	762	878
1954	—	—	—	—	1,082	1,418	1,082	1,418
1955	—	—	—	—	1,309	1,755	1,309	1,755
1956	—	—	—	—	1,650	2,123	1,650	2,123
1957	—	—	—	—	1,655	2,460	1,655	2,460
1958	—	—	—	—	1,769	2,610	1,769	2,610
1959	6	6	—	—	1,854	2,490	1,860	2,496
1960	12	12	32	32	1,735	3,083	1,779	3,127
1961	57	57	80	80	1,526	3,016	1,663	3,153
1962	203	203	144	144	1,595	3,104	1,942	3,451
1963	597	597	160	160	1,335	3,293	2,092	4,050
1964	907	907	320	384	1,111	3,427	2,338	4,718
1965	854	854	384	736	807	3,465	2,045	5,055
1966	1,004	1,004	560	1,264	674	3,476	2,238	5,744
1967	1,054	1,044	656	1,552	669	3,630	2,379	6,226
1968	1,054	1,044	656	1,552	655	3,521	2,365	6,117

Large and graceful, in the 1950s the B-36 Peacemaker symbolized the strength and global reach of U.S. strategic striking power. (U.S. Air Force)

1969	1,054	1,044	656	1,552	549	3,366	2,259	5,962
1970	1,054	1,244	656	1,552	501	3,339	2,211	6,135
1971	1,054	1,444	656	2,464	478	3,232	2,188	7,140
1972	1,054	1,644	656	3,120	462	3,845	2,172	8,609
1973	1,054	1,844	656	4,112	493	3,776	2,203	9,732
1974	1,054	1,944	656	4,432	494	3,819	2,204	10,195
1975	1,054	2,144	656	4,544	489	3,978	2,199	10,666
1976	1,054	2,144	656	5,104	487	3,850	2,197	11,098
1977	1,054	2,144	656	5,216	483	3,834	2,193	11,194
1978	1,054	2,144	656	5,440	410	3,568	2,120	11,152
1979	1,054	2,144	656	5,376	408	3,568	2,118	11,088
1980	1,054	2,144	592	5,056	406	3,568	2,052	10,768
1981	1,054	2,144	512	4,752	406	3,568	1,972	10,464
1982	1,049	2,139	520	4,768	362	3,384	1,931	10,291
1983	1,040	2,130	544	4,960	323	3,520	1,907	10,610
1984	1,030	2,120	592	5,344	322	3,844	1,944	11,308
1985	1,020	2,110	600	5,376	321	4,104	1,941	11,590
1986	1,005	2,165	616	5,440	286	4,709	1,907	12,314
1987	1,000	2,300	640	5,632	396	5,753	2,016	13,685
1988	1,000	2,440	608	5,312	411	5,328	2,019	13,080
1989	1,000	2,440	592	5,152	400	5,188	1,992	12,780
1990	1,000	2,440	608	5,216	346	4,648	1,954	12,304

1991	550	2,000	480	3,456	287	3,844	1,317	9,300
1992	550	2,000	488	3,456	190	2,824	1,228	8,280
1993	550	2,000	336	2,688	191	2,840	1,077	7,528
1994	580	2,000	360	2,880	194	2,808	1,134	7,688
1995	575	2,000	384	3,072	197	2,176	1,156	7,248
1996	575	1,950	408	2,880	202	2,032	1,185	6,862
1997	550	1,950	432	3,072	115	1,264	1,097	6,286
1998	550	1,900	432	3,072	115	1,264	1,097	6,236
1999	550	1,850	432	3,072	115	1,376	1,097	6,298
2000	550	1,850	432	3,072	115	1,376	1,097	6,298
2001	550	1,700	432	2,304	115	1,376	1,097	5,380
2002	550	1,700	384	2,016	115	1,376	1,049	5,092
2003	532	1,600	360	1,872	115	1,376	1,007	4,848
2004	516	1,490	336	1,728	115	1,376	967	4,594
2005	510	1,100	288	1,584	115	1,376	913	4,060
2006	500	1,050	288	1,728	115	1,376	903	4,154
2007	488	764	288	1,728	115	1,376	891	3,868
2008	450	650	288	1,728	113	1,376	851	3,754

* Total inventory. Not every bomber was certified for nuclear missions or was combat ready.

Recommended Reading

BOOKS

Bacevich, Lt. Col. A. J., USA. *The Pentomic Era.* Washington, D.C.: National Defense University Press, 1986.

Ball, Desmond. *Politics and Force Levels: The Strategic Missile Program of the Kennedy Administration.* Berkeley, Calif.: University of California Press, 1980.

Bowers, Peter M. *Boeing Aircraft since 1916.* London: Putnam, 1989.

Boyne, Walter. *Boeing B-52: A Documentary History.* London: Jane's, 1981.

Brodie, Bernard. *The Absolute Weapon.* New York: Harcourt, Brace, 1946.

Bruins, Berend Derk. "U.S. Naval Bombardment Missiles, 1940–1958: A Study of the Weapons Innovation Process." PhD diss., Columbia University, 1981.

Cochran, Thomas B., William M. Arkin, and Milton M. Hoenig. *Nuclear Weapons Databook: U.S. Nuclear Forces and Capabilities.* Vol. I. Cambridge, Mass.: Ballinger, 1984.

Cochran, Thomas B., William M. Arkin, Robert S. Norris, and Milton M. Hoenig. *Nuclear Weapons Databook: U.S. Nuclear Warhead Production.* Vol. II. Cambridge, Mass.: Ballinger, 1987.

Douhet, Gen. Giulio, Italian Navy. *The Command of the Air.* Washington, D.C.: Office of Air Force History, 1983.

Eliot, George Fielding. *Victory without War 1958–1961.* Annapolis, Md.: U.S. Naval Institute, 1958.

Francillon, Rene J. *Grumman Aircraft since 1929.* London: Putnam, 1989.

———. *Lockheed Aircraft since 1913.* London: Putnam, 1987.

———. *McDonnell Douglas Aircraft since 1920.* London: Putnam, 1979.

Friedman, Norman. *U.S. Naval Weapons.* London: Conway, 1983.

Futrell, Robert F. *The United States Air Force in Korea 1950–1953.* New York: Duell, Sloan and Pearce, 1961.

Gibson, James N. *Nuclear Weapons of the United States: An Illustrated History.* Atglen, Pa.: Schiffer, 1996.

Glennon, John P., Edward C. Keefer, and David W. Mabon, eds. *Foreign Relations of the United States, 1958–1960,* vol. VIII in *National Security Policy.* Washington, D.C.: Government Printing Office, 1996.

Groves, Lt. Gen. Leslie R., USA (Ret.). *Now It Can Be Told: The Story of the Manhattan Project.* New York: Harper, 1962.

Gunston, Bill. *Attack Aircraft of the West.* New York: Charles Scribner's Sons, 1974.

———. *Bombers of the West.* New York: Charles Scribner's Sons, 1973.

———. *The Illustrated Encyclopedia of Aircraft Armament.* New York: Orion Books, 1988.

———. *The Illustrated Encyclopedia of the World's Missiles and Rockets.* London: Salamander Books, 1979.

Hansen, Chuck. *U.S. Nuclear Weapons: The Secret History.* New York: Orion, 1988.

Hartt, Julian. *The Mighty Thor: Missile in Readiness.* New York: Duell, Sloan, and Pearce, 1961.

Hopkins, J. C., and Sheldon A. Goldberg. *The Development of Strategic Air Command 1946–1986.* Offutt AFB, Neb.: Office of the Historian, Headquarters Strategic Air Command, 1986.

Knaack, Marcelle Size. *Post–World War II Bombers 1945–1973,* Vol. II. *Encyclopedia of U.S. Air Force Aircraft and Missile Systems.* Washington, D.C.: Office of Air Force History, 1988.

———. *Post–World War II Fighters 1945–1973,* Vol. I. *Encyclopedia of U.S. Air Force Aircraft and Missile Systems.* Washington, D.C.: Office of Air Force History, 1978.

Lonnquest, John C., and David F. Winkler. *To Defend and Deter: The Legacy of the United States Cold War Missile Program.* Washington, D.C.: Department of Defense Legacy Resource Management Program Cold War Project, November 1996.

McLucas, John L. *Reflections of a Technocrat: Managing Defense, Air, and Space Programs during the Cold War.* Maxwell AFB, Ala.: Air University Press, August 2006.

Miller, Vice Adm. Jerry, USN (Ret). *Nuclear Weapons and Aircraft Carriers: How the Bomb Saved Naval Aviation.* Washington, D.C.: Smithsonian Institution Press, 2001.

Moody, Walter S. *Building a Strategic Air Force.* Washington, D.C.: Air Force History and Museum Program, 1955.

Muir, Malcolm, Jr. *Black Shoes and Blue Water: Surface Warfare in the United States Navy, 1945–1975.* Washington, D.C.: Naval Historical Center, 1996.

Neufeld, Jacob. *The Development of Ballistic Missiles in the United States Air Force 1945–1960.* Washington, D.C.: Office of Air Force History, 1990.

Peacock, Lindsay T. *Strategic Air Command.* London: Arms & Armour Press, 1988.

Polmar, Norman. *Aircraft Carriers: A History of Carrier Aviation and Its Influence on World Events, 1946–2006.* Dulles, Va.: Potomac Books, 2008.

Polmar, Norman, and Kenneth J. Moore. *Cold War Submarines: The Design and Construction of U.S. and Soviet Submarines.* Dulles, Va.: Brassey's/Potomac Books, 2004.

Rausa, Capt. Rosario, USNR (Ret.). *The Douglas A-1 "Flying Dump Truck."* Baltimore, Md.: Nautical & Aviation, 1982.

Reed, Thomas C. *At the Abyss: An Insider's History of the Cold War.* New York: Random House, 2004.

Rosenberg, Max. *The Air Force and the National Guided Missile Program 1944–1950.* Washington, D.C.: USAF Historical Division Liaison Office, June 1964.

Schaffel, Kenneth. *The Emerging Shield: The Air Force and the Evolution of Continental Air Defense, 1945–1960.* Washington, D.C.: Office of Air Force History, 1991.

Schwiebert, Ernest G. *A History of the U.S. Air Force Ballistic Missiles.* New York: Praeger, 1964.

Spinardi, Graham. *From Polaris to Trident: The Development of US Fleet Ballistic Missile Technology.* Cambridge, U.K.: Cambridge University Press, 1994.

Stumpf, David K. *Regulus: The Forgotten Weapon.* Paducah, Ky.: Turner, 1996.

Swanborough, Gordon, and Peter M. Bowers. *United States Military Aircraft since 1909,* 3rd ed. Washington, D.C.: Smithsonian Institution Press, 1989.

———. *United States Navy Aircraft since 1911.* 3rd ed. Annapolis, Md.: Naval Institute Press, 1990.

Wegg, John. *General Dynamics Aircraft and their Predecessors.* London: Putnam, 1990.

Werrell, Kenneth P. *The Evolution of the Cruise Missile.* Maxwell AFB, Ala.: Air University Press, September 1985.

Wynn, Humphrey. *The RAF Strategic Nuclear Deterrent Forces: Their Origins, Roles and Deployment 1946–1969: A Documentary History.* London: Her Majesty's Stationery Office, 1994.

Zaloga, Steven. *Soviet Air Defense Missles: Design, Development, and Tactics.* London: Jane's, 1989.

OFFICIAL PUBLICATIONS

Aircraft data are based primarily on the Standard Aircraft Characteristics sheets produced by the Air Force and Navy.

U.S. Army Command and General Staff College. "Conventional-Nuclear Operations," RB 100-30. Fort Leavenworth, Kan., 6 August 1976.

U.S. Department of Energy, Nevada Operations Office. "Announced United States Nuclear Tests: July 1945 through December 1962." Las Vegas, Nev., January 1983.

U.S. Navy Department. Strategic Systems Program

Office. "FBM Facts/Chronology: Polaris, Poseidon, Trident." Washington, D.C., 1990.

U.S. Strategic Air Command, Office of the Historian. "Alert Operations and the Strategic Air Command, 1957–1991." Offutt AFB, Neb.: Office of the Historian, 7 December 1991.

———. "From Snark to SRAM: A Pictorial History of Strategic Air Command Missiles." Offutt AFB, Neb.: Office of the Historian, 1 May 1990.

———. "SAC Missile Chronology, 1939–1988." Offutt AFB, Neb.: Office of the Historian, 1 May 1990.

COMPACT DISKS

Hansen, Chuck. *The Swords of Armageddon.* 2007.

Sandia National Laboratories. *U.S. Strategic Nuclear Policy: An Oral History 1942–2004.* 2006.

PERIODICALS

The following periodicals have carried significant articles related to nuclear weapons and systems.

Air Enthusiast
Air Force
Air International
Aviation Week & Space Technology
The Bulletin of the Atomic Scientists
U.S. Naval Institute *Proceedings*
World Airpower Journal

Index

Notes: Ranks listed are the highest used in this book. Aircraft data are listed by their designations, with aircraft designations at the beginning of each letter group. The prefixes B, EC, Mark, TX, and W were often used interchangeably for nuclear warheads.

About the Authors

Norman Polmar, a well-known defense analyst and writer, has participated in several studies and authored several articles and books directly related to nuclear weapons development, including studies related to naval nuclear weapons for the Defense Advanced Research Projects Agency (DARPA), the U.S. Navy, and the Defense Nuclear Agency (DNA). He was head of the contractor team for Navy participation in the 1974–1975 Department of Defense project "History of the US-USSR Strategic Arms Competition, 1945–1972." He is also an author or coauthor of several papers on nuclear forces, including "Measures and Trends—U.S. and USSR Strategic Force Effectiveness" (Defense Nuclear Agency, 1978), "U.S. Strategic Weapon Programs" (General Electric, 1985), and several classified papers. For several years in the 1980s Mr. Polmar was a consultant to the director of the Los Alamos National Laboratory.

Mr. Polmar is also author of *Strategic Weapons: An Introduction* (Crane, Russell, 1975), which, in several printings, was used as a text at U.S. military academies and several universities. Nuclear weapons are also discussed in his several editions of *Guide to the Soviet Navy* and *Ships and Aircraft of the U.S. Fleet* (both Naval Institute Press) as well as in several of his articles in the Naval Institute *Proceedings* and other professional journals.

He has served on the advisory boards for several of the nuclear weapons studies produced by the Natural Resources Defense Council and has lectured on nuclear weapons development at the Department of Energy and the Soviet Institute of U.S. Studies (Moscow). Mr. Polmar served on the Secretary of the Navy's Research Advisory Committee (NRAC) for almost ten years and in 2007 was named chairman of the Department of Homeland Security's Science and Technology Advisory Committee.

Robert S. (Stan) Norris is senior research associate at the Natural Resources Defense Council in Washington, D.C. He has been a primary contributor to the NRDC's series of Nuclear Weapon Databooks and is coauthor of a column on nuclear weapons in *The Bulletin of the Atomic Scientists*. Other writings by Dr. Norris in this field include coauthoring the chapter on nuclear weapons in five editions of the *SIPRI Yearbook* and the entry on Nuclear Weapons in the latest edition of *Encyclopaedia Britannica*. He is author of the award-winning *Racing for the Bomb: General Leslie R. Groves, the Manhattan Project's Indispensable Man* (Steerforth Press, 2002), and is coauthor with Thomas B. Cochran and Oleg A. Bukharin of *Making of the Russian Bomb: From Stalin to Yeltsin* (Westview Press, 1994).

He has taught political science and international relations at the university level in the United States and abroad. He holds a Ph.D. in political science from New York University.

The Naval Institute Press is the book-publishing arm of the U.S. Naval Institute, a private, nonprofit, membership society for sea service professionals and others who share an interest in naval and maritime affairs. Established in 1873 at the U.S. Naval Academy in Annapolis, Maryland, where its offices remain today, the Naval Institute has members worldwide.

Members of the Naval Institute support the education programs of the society and receive the influential monthly magazine *Proceedings* or the colorful bimonthly magazine *Naval History* and discounts on fine nautical prints and on ship and aircraft photos. They also have access to the transcripts of the Institute's Oral History Program and get discounted admission to any of the Institute-sponsored seminars offered around the country.

The Naval Institute's book-publishing program, begun in 1898 with basic guides to naval practices, has broadened its scope to include books of more general interest. Now the Naval Institute Press publishes about seventy titles each year, ranging from how-to books on boating and navigation to battle histories, biographies, ship and aircraft guides, and novels. Institute members receive significant discounts on the Press's more than eight hundred books in print.

Full-time students are eligible for special half-price membership rates. Life memberships are also available.

For a free catalog describing Naval Institute Press books currently available, and for further information about joining the U.S. Naval Institute, please write to:

Member Services
U.S. Naval Institute
291 Wood Road
Annapolis, MD 21402-5034
Telephone: (800) 233-8764
Fax: (410) 571-1703
Web address: www.usni.org